1995
NATIONAL BUILDING COST MANUAL

Nineteenth Edition

Edited by
Martin D. Kiley
and
Marques Allyn

Craftsman Book Company
6058 Corte del Cedro / P.O. Box 6500 / Carlsbad, CA 92018

Looking for other construction reference manuals?

Craftsman has the books to fill your needs. Call **toll-free 1-800-829-8123** or write to Craftsman Book Company, P.O. Box 6500, Carlsbad, CA 92018 for a **FREE CATALOG** of books, videos and audios.

Cover photo by Ed Kessler
Illustrations by Laura Knight
© 1994 Craftsman Book Company
ISBN 1-57218-001-3

Contents Of This Manual

Explanation of the Cost Tables

This manual provides construction or replacement costs for a wide variety of residential, commercial, industrial, agricultural and military buildings. For your convenience and to avoid possible errors, all the cost and reference information you need for each building type is listed with the primary cost figures for that building. After reading this and the following two pages you should be able to turn directly to any building type and make an error-free estimate or appraisal.

The costs are per square foot of floor area for the basic building and additional costs for optional or extra components that differ from building to building. Building shape, floor area, design elements, materials used, and overall quality influence the basic structure cost. These and other cost variables are isolated for the building types. Components included in the basic square foot cost are listed with each building type. Instructions for using the basic building costs are included above the cost tables. These instructions include a list of components that may have to be added to the basic cost to find the total cost for your structure.

The figures in this manual are intended to reflect the amount that would be paid by the end user of a building as of mid 1995.

They show the total construction cost including all design fees, permits, and the builder's supervision, overhead, and profit. These figures do not include land value, site development costs, or the cost of modifying unusual soil conditions or grades.

Building Quality

Structures vary widely in quality and the quality of construction is the most significant variable in the finished cost. For estimating purposes the structure should be placed in one or more quality classes. These classes are numbered from 1 which is the highest quality generally encountered. Each section of this manual has a page describing typical specifications which define the quality class. Each number class has been assigned a word description (such as best, good, average or low) for convenience and to help avoid possible errors.

The quality specifications do not reflect some design features and construction details that can make a building both more desirable and more costly. When substantially more than basic design elements are present, and when these elements add significantly to the cost, it is appropriate to classify the quality of the building as higher than would be warranted by the materials used in construction.

Many structures do not fall into a single class and have features of two quality classes. The tables have "half classes" which apply to structures which have some features of one class and some features of a higher or lower class. Classify a building into a "half class" when the quality elements are fairly evenly divided between two classes. Generally quality elements do not vary widely in a single building. For example, it would be unusual to find a top quality single family residence with minimum quality roof cover. The most weight should be given to quality elements that have the greatest cost. For example, the type of wall and roof framing or the quality of interior finish are more significant than the roof cover or bathroom wall finish. Careful evaluation may determine that certain structures fall into two distinct classes. In this case the cost of each part of the building should be evaluated separately.

Building Shapes

Shape classification considers any cost differences that arise from variations in building outline. Shape classification considerations vary somewhat with different building types. Where the building shape often varies widely between buildings and shape has a significant effect on the building cost, basic building costs are given for several shapes. Use the table that most closely matches the shape of the building you are evaluating. If the shape falls near the division between two basic building cost tables, it is appropriate to average the square foot cost from those two tables.

Area of Buildings

The basic building cost tables reflects the fact that larger buildings generally cost less per square foot than smaller buildings. The cost tables are based on square foot areas which include the following:

1. All floor area within and including the exterior walls of the main building.
2. Inset areas such as vestibules, entrances or porches outside of the exterior wall but under the main roof.
3. Any enclosed additions, annexes or lean-tos with a square foot cost greater than three-fourths of the square foot cost of the main building.

Select the basic building cost listed below the area which falls closest to the actual area of your building. If the area of your building falls nearly mid-way between two listed building areas, it is appropriate to average the square foot costs for the listed areas.

4

Wall Heights

Building costs are based on the wall heights given in the instructions for each building cost table. Wall height for the various floors of a building are computed as follows: The basement is measured from the bottom of floor slab to the bottom of the first floor slab or joist. The main or first floor extends from the bottom of the first floor slab or joist to the top of the roof slab or ceiling joist. Upper floors are measured from the top of the floor slab or floor joist to the top of the roof slab or ceiling joist. These measurements may be illustrated as follows:

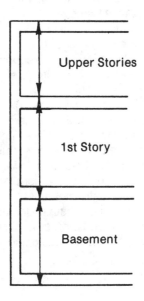

Square foot costs of most building design types must be adjusted if the actual wall height differs from the listed wall height. Wall height adjustment tables are included for building requiring this adjustment. Wall height adjustment tables list square foot costs for a foot of difference in perimeter wall height of buildings of various areas. The amount applicable to the actual building area is added or deducted for each foot of difference from the basic wall height.

Buildings such as residences, medical-dental buildings, funeral homes and convalescent hospitals usually have a standard eight foot ceiling height except in chapels or day room areas. If a significant cost difference exists due to a wall height variation, this factor should be considered in establishing the quality class.

Other Adjustments

A common wall exists when two buildings share one wall. Common wall adjustments are made by deducting the in-place cost of the exterior wall finish plus one-half of the in-place cost of the structural portion of the common wall area.

If an owner has no ownership in a wall, the in-place cost of the exterior wall finish plus the in-place cost of the structural portion of the wall should be deducted from the total building costs. Suggested common wall and no wall ownership costs are included for many of the building types.

Some square foot costs include the cost of expensive veneer finishes on the entire perimeter wall. When these buildings butt against other buildings, adjustments should be made for the lack of this finish. Where applicable, linear foot cost deductions are provided.

The square foot costs in this manual are based on composite costs of total buildings including usual work room or storage areas. They are intended to be applied on a 100% basis to the total building area even though certain areas may or may not have interior finish. Only in rare instances will it be necessary to modify the square foot cost of a portion of a building.

Multiple story buildings usually share a common roof structure and cover, a common foundation and common floor or ceiling structures. The costs of these components are included in the various floor levels as follows:

The first or main floor includes the cost of a floor structure built at ground level, foundation costs for a one-story building, a complete ceiling and roof structure, and a roof cover. The basement includes the basement floor structure and the difference between the cost of the first floor structure built at ground level and its cost built over a basement. The second floor includes the difference between the cost of a foundation for a one-story building and the cost of a foundation for a two story building and the cost of the second story floor structure.

Location Adjustments

The figures in this manual are intended as national averages for metropolitan areas of the United States. Use the information on page 7 to adapt the basic building costs to any area listed. Frequently building costs outside metropolitan areas are 2 to 6 percent lower if skilled, productive, lower cost labor is available in the area. The factors on page 7 can be applied to nearly all the square foot costs and some of the "additional" costs in this book.

Depreciation

Depreciation is the loss in value of a structure from all causes and is caused primarily by three forms of obsolescence: (1) physical (2) functional, and (3) economic.

Physical obsolescence is the deterioration of building components such as paint, carpets or roofing. This deterioration may be partially curable. The tables on pages 40, 200 and 233 consider only typical physical obsolescence. Individual judgements will have to be made of functional and economic obsolescence.

Functional obsolescence is due to a deficiency or inadequacy in some characteristic of the building, such as too few bathrooms for the number of bedrooms, or some excess, such as a 10 foot ceiling in a residence. This obsolescence may be curable. The tables do not include functional obsolescence considerations.

Economic obsolescence is caused by factors not directly concerning the structure, but rather, by adverse environmental factors, resulting in loss of desirability. Examples include the obsolescence of a store in an area of declining economic activity, or obsolescence resulting from governmental regulation changing the zone of an area. Because this kind of obsolescence is particularly difficult to measure, it is not included in the tables.

"Effective age" considers all forms of depreciation. It may be less than chronological age, if recently remodeled or improved, or more than the actual age, if deterioration is particularly bad. Though effective age is not considered in the physical life tables, it may yield a better picture of a structure's life than the actual physical age.

Once the effective age is determined, considering physical, functional and economic deterioration, use the percent good tables on pages 40 or 200 to determine the present value of a depreciated building. Present value is the result of multiplying the replacement cost (found by using the cost tables) by the appropriate percent good.

Limitations

This manual will be a useful reference for anyone who has to develop budget estimates or replacement costs for buildings. Anyone familiar with construction estimating understands that even very competent estimators with complete working drawings, full specifications and precise labor and material costs can disagree on the cost of a building. Frequently exhaustive estimates for even relatively simple structures can vary 5% or more. The range of competitive bids on some building projects is as much as 10%. Estimating costs is not an exact science and there is room for legitimate disagreement on what the "right" cost is. This manual can not help you do in a few minutes what skilled estimators may not be able to do in several hours. This manual will help you determine a reasonable replacement or construction cost for most buildings. It is not intended as a substitute for judgement or as a replacement for sound professional practice, but should prove a valuable aid to developing an informed opinion of value.

Area Modification Factors

Construction costs are higher in some cities than in other cities. Use the factors on this page or page 8 to adapt the costs listed in this book to your job site. Multiply your estimated total project cost by the factor listed for the appropriate city in this table to find your estimated building cost.

These factors were compiled by comparing the actual construction cost of residential, institutional and commercial buildings in 402 communities throughout the United States. Because these factors are based on completed project costs, they consider all construction cost variables, including labor,

equipment and material cost, labor productivity, climate, job conditions and markup.

Use the factor for the nearest or most comparable city. If the city you need is not listed in the table, use the factor for the appropriate state.

Note that these location factors are composites of many costs and will not necessarily be accurate when estimating the cost of any particular part of a building. But when used to modify all estimated costs on a job, they should improve the accuracy of your estimates.

Alabama	**0.82**	Ridgecrest	1.11	**Guam**	**1.45**	Middlesboro	0.92
Birmingham	0.84	Sacramento	1.03	Agana	1.45	Owensboro	0.89
Dothan	0.78	San Bernardino	1.17			Paducah	0.88
Florence	0.80	San Diego	1.13	**Hawaii**	**1.46**		
Huntsville	0.82	San Francisco	1.39	Hilo	1.47	**Louisiana**	**0.87**
Mobile	0.85	San Jose	1.25	Honolulu	1.40	Alexandria	0.82
Montgomery	0.83	Santa Ana	1.23	Kauai	1.50	Baton Rouge	0.91
Tuscaloosa	0.81	Santa Barbara	1.20	Kona	1.49	Houma	0.89
		Santa Cruz	1.24	Maui	1.46	Lafayette	0.87
Alaska	**1.45**	Santa Maria	1.18			Lake Charles	0.90
Anchorage	1.43	Santa Rosa	1.13	**Idaho**	**0.95**	Marshall	0.90
Fairbanks	1.53	South Lake Tahoe	1.05	Boise	0.93	Monroe	0.83
Juneau	1.47	Tehachapi	1.16	Coeur D'Alene	0.98	New Orleans	0.90
Kenai	1.31	Ventura	1.19	Idaho Falls	0.93	Shreveport	0.85
Ketchikan	1.47	Victorville	1.16	Pocatello	0.97		
Sitka	1.48					**Maine**	**1.02**
		Colorado	**0.95**	**Illinois**	**1.00**	Augusta	1.03
Arizona	**0.97**	Aspen–Vail	1.03	Bellville	0.98	Bangor	0.97
Casa Grande	1.01	Denver	0.90	Chicago	1.07	Brunswick	1.06
Flagstaff	0.94	Grand Junction	0.92	East St. Louis	1.00	Lewiston	1.04
Kingman	0.94			Moline	1.00	Portland	1.04
Phoenix	1.03	**Connecticut**	**1.09**	Springfield	0.97	Waterville	0.99
Sierra Vista	0.92	Bridgeport	1.12				
Tucson	1.00	Hartford	1.10	**Indiana**	**0.96**	**Maryland**	**0.99**
Yuma	0.94	New Haven	1.08	Bloomington	0.95	Baltimore	1.01
		New London	1.06	Evansville	0.93	Baltimore City	1.04
Arkansas	**0.84**	New Milford	1.09	Fort Wayne	0.93	Hagerstown	0.94
Fayetteville	0.83	Norwich	1.05	Gary	1.03	Salisbury	0.97
Fort Smith	0.83	Ridgefield	1.13	Hammond	1.01	Waldorf	0.99
Jonesboro	0.82	Windam	1.05	Indianapolis	0.94		
Little Rock	0.87			Lafayette	0.95	**Massachusetts**	**1.14**
Texarkana	0.83	**Delaware**	**1.03**	South Bend	0.96	Boston	1.21
		Dover	1.02	Terre Haute	0.97	Fall River	1.11
California	**1.15**	Wilmington	1.04			Worcester	1.09
Arrowhead	1.17			**Iowa**	**0.92**		
Bakersfield	1.12	**District of Columbia**	**1.04**	Bettendorf	0.97	**Michigan**	**0.98**
Barstow	1.16	Washington D.C.	1.04	Cedar Rapids	0.93	Ann Arbor	1.02
Big Bear	1.18			Council Bluffs	0.89	Battle Creek	0.94
Desert Center	1.23	**Florida**	**0.87**	Davenport	0.95	Benton Harbor	0.97
El Cajon	1.14	Jacksonville	0.81	Des Moines	0.92	Detroit	1.04
Eureka	1.11	Key West	0.90	Dubuque	0.92	Flint	1.01
Fresno	1.04	Miami	0.89	Mason City	0.91	Grand Rapids	0.93
Inyokern	1.14	Orlando	0.95	Sioux City	0.90	Jackson	0.99
Lancaster	1.22	Pensacola	0.80	Waterloo	0.92	Lansing	1.02
Los Angeles	1.26	Tampa	0.87			Marquette	0.98
Modesto	1.02			**Kansas**	**0.91**	Mt. Pleasant	0.97
Monterey	1.22	**Georgia**	**0.82**	Garden City	0.87	Muskegon	0.92
Mojave	1.17	Albany	0.82	Kansas City	1.03	Saginaw	0.96
Needles	1.16	Atlanta	0.85	Pittsburg	0.86	Ypsilanti	1.05
Oakland	1.25	Augusta	0.82	Salina	0.88		
Ojai	1.16	Brunswick	0.81	Topeka	0.94	**Minnesota**	**0.98**
Oxnard	1.19	Columbus	0.81	Wichita	0.88	Duluth	1.00
Paso Robles	1.16	Macon	0.81			Mankato	0.96
Piru	1.15	Rome	0.80	**Kentucky**	**0.92**	Minneapolis	1.03
Placerville	1.04	Savannah	0.81	Ashland	0.95	Rochester	0.98
Redding	1.00	Valdosta	0.81	Covington	0.97	St. Cloud	0.97
				Louisville	0.88	Worthington	0.91

Mississippi **0.84**
Corinth 0.84
Greenville 0.85
Greenwood 0.84
Gulfport 0.86
Hattisburg 0.83
Jackson 0.84
Southaven 0.84

Missouri **0.97**
Cape Girardeau 0.97
Columbia 0.96
Joplin 0.91
Kansas City 1.02
Kirksville 0.97
Rolla 0.91
St. Joseph 0.97
St. Louis 1.06
Sedalia 0.95
Springfield 0.94

Montana **0.92**
Billings 0.92
Great Falls 0.93
Helena 0.92
Missoula 0.92

Nebraska **0.90**
Grand Island 0.92
Lincoln 0.89
Omaha 0.92
Norfolk 0.92
North Platte 0.88
Scottsbluff 0.88

Nevada **1.10**
Las Vegas 1.10
Reno 1.09

New Hampshire **1.03**
Concord 1.03
Dover 1.03
Keene 1.01
Manchester 1.03
Nashua 1.05
Portsmouth 1.05

New Jersey **1.16**
Ashbury Park 1.15
Atlantic City 1.17
Burlington 1.14
Camden 1.15
Freehold 1.15
Gloucester 1.14
Newark 1.22
North Bergen 1.22
Trenton 1.15
Vineland 1.13

New Mexico **0.92**
Albuquerque 0.90
Clovis 0.89
Santa Fe 0.93
Silver City 0.91
Taos 0.99

New York **1.15**
Albany 1.02
Binghamton 1.00
Buffalo 1.01
Elmira 0.97

Jamestown 0.93
Nassau County 1.34
N.Y. City (Metro) 1.39
N.Y. City (Inner) 1.44
Orange County 1.28
Plattsburgh 0.96
Poughkeepsie 1.15
Rochester 0.99
Rockland County 1.29
Suffolk County 1.20
Syracuse 1.00
Westchester County 1.35

North Carolina **0.87**
Asheville 0.88
Charlotte 0.88
Durham 0.89
Elizabeth City 0.89
Fayetteville 0.85
Greensboro 0.86
Greenville 0.85
Raleigh 0.90
Wilmington 0.86
Winston-Salem 0.86

North Dakota **0.92**
Bismarck 0.94
Dickinson 0.93
Fargo 0.90

Ohio **1.03**
Akron 1.09
Cincinnati 0.98
Cleveland 1.12
Columbus 0.97
Dayton 0.97
Findlay 1.00
Lorain 1.08
Mansfield 1.00
Toledo 1.09
Youngstown 1.03

Oklahoma **0.88**
Ada 0.89
Ardmore 0.88
Bartlesville 0.88
Enid 0.87
Guymon 0.89
Lawton 0.87
McAlester 0.89
Muskogee 0.90
Oklahoma City 0.88
Shawnee 0.89
Stillwater 0.89
Tulsa 0.88
Woodward 0.88

Oregon **0.99**
Bend 0.98
Coos Bay 1.01
Eugene 0.97
Portland 1.00

Pennsylvania **1.05**
Allentown 1.13
Altoona 1.02
Bellefonte 1.04
Erie 1.05
Harrisburg 1.00
Johnstown 1.02

Lancaster 1.04
Philadelphia 1.17
Pittsburgh 1.06
Reading 1.02
Scranton 1.04
Wellsboro 1.03
York 0.99

Puerto Rico **1.12**
Arecibo 1.12
Mayaguez 1.12
Old San Juan 1.13
Ponce 1.12
San Juan 1.11

Rhode Island **1.07**
Providence 1.07

South Carolina **0.86**
Aiken 0.85
Anderson 0.85
Beaufort 0.87
Charleston 0.89
Columbia 0.85
Florence 0.84
Greenville 0.86
Greenwood 0.85
Myrtle Beach 0.87
North Augusta 0.87
Orangeburg 0.85
Rockhill 0.85
Spartanburg 0.86

South Dakota **0.90**
Pierre 0.91
Rapid City 0.88
Sioux Falls 0.91

Tennessee **0.88**
Chattanooga 0.90
Clarksville 0.86
Columbia 0.88
Jackson 0.89
Johnson City 0.86
Kingsport 0.87
Knoxville 0.89
Memphis 0.89
Nashville 0.87
Oak Ridge 0.88

Texas **0.91**
Abilene 0.93
Amarillo 0.95
Austin 0.89
Beaumont 0.93
Bryan 0.93
Corpus Christi 0.84
Dallas 0.84
Eagle Pass 0.86
El Campo 0.93
El Paso 0.90
Fort Worth 0.86
Harlingen 0.84
Houston 0.92
Junction 0.87
Laredo 0.84
Lubbock 0.96
Lufkin 0.92
Midland 0.99
Odessa 0.99

San Angelo 0.92
San Antonio 0.97
Sherman 0.92
Texas City 0.93
Tyler 0.90
Waco 0.89
Wichita Falls 0.93
Victoria 0.87

Utah **0.93**
Cedar City 0.91
Salt Lake City 0.94
Vernal 0.93

Vermont **1.01**
Bennington 1.03
Brattleboro 1.03
Burlington 1.00
Montpelier 1.00
Rutland 1.06

Virginia **0.97**
Charlottesville 0.97
Harrisonburg 0.93
Newport News 0.96
Norfolk 0.98
Norton 0.97
Richmond 0.99

Virgin Islands **1.16**
St. Thomas 1.17
St. Croix 1.15

Washington **1.02**
Aberdeen 1.02
Bellingham 1.02
Cheney 1.01
Kennewick 1.03
Longview 1.01
Olympia 1.03
Port Angeles 1.03
Pullman 1.02
Seattle 1.03
Spokane 1.00
Yakima 1.02

West Virginia **0.98**
Bluefield 0.93
Charleston 0.99
Fairmont 1.00
Huntington 0.96
Martinsburg 0.93
Parkersburg 0.98
Point Pleasant 1.00
Wheeling 1.05

Wisconsin **0.98**
Eau Claire 0.99
Green Bay 0.95
Madison 0.99
Milwaukee 1.03
Reedsville 0.97
Superior 1.01
Wausau 0.96

Wyoming **0.92**
Casper 0.92
Cheyenne 0.90
Cody 0.93

Building Cost Historical Index

Use this table to find the approximate current dollar building cost when the actual cost is known for any year since 1928. Multiply the figure listed below for the building type and year of construction by the known cost. The result is the estimated 1995 construction cost.

Year	Masonry Buildings	Concrete Buildings	Steel Buildings	Wood-Frame Buildings	Agricultural Buildings	Year of Construction
1928	17.71	18.55	19.41	16.56	16.74	1928
1929	17.53	18.33	19.64	16.62	16.22	1929
1930	17.94	19.34	18.94	17.19	16.94	1930
1931	18.94	21.00	20.28	18.45	18.80	1931
1932	21.47	23.10	22.73	19.88	20.52	1932
1933	20.63	22.57	20.61	19.28	19.72	1933
1934	19.20	20.49	19.86	18.44	19.01	1934
1935	19.38	20.00	19.63	17.99	18.80	1935
1936	19.02	19.80	19.63	17.38	17.68	1936
1937	17.00	18.20	17.03	15.75	16.09	1937
1938	16.48	18.14	16.92	15.75	15.91	1938
1939	16.32	18.01	16.71	15.29	15.89	1939
1940	16.13	17.81	16.29	15.12	15.44	1940
1941	15.09	16.50	15.59	14.84	14.30	1941
1942	13.76	15.36	14.65	12.72	13.48	1942
1943	13.56	14.71	14.01	12.12	12.08	1943
1944	12.84	14.25	13.48	11.10	11.53	1944
1945	12.15	13.72	13.06	10.52	10.85	1945
1946	10.42	12.04	11.61	9.40	9.37	1946
1947	8.88	9.94	10.15	7.56	7.60	1947
1948	7.79	8.51	9.05	6.93	6.90	1948
1949	7.85	8.42	9.01	7.01	7.17	1949
1950	7.45	8.05	8.84	6.70	6.67	1950
1951	6.97	7.58	8.02	6.26	6.19	1951
1952	6.73	7.41	7.86	6.15	6.13	1952
1953	6.64	7.16	7.50	6.01	5.99	1953
1954	6.51	6.91	7.50	6.01	5.99	1954
1955	6.24	6.59	7.10	5.68	5.73	1955
1956	5.93	6.30	6.53	5.44	5.48	1956
1957	5.75	6.06	6.28	5.40	5.37	1957
1958	5.60	5.82	5.98	5.39	5.46	1958
1959	5.43	5.64	5.84	5.18	5.12	1959
1960	5.30	5.55	5.72	5.08	5.02	1960
1961	5.17	5.52	5.64	5.00	5.00	1961
1962	5.06	5.36	5.53	4.94	4.92	1962
1963	4.98	5.23	5.44	4.83	4.47	1963
1964	4.85	5.14	5.37	4.67	4.71	1964
1965	4.70	5.03	5.17	4.58	4.56	1965
1966	4.48	4.88	4.99	4.37	4.44	1966
1967	4.38	4.66	4.67	4.17	4.26	1967
1968	4.19	4.40	4.45	3.92	4.07	1968
1969	3.97	4.20	4.31	3.78	3.84	1969
1970	3.81	4.01	4.08	3.60	3.66	1970
1971	3.57	3.69	3.78	3.11	3.42	1971
1972	3.31	3.40	3.54	3.12	3.16	1972
1973	3.03	3.23	3.13	2.87	2.98	1973
1974	2.70	2.95	2.94	2.69	2.77	1974
1975	2.46	2.61	2.65	2.53	2.45	1975
1976	2.30	2.49	2.52	2.43	2.33	1976
1977	2.14	2.33	2.39	2.25	2.18	1977
1978	2.00	2.18	2.20	2.07	1.99	1978
1979	1.83	1.94	1.96	1.89	1.90	1979
1980	1.65	1.77	1.76	1.71	1.71	1980
1981	1.56	1.66	1.61	1.63	1.59	1981
1982	1.50	1.59	1.57	1.58	1.52	1982
1983	1.45	1.53	1.53	1.49	1.46	1983
1984	1.35	1.45	1.46	1.40	1.40	1984
1985	1.32	1.38	1.43	1.34	1.37	1985
1986	1.29	1.36	1.40	1.31	1.35	1986
1987	1.28	1.34	1.38	1.29	1.33	1987
1988	1.25	1.28	1.34	1.27	1.30	1988
1989	1.21	1.26	1.28	1.24	1.26	1989
1990	1.14	1.23	1.23	1.18	1.22	1990
1991	1.24	1.20	1.17	1.12	1.16	1991
1992	1.10	1.18	1.14	1.11	1.15	1992
1993	1.07	1.16	1.11	1.09	1.13	1993
1994	1.04	1.08	1.06	1.05	1.06	1994
1995 (estimated)	1.00	1.00	1.00	1.00	1.00	1995

Residential Structures Section

Section Contents

The figures in this section include all costs associated with normal construction:

Foundations as required for normal soil conditions. Excavation for foundations, piers, and other foundation components given a fairly level construction site. Floor, wall, and roof structures. Interior floor, wall, and ceiling finishes. Exterior wall finish and roof cover. Interior partitions as described in the quality class. Finish carpentry, doors, windows, trim, etc. Electric wiring and fixtures. Rough and finish plumbing as described in applicable building specifications. Built-in appliances as described in applicable building specifications. All labor and materials including supervision. All design and engineering fees, if necessary. Permits and fees. Utility hook-ups. Contractors' contingency, overhead and profit.

The square foot costs do not include heating and cooling equipment or the items listed in the section "Additional Costs for Residential Structures" which appear on pages 24 to 28. The costs of the following should be figured separately and added to the basic structure cost: porches, basements, balconies, exterior stairways, built-in equipment beyond that listed in the quality classifications, fireplaces, garages and carports.

Single Family Residences

Single family residential structures vary widely in quality and the quality of construction is a significant factor in the finished cost. For estimating purposes the structure should be placed in one or more classes from 1 to 4. Class 1 construction is top quality work and Class 4 is the minimum required under most building codes. Many structures do not fall into a single class and have features of two quality classes. The tables have 3 "half classes" which apply to structures which have some features of two quality classes. Some structures may fall into two distinct classes and the square foot area of each part should be computed separately.

The shape of the outside perimeter is an important consideration in estimating the total construction cost. Generally, the more complex the shape, the more expensive the structure per square foot of floor area. The shape classification of multiple story or split level homes is based on the outline formed by the outer-most exterior walls, including the garage area, regardless of the varying level. Most structures have 4, 6, 8 or 10 corners, as illustrated in the example. Small insets not requiring a change in the roof shape can be ignored when determining the shape.

These figures can be applied to nearly all single family dwellings built using conventional methods and common materials. There is no simplified method of evaluating the cost of the relatively small number of highly decorative, starkly original or exceptionally well-appointed residences. It is safe to assume, however, that the cost of these highly customized homes will exceed the figures that appear on the following pages.

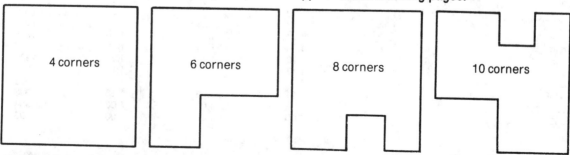

4 corners   6 corners   8 corners   10 corners

Single Family Residences

Quality Classification

	Class 1 Best Quality	Class 2 Good Quality	Class 3 Average Quality	Class 4 Low Quality
Exterior Finish	Good quality wood siding, stucco or masonry with good trim.	Good quality hardboard or plywood siding, some masonry veneer stucco.	Hardwood or shingle or minimum quality stucco.	Composition siding or minimum quality stucco.
Windows	Good quality wood casement or aluminum.	Average quality alunimum or wood.	Minimum quality wood or aluminum	Aluminum sliding.
Roofing	Heavy shake, slate or tile with a 24" to 36" overhang.	Shingle or built-up with rock cover. Less than 24" overhang.	Minimum shingle or built-up with rock cover or good quality composition roof.	Minimum quality built-up with gravel surface or composition small open cornice.
Flooring	Terrazzo or tile entry. Carpet or good quality hardwood in most rooms. Sheet vinyl in kitchen, family and breakfast rooms.	Average quality tile or hardwood entry. Average quality hard Average quality hardwood or carpet in most rooms. Resilient tile in kitchen, family and breakfast rooms.	3/8" hardwood or low cost carpet in most ooms, vinyl or resilient tile in remaining rooms.	Asphalt or composition tile in most rooms, some low cost carpet.
Interior Finish	Gypsum board with heavy texture or plaster. Some wallpaper or vinyl covered walls or wood paneling.	Taped and textured gypsum wallboard, some wallpaper and plywood paneling in family room.	Taped and textured gypsum wallboard, some wallpaper.	Minimum quality wallboard.
Bathrooms	2-1/2 bathrooms, tile floor, ceramic tile shower over tub, glass shower door, 4' to 6' tile or cultured marble pullman in 2 baths.	2 bathrooms, sheet vinyl floor, some average quality tile, 4' to 6' tile or laminated plastic pullman in each bath.	2 bathrooms, usually back to back, resilient tile floor, plastic hardboard or fiberglass around shower and over tub. 4' laminated plastic pullman in each bath.	2 bathrooms back to back, composition tile, plastic shower door, fiberglass enclosures.
Kitchen	20 linear feet of hardwood veneer base cabinets, 20 linear feet of ceramic tile or good quality laminated plastic counter. Built in oven, range, dishwasher, garbage disposer, range hood with fan, all good quality.	16 linear feet of average quality hardwood veneer cabinets and ceramic tile or good quality laminated plastic counter. Built-in oven, range, dishwasher, garbage disposer, range hood and fan, all average quality.	12 linear feet of low cost hardwood veneer cabinets and laminated plastic counter. Low cost built-in oven, appliances, garbage disposer and range hood.	8 linear feet of low cost cabinets with laminated plastic. Few built-in appliances.
Plumbing	10 to 12 good quality fixtures, copper tubing supply and waste.	8 good quality fixtures.	7 average quality fixtures.	7 low cost fixtures, plastic supply

Single Family Residences

4 Corners

Estimating Procedure

1. Establish the structure quality class by applying the information on page 11.
2. Multiply the structure floor area (excluding the garage) by the appropriate square foot cost below.
3. Multiply the total from step 2 by the correct location factor listed on page 7.
4. Add, when appropriate, the cost of a porch, garage, heating and cooling equipment, basement, fireplace, carport, appliances and plumbing fixtures beyond that listed in the quality classification. See the cost of these items on pages 24 to 26.

Single Family Residence, Class 3

Single Family Residence, Class 4

Square Foot Area

Quality Class	700	800	900	1,000	1,100	1,200	1,300	1,400	1,500	1,600	1,700
1, Best	82.09	78.65	75.83	73.44	71.40	69.61	68.03	66.63	65.40	64.25	63.21
1 & 2	75.27	72.12	69.52	67.34	65.45	63.82	62.38	61.10	59.96	58.90	57.96
2, Good	69.92	66.99	64.58	62.56	60.80	59.29	57.95	56.76	55.68	54.73	53.84
2 & 3	64.45	61.76	59.54	57.66	56.06	54.64	53.41	52.32	51.33	50.45	49.64
3, Average	59.03	56.57	54.54	52.82	51.35	50.04	48.92	47.93	47.01	46.21	45.47
3 & 4	53.61	51.36	49.52	47.96	46.61	45.44	44.43	43.50	42.69	41.95	41.27
4, Low	48.59	46.57	44.89	43.48	42.25	41.20	40.29	39.44	38.71	38.02	37.42

Square Foot Area

Quality Class	1,800	2,000	2,200	2,400	2,600	2,800	3,000	3,200	3,400	3,600	4,000
1, Best	62.08	61.76	61.31	60.74	60.15	59.52	58.89	58.25	57.63	57.02	55.84
1 & 2	56.05	55.78	55.34	54.83	54.31	53.75	53.16	52.59	52.02	51.47	50.41
2, Good	50.50	50.25	49.88	49.42	48.92	48.42	47.90	47.39	46.88	46.37	45.43
2 & 3	46.39	46.16	45.80	45.38	44.95	44.48	44.00	43.54	43.06	42.61	41.73
3, Average	42.86	42.63	42.32	41.94	41.51	41.09	40.66	40.21	39.79	39.35	38.55
3 & 4	39.31	39.10	38.81	38.46	38.06	37.69	37.28	36.87	36.49	36.09	35.35
4, Low	36.10	35.89	35.64	35.31	34.97	34.60	34.23	33.86	33.51	33.14	32.46

Note: Tract work and highly repetitive jobs may reduce the cost 8 to 12%. Add 4% to the square foot cost of floors above the second floor level. Work outside metropolitan areas may cost 2 to 6% less. When the exterior walls are masonry, add 8 to 9% for class 2 and 1 structures and 6 to 7% for class 4 and 3 structures. Deduct 2% for area built on a concrete slab.

The building area includes all full story (7'6" to 8' high) areas within and including the exterior walls of all floor areas of the building, including small inset areas such as entrances outside the exterior wall but under the main roof. For areas with a ceiling height of less than 80", see the section on half-story areas on page 27.

Single Family Residences

6 Corners

Estimating Procedure

1. Establish the structure quality class by applying the information on page 11.
2. Multiply the structure floor area (excluding the garage) by the appropriate square foot cost below.
3. Multiply the total from step 2 by the correct location factor listed on page 7.
4. Add, when appropriate, the cost of a porch, garage, heating and cooling equipment, basement, fireplace, carport, appliances and plumbing fixtures beyond that listed in the quality classification. See the cost of these items on pages 24 to 26.

Single Family Residence, Class 4

Single Family Residence, Class 2

Square Foot Area

Quality Class	700	800	900	1,000	1,100	1,200	1,300	1,400	1,500	1,600	1,700
1, Best	84.21	80.20	77.00	74.44	72.33	70.55	69.03	67.71	66.55	65.55	64.65
1 & 2	77.86	74.13	71.20	68.83	66.87	65.23	63.82	62.60	61.54	60.61	59.76
2, Good	72.01	68.58	65.85	63.66	61.86	60.32	59.02	57.91	56.92	56.06	55.28
2 & 3	66.30	63.13	60.64	58.61	56.94	55.54	54.34	53.31	52.40	51.61	50.90
3, Average	60.85	57.93	55.63	53.78	52.25	50.97	49.87	48.90	48.08	47.36	46.72
3 & 4	55.18	52.53	50.45	48.76	47.37	46.21	45.22	44.35	43.61	42.94	42.35
4, Low	50.01	47.63	45.74	44.21	42.96	41.90	41.00	40.21	39.52	38.93	38.39

Square Foot Area

Quality Class	1,800	2,000	2,200	2,400	2,600	2,800	3,000	3,200	3,400	3,600	4,000
1, Best	64.14	63.85	63.38	62.80	62.16	61.50	60.84	60.16	59.49	58.83	57.57
1 & 2	57.07	56.80	56.40	55.88	55.31	54.71	54.11	53.51	52.93	52.35	51.21
2, Good	51.51	51.29	50.90	50.44	49.94	49.41	48.86	48.32	47.78	47.26	46.24
2 & 3	47.73	47.52	47.16	46.74	46.27	45.77	45.26	44.78	44.28	43.78	42.85
3, Average	43.96	43.76	43.45	43.06	42.62	42.18	41.71	41.23	40.78	40.33	39.47
3 & 4	40.32	40.12	39.84	39.47	39.08	38.66	38.23	37.81	37.39	36.97	36.18
4, Low	37.39	37.23	36.94	36.61	36.25	35.85	35.48	35.08	34.68	34.29	33.56

Note: Tract work and highly repetitive jobs may reduce the cost 8 to 12%. Add 4% to the square foot cost of floors above the second floor level. Work outside metropolitan areas may cost 2 to 6% less. When the exterior walls are masonry, add 8 to 9% for class 2 and 1 structures and 6 to 7% for class 4 and 3 structures. Deduct 2% for area built on a concrete slab.

The building area includes all full story (7'6" to 8' high) areas within and including the exterior walls of all floor areas of the building, including small inset areas such as entrances outside the exterior wall but under the main roof. For areas with a ceiling height of less than 80", see the section on half-story areas on page 27.

Single Family Residences

8 Corners

Estimating Procedure

1. Establish the structure quality class by applying the information on page 11.
2. Multiply the structure floor area (excluding the garage) by the appropriate square foot cost below.
3. Multiply the total from step 2 by the correct location factor listed on page 7.
4. Add, when appropriate, the cost of a porch, garage, heating and cooling equipment, basement, fireplace, carport, appliances and plumbing fixtures beyond that listed in the quality classification. See the cost of these items on pages 24 to 26.

Single Family Residence, Class 1

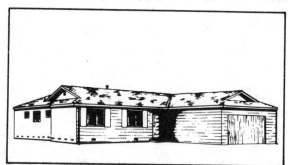

Single Family Residence, Class 3

Square Foot Area

Quality Class	700	800	900	1,000	1,100	1,200	1,300	1,400	1,500	1,600	1,700
1, Best	84.97	81.09	78.02	75.57	73.54	71.84	70.39	69.13	68.03	67.07	66.23
1 & 2	78.63	75.04	72.20	69.94	68.05	66.48	65.14	63.97	62.96	62.08	61.28
2, Good	72.79	69.47	66.86	64.74	63.01	61.55	60.30	59.24	58.29	57.47	56.75
2 & 3	67.10	64.03	61.63	59.69	58.09	56.73	55.59	54.60	53.73	52.97	52.30
3, Average	61.53	58.71	56.51	54.73	53.27	52.04	50.98	50.06	49.28	48.58	47.96
3 & 4	55.81	53.25	51.26	49.65	48.30	47.18	46.24	45.41	44.69	44.06	43.49
4, Low	50.57	44.89	46.45	44.99	43.77	42.76	41.91	41.15	40.50	39.93	39.42

Square Foot Area

Quality Class	1,800	2,000	2,200	2,400	2,600	2,800	3,000	3,200	3,400	3,600	4,000
1, Best	65.67	64.87	64.07	63.30	62.57	61.88	61.22	60.60	60.01	59.45	58.42
1 & 2	59.20	58.47	57.73	57.06	56.40	55.78	55.18	54.62	54.08	53.58	52.66
2, Good	53.54	52.85	52.21	51.58	50.99	50.42	49.90	49.39	48.90	48.44	47.61
2 & 3	49.52	48.91	48.30	47.72	47.19	46.66	46.17	45.70	45.25	44.84	44.06
3, Average	45.65	45.09	44.53	44.00	43.49	43.01	42.55	42.13	41.72	41.33	40.60
3 & 4	41.85	41.32	40.83	40.34	39.88	39.44	39.01	38.61	38.24	37.89	37.23
4, Low	38.48	37.99	37.52	37.09	36.66	36.25	35.86	35.51	35.15	34.82	34.22

Note: Tract work and highly repetitive jobs may reduce the cost 8 to 12%. Add 4% to the square foot cost of floors above the second floor level. Work outside metropolitan areas may cost 2 to 6% less. When the exterior walls are masonry, add 8 to 9% for class 2 and 1 structures and 6 to 7% for class 4 and 3 structures.

The building area includes all full story (7'6" to 8' high) areas within and including the exterior walls of all floor areas of the building, including small inset areas such as entrances outside the exterior wall but under the main roof. For areas with a ceiling height of less than 80", see the section on half-story areas on page 27.

Single Family Residences

10 Corners

Estimating Procedure

1. Establish the structure quality class by applying the information on page 11.
2. Multiply the structure floor area (excluding the garage) by the appropriate square foot cost below.
3. Multiply the total from step 2 by the correct location factor listed on page 7.
4. Add, when appropriate, the cost of a porch, garage, heating and cooling equipment, basement, fireplace, carport, appliances and plumbing fixtures beyond that listed in the quality classification. See the cost of these items on pages 24 to 26.

Single Family Residence, Class 1

Single Family Residence, Class 2

Square Foot Area

Quality Class	700	800	900	1,000	1,100	1,200	1,300	1,400	1,500	1,600	1,700
Best	86.31	82.41	79.29	76.77	74.64	72.83	71.29	69.93	68.73	67.67	66.71
1 & 2	80.49	76.85	73.96	71.57	69.58	67.92	66.47	65.21	64.10	63.10	62.21
2, Good	74.53	71.16	68.46	66.27	64.45	62.89	61.53	60.37	59.34	58.43	57.61
2 & 3	68.63	65.52	63.06	61.03	59.34	57.90	56.67	55.59	54.63	53.81	53.04
3, Average	62.84	60.00	57.74	55.88	54.34	53.03	51.88	50.90	50.03	49.27	48.57
3 & 4	57.08	54.48	52.44	50.76	49.34	48.15	47.13	46.24	45.44	44.74	44.12
4, Low	51.72	49.36	47.51	45.99	44.72	43.63	42.70	41.90	41.17	40.55	39.97

Square Foot Area

Quality Class	1,800	2,000	2,200	2,400	2,600	2,800	3,000	3,200	3,400	3,600	4,000
1, Best	66.30	65.93	65.44	64.85	64.20	63.54	62.87	62.20	61.55	60.91	59.67
1 & 2	59.78	59.46	59.02	58.49	57.91	57.31	56.71	56.10	55.51	54.93	53.83
2, Good	54.05	53.75	53.33	52.85	52.34	51.80	51.26	50.71	50.18	49.66	48.65
2 & 3	50.09	49.83	49.44	48.99	48.52	48.02	47.51	47.00	46.50	46.03	45.09
3, Average	46.16	45.91	45.56	45.14	44.70	44.25	43.78	43.31	42.86	42.41	41.55
3 & 4	42.31	42.07	41.76	41.38	40.96	40.55	40.11	39.69	39.28	38.87	38.09
4, Low	38.93	38.70	38.39	38.03	37.68	37.28	36.89	36.51	36.13	35.74	35.02

Note: Tract work and highly repetitive jobs may reduce the cost 8 to 12%. Add 4% to the square foot cost of floors above the second floor level. Work outside metropolitan areas may cost 2 to 6% less. When the exterior walls are masonry, add 8 to 9% for class 2 and 1 structures and 6 to 7% for class 2 and 1 structures and 6 to 7% for class 4 and 3 structures.

The building area includes all full story (7'6" to 8' high) areas within and including the exterior walls of all floor areas of the building, including small inset areas such as entrances outside the exterior wall but under the main roof. For areas with a ceiling height of less than 80", see the section on half-story areas on page 27.

Multi-Family Residences
Quality Classification

	Class 1 Best Quality	Class 2 Good Quality	Class 3 Average Quality	Class 4 Low Quality
Framing*	Wood frame.	Wood frame.	Wood frame.	Wood frame.
Exterior Walls	Good wood siding or stucco. Masonry veneer on front.	Good quality stucco, good hardboard, plywood or board, masonry veneer on front.	Stucco, hardboard, low cost plywood or low cost shingle.	Light stucco or hardboard.
Windows	Large top quality wood or aluminum.	Average quality wood or aluminum.	Average quality wood or aluminum.	Few & smaller low cost aluminum or wood.
Roofing	Heavy shake or tile.	Good shingle or shake or built-up with heavy rock dress.	Wood or composition shingle or built-up with rock.	Asphalt shingles or built-up, light weight.
Overhang	36" open or 24" closed.	30", open.	16", open.	12" to 16" open.
Gutters	Good quality at all eaves.	6", at all eaves.	Over entrances.	Minimum or none.
Flooring	Terrazzo or tile entry. Good hardwood or carpet. Good sheet vinyl in kitchen.	Hardwood or average quality carpet. Sheet vinyl in kitchen.	Average quality hardwood or low cost carpet. Vinyl tile in kitchen.	Composition tile.
Interior Walls	Heavy textured wallboard or plaster with putty finish. Good wallpaper or vinyl. Veneer hardwood paneling.	Gypsum board or plaster with good finish. Some wallpaper or paneling.	Gypsum board taped and textured. Some acoustical treatment or wallpaper	Gypsum board taped and textured.
Trim	Good quality trim with some hardwood.	Painted softwood with some hardwood.	Painted softwood.	Painted softwood.
Closets	Ample, good doors.	Average space and doors.	Average space, low cost doors.	Minimum.
Kitchens	10' good cabinets 10' tile drainboard.	8' average quality cabinets, 8' ceramic tile or good plastic drainboard.	6' low cost cabinets and plastic or ceramic drainboard.	5' low cost cabinets and plastic drainboard.
Bathrooms	Sheet vinyl floor, tile over tub with glass shower door. 6' marble top vanity.	Sheet vinyl or inlaid tile, tile over tub with glass shower door. 4' plastic vanity	Tile floor, plastic coated hardboard over tub. Glass or plastic shower door. 3' plastic vanity.	Tile floor, plastic coated hardboard over tub.
Plumbing	Copper tubing, good quality fixtures.	Galvanized pipe, good quality fixtures.	Average fixtures.	Plastic pipe, low cost fixtures.
Special Features	8' sliding glass door, good quality built-in oven, range with hood, dishwasher.	Average quality disposer and range hood. Low cost dishwasher.	One or two low cost built-in appliances.	None.
*For Masonry Walls	Textured block, tile or brick with masonry facing.	Colored or detailed concrete block, brick or tile with facing.	Colored reinforced concrete block, clay tile or common brick.	Reinforced concrete block or clay tile.

Plumbing costs assume 1 bathroom per unit. See page 27 for the costs of additional bathrooms.
When masonry walls are used in lieu of wood frame walls, add 5% to the appropriate S.F. cost.

Multi-Family Residences

2 or 3 Units

Estimating Procedure

1. Establish the structure quality class by applying the information on page 16.
2. Multiply the average unit area by the appropriate square foot cost below. The average unit area is found by dividing the building area on all floors by the number of units in the building. The building area should include office and utility rooms, interior hallways and interior stairways.
3. Multiply the total from step 2 by the correct location factor listed on page 7.
4. Add, when appropriate, the cost of balconies, porches, garages, heating and cooling equipment, basements, fireplaces, carports, appliances and plumbing fixtures beyond that listed in the quality classification. See the cost of these items on pages 24 to 28.
5. Costs assume one bathroom per unit. Add the cost of additional bathrooms from page 27.

Multi-Family, Class 1

Multi-Family, Class 3 & 4

Average Unit Area in Square Feet

Quality Class	400	450	500	550	600	650	700	750	800	900	1,000
1, Best	78.25	75.46	73.30	71.56	70.17	69.03	68.05	67.25	66.55	65.45	64.59
1 & 2	72.85	70.27	68.27	66.66	65.35	64.28	63.39	62.63	61.99	60.93	60.15
2, Good	67.85	65.44	63.57	62.06	60.85	59.85	59.01	58.30	57.70	56.75	56.01
2 & 3	63.19	60.96	59.21	57.82	56.69	55.75	54.99	54.32	53.76	52.87	52.18
3, Average	58.86	56.76	55.14	53.83	52.79	51.91	51.19	50.58	50.05	49.22	48.59
3 & 4	54.44	52.52	51.01	49.81	48.83	48.03	47.36	46.79	46.33	45.53	44.94
4, Low	50.38	48.58	47.19	46.09	45.20	44.44	43.82	43.30	42.86	42.14	41.58

Average Unit Area in Square Feet

Quality Class	1,100	1,200	1,300	1,400	1,500	1,600	1,700	1,800	1,900	2,000	2,200
1, Best	63.70	63.00	62.42	61.94	61.53	61.18	60.88	60.62	60.40	60.19	59.86
1 & 2	59.35	58.70	58.15	57.70	57.33	56.99	56.72	56.48	56.25	56.08	55.79
2, Good	55.26	54.64	54.15	53.72	53.36	53.07	52.81	52.58	52.39	52.22	51.93
2 & 3	51.46	50.89	50.42	50.02	49.71	49.41	49.18	48.96	48.78	48.62	48.34
3, Average	47.93	47.40	46.96	46.60	46.28	46.03	45.79	45.61	45.43	45.29	45.04
3 & 4	44.33	43.86	43.45	43.11	42.82	42.58	42.37	42.19	42.03	41.89	41.67
4, Low	41.04	40.58	40.21	39.90	39.63	39.41	39.23	39.05	38.91	38.78	38.55

Note: Work outside metropolitan areas may cost 2 to 6% less. Add 2% to the costs for second floor areas and 4% for third floor areas. Add 5% when the exterior walls are masonry.

Multi-Family Residences

4 to 9 Units

Estimating Procedure

1. Establish the structure quality class by applying the information on page 16.
2. Multiply the average unit area by the appropriate square foot cost below. The average unit area is found by dividing the building area on all floors by the number of units in the building. The building area should include office and utility rooms, interior hallways and interior stairways.
3. Multiply the total from step 2 by the correct location factor listed on page 7.
4. Add, when appropriate, the cost of balconies, porches, garages, heating and cooling equipment, basements, fireplaces, carports, appliances and plumbing fixtures beyond that listed in the quality classification. See the cost of these items on pages 24 to 28.
5. Costs assume one bathroom per unit. Add the cost of additional bathrooms from page 27.

Multi-Family, Class 2 & 3

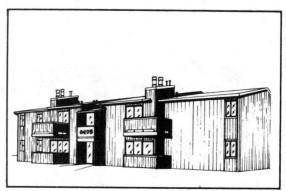

Multi-Family, Class 3

Average Unit Area in Square Feet

Quality Class	400	450	500	550	600	650	700	750	800	900	1,000
1, Best	73.71	71.04	68.98	67.34	66.02	64.94	64.05	63.30	62.66	61.64	60.87
1 & 2	68.79	66.30	64.38	62.86	61.62	60.62	59.77	59.08	58.49	57.53	56.81
2, Good	64.05	61.76	59.96	58.54	57.40	56.45	55.67	55.01	54.44	53.57	52.92
2 & 3	59.65	57.51	55.83	54.52	53.45	52.58	51.84	51.23	50.70	49.89	49.27
3, Average	55.82	53.79	52.23	51.00	50.01	49.17	48.49	47.92	47.45	46.66	46.09
3 & 4	51.19	49.34	47.91	46.77	45.86	45.11	44.49	43.96	43.51	42.81	42.27
4, Low	47.58	45.86	44.53	43.47	42.63	41.93	41.36	40.86	40.45	39.80	39.31

Average Unit Area in Square Feet

Quality Class	1,100	1,200	1,300	1,400	1,500	1,600	1,700	1,800	1,900	2,000	2,200
1, Best	60.12	59.45	58.91	58.44	58.05	57.72	57.44	57.20	56.98	56.78	56.49
1 & 2	55.99	55.37	54.86	54.44	54.07	53.78	53.49	53.27	53.08	52.90	52.60
2, Good	52.08	51.50	51.01	50.64	50.28	50.00	49.75	49.55	49.37	49.20	48.92
2 & 3	48.56	48.02	47.56	47.20	46.88	46.62	46.38	46.19	46.02	45.86	45.61
3, Average	45.21	44.72	44.31	43.96	43.67	43.43	43.21	43.03	42.87	42.73	42.49
3 & 4	41.83	41.36	40.98	40.66	40.40	40.16	39.96	39.81	39.64	39.51	39.31
4, Low	38.71	38.28	37.92	37.63	37.37	37.16	36.97	36.82	36.68	36.57	36.36

Note: Work outside metropolitan areas may cost 2 to 6% less. Add 2% to the costs for second floor areas and 4% for third floor areas. Add 5% when the exterior walls are masonry.

Multi-Family Residences

10 or More Units

Estimating Procedure

1. Establish the structure quality class by applying the information on page 16.
2. Multiply the average unit area by the appropriate square foot cost below. The average unit area is found by dividing the building area on all floors by the number of units in the building. The building area should include office and utility rooms, interior hallways and interior stairways.
3. Multiply the total from step 2 by the correct location factor listed on page 7.
4. Add, when appropriate, the cost of balconies, porches, garages, heating and cooling equipment, basements, fireplaces, carports, appliances and plumbing fixtures beyond that listed in the quality classification. See the cost of these items on pages 24 to 28.
5. Costs assume one bathroom per unit. Add the cost of additional bathrooms from page 27.

Multi-Family, Class 3

Multi-Family, Class 3 & 4

Average Unit in Square Feet

Quality Class	400	450	500	550	600	650	700	750	800	900	1,000
1, Best	69.62	67.15	65.21	63.69	62.45	61.43	60.57	59.85	59.23	58.23	57.47
1 & 2	64.83	62.55	60.74	59.32	58.16	57.22	56.43	55.75	55.17	54.23	53.52
2, Good	60.37	58.22	56.57	55.23	54.16	53.27	52.53	51.91	51.37	50.50	49.85
2 & 3	56.23	54.24	52.68	51.45	50.44	49.62	48.91	48.32	47.84	47.03	46.42
3, Average	52.36	50.52	49.08	47.93	46.98	46.22	45.58	45.04	44.56	43.81	43.24
3 & 4	48.47	46.75	45.41	44.35	43.48	42.77	42.18	41.67	41.24	40.55	40.03
4, Low	44.84	43.26	42.02	41.03	40.23	39.56	39.02	38.54	38.16	37.51	37.03

Average Unit in Square Feet

Quality Class	1,100	1,200	1,300	1,400	1,500	1,600	1,700	1,800	1,900	2,000	2,200
1, Best	56.69	56.07	55.58	55.14	54.77	54.46	54.19	53.94	53.74	53.56	53.24
1 & 2	52.83	52.26	51.79	51.38	51.05	50.72	50.53	50.27	50.09	49.91	49.63
2, Good	49.19	48.65	48.20	47.83	47.52	47.25	47.00	46.80	46.62	46.47	46.19
2 & 3	45.80	45.30	44.89	44.54	44.26	44.00	43.78	43.60	43.41	43.27	43.02
3, Average	42.64	42.19	41.80	41.47	41.20	40.97	40.76	40.58	40.44	40.28	40.07
3 & 4	39.47	39.04	38.68	38.38	38.12	37.91	37.73	37.55	37.40	37.28	37.08
4, Low	36.51	36.11	35.78	35.51	35.26	35.07	34.90	34.74	34.60	34.49	34.28

Note: Work outside metropolitan areas may cost 2 to 6% less. Add 2% to the costs for second floor areas and 4% for third floor areas. Add 5% when the exterior walls are masonry.

Motels

Quality Classification

	Class 1 Best Quality	Class 2 Good Quality	Class 3 Average Quality	Class 4 Low Quality
Framing*	Wood frame.	Wood frame.	Wood frame.	Wood frame.
Windows	Large, good quality.	Average number and quality.	Average number and quality.	Small, few, low cost.
Roofing	Heavy, shake, tile or slate.	Medium shake or good built-up with large rock, inexpensive tile.	Wood or good composition shingle, light shake, or good built-up with rock.	Inexpensive shingles or built-up with rock.
Overhang	36" open or 24" closed.	30" open or small closed.	16" open.	12" to 16" open.
Gutters	Good quality at all eaves.	6" at all eaves.	4" at all eaves.	Over entrances.
Exterior Walls	Good wood or stucco, masonry veneer on front.	Good wood siding or stucco with some veneer.	Hardboard, wood shingles, plywood or stucco.	Low cost stucco, hardboard or plywood
Flooring	Good carpet, good sheet vinyl.	Good carpet, sheet vinyl or inlaid resilient.	Average carpet, average resilient tile in bath.	Minimum tile or low cost carpet.
Interior Walls	Gypsum board with heavy texture or plaster with putty coat. Some good sheet wall cover or paneling.	Gypsum board, taped, textured and painted or plaster. Some wall paper.	Gypsum board taped and textured or colored interior stucco.	Minimum gypsum board.
Trim	Hardboard with some softwood.	Softwood with some hardwood.	Painted softwood.	Painted softwood.
Closets	Ample space with good wood doors.	Adequate, with good wood doors.	Moderate size, low cost plastic, wood or metal doors.	Small, low cost or no doors.
Baths	Vinyl or foil wall cover, ceramic tile over tub with glass shower door, ample mirrors.	Ceramic tile over tub with glass shower door.	Plastic coated hardboard with low cost glasss shower door.	Plastic coated hardboard with one small mirror.
Plumbing	Copper tube, good quality fixtures.	Galvanized pipe, good fixtures.	Average cost fixtures.	Plastic pipe, low cost fixtures.
Special Features	8' sliding glass door, 8' to 10' tile pullman in bath.	8' sliding glass door, good tile or plastic top pullman in bath.	Small tile or plastic pullman in bath.	None.
***For Masonry Walls**	8" textured face reinforced masonry.	8" colored or detailed reinforced masonry.	8" colored block or common brick, reinforced.	8" painted concrete block.

Note: When masonry walls are used in lieu of wood frame walls add 8% to the appropriate cost

Add the Following Amounts per Kitchen Unit			
Kitchens	Good sink, 8' to 10' of good cabinets and drainboard - $2,700.	Average sink and 6' to 8' average cabinet and drainboard - $2,500.	Low cost sink, and 5' of cabinets and drainboard - $1,800.	Minimum sink, cabinets and drainboard - $1,500.

Note: Add the cost of built-in kitchen fixtures from the table of costs for built-in appliances on page 26.

Motels

9 Units or Less

Estimating Procedure

1. Establish the structure quality class by applying the information on page 20.
2. Multiply the average unit area by the appropriate cost below. The average unit area is found by dividing the total building area on all floors (including office and manager's area, utility rooms, interior hallways and stairway area) by the number of units in the building.
3. Multiply the total from step 2 by the correct location factor listed on page 7.
4. Add, when appropriate, the cost of heating and cooling equipment, porches, balconies, exterior stairs, garages, kitchens, built-in kitchen appliances and fireplaces. See pages 20 and 24 to 28.

Motel, Class 3 & 4

Average Unit in Square Feet

Quality Class	200	225	250	275	300	330	375	425	500	600	720
1, Best	86.82	83.73	81.26	79.22	77.52	75.83	73.78	72.02	70.01	68.10	66.52
1 & 2	79.36	76.53	74.27	72.42	70.87	69.30	67.43	65.82	63.98	62.26	60.81
2, Good	72.60	70.01	67.94	66.24	64.82	63.40	61.69	60.22	58.54	56.94	55.63
2 & 3	66.39	64.03	62.13	60.58	59.29	57.98	56.41	55.06	53.53	52.09	50.88
3, Average	60.71	58.54	56.83	55.40	54.21	53.02	51.59	50.36	48.95	47.63	46.52
3 & 4	55.47	53.49	51.91	50.61	49.53	48.44	47.12	46.01	44.72	43.52	42.50
4, Low	50.68	48.88	47.44	46.25	45.25	44.28	43.07	42.04	40.86	39.76	38.83

Note: Add 2% for work above the first floor. Work outside metropolitan areas may cost 2 to 6% less. Add 8% when the exterior walls are masonry. Deduct 2% for area built on a concrete slab.

10 to 24 Units

Estimating Procedure

1. Establish the structure quality class by applying the information on page 20.
2. Multiply the average unit area by the appropriate cost below. The average unit area is found by dividing the total building area on all floors (including office and manager's area, utility rooms, interior hallways and stairway area) by the number of units in the building.
3. Multiply the total from step 2 by the correct location factor listed on page 7.
4. Add, when appropriate, the cost of heating and cooling equipment, porches, balconies, exterior stairs, garages, kitchens, built-in kitchen appliances and fireplaces. See pages 20 and 24 to 28.

Motel, Class 3

Average Unit Area in Square Feet

Quality Class	200	225	250	275	300	330	375	425	500	600	720
1, Best	84.14	81.14	78.73	76.78	75.12	73.47	71.49	69.77	67.84	65.99	64.48
1 & 2	76.91	74.18	71.96	70.17	68.67	67.16	65.35	63.78	62.01	60.33	58.92
2, Good	70.39	67.88	65.86	64.21	62.85	61.46	59.80	58.37	56.75	55.21	53.95
2 & 3	64.33	62.03	60.18	58.68	57.43	56.15	54.66	53.34	51.84	50.45	49.28
3, Average	58.83	56.73	55.06	53.67	52.51	51.37	49.98	48.78	47.43	46.15	45.08
3 & 4	53.74	51.82	50.29	49.03	47.98	46.93	45.65	44.56	43.31	42.16	41.18
4, Low	49.12	47.35	45.96	44.82	43.85	42.90	41.73	40.73	39.61	38.53	37.64

Note: Add 2% for work above the first floor. Work outside metropolitan areas may cost 2 to 6% less. Add 8% when the exterior walls are masonry. Deduct 2% for area built on a concrete slab.

Motels
Over 24 Units

Estimating Procedure

1. Establish the structure quality class by applying the information on page 20.
2. Multiply the average unit area by the appropriate cost below. The average unit area is found by dividing the total building area on all floors (including office and manager's area, utility rooms, interior hallways and stairway area) by the number of units in the building.
3. Multiply the total from step 2 by the correct location factor listed on page 7.
4. Add, when appropriate, the cost of heating and cooling equipment, porches, balconies, exterior stairs, garages, kitchens, built-in kitchen appliances and fireplaces. See pages 20 and 24 to 28.

Motel, Class 2 & 3

Average Unit Area in Square Feet

Quality Class	200	225	250	275	300	330	375	425	500	600	720
1, Best	81.37	78.46	76.13	74.23	72.63	71.04	69.12	67.48	65.61	63.84	62.37
1 & 2	74.39	71.73	69.60	67.86	66.41	64.95	63.20	61.69	59.98	58.37	57.02
2, Good	68.08	65.66	63.71	62.12	60.79	59.45	57.86	56.46	54.90	53.41	52.19
2 & 3	62.22	60.00	58.22	56.76	55.54	54.32	52.88	51.58	50.17	48.82	47.69
3, Average	56.88	54.85	53.24	51.89	50.78	49.67	48.34	47.19	45.88	44.63	43.61
3 & 4	51.96	50.10	48.60	47.39	46.38	45.37	44.13	43.08	41.88	40.78	39.82
4, Low	47.50	45.79	44.43	43.32	42.39	41.48	40.34	39.38	38.29	37.26	36.40

Note: Add 2% for work above the first floor. Work outside metropolitan areas may cost 2 to 6% less. Add 8% when the exterior walls are masonry. Deduct 2% for area built on a concrete slab.

Additional Costs for Residential Structures

Covered Porches
Estimate covered porches by applying a fraction of the main building square foot cost.

Porch Description	Suggested Fraction
Ground level floor (usually concrete) without banister, with no ceiling and shed-type roof.	1/4 to 1/3
High (house floor level) floor (concrete or wood) with light banister, no ceiling and shed-type roof.	1/3 to 1/2
Same as above with a finished ceiling and roof like the residence (most typical).	1/2
Same as above but partially enclosed with screen or glass.	1/2 to 2/3
Enclosed lean-to (sleeping porch, etc.) with lighter foundation, wall structure, interior finish or roof than that of house to which it is attached.	1/2 to 3/4
Roofed, enclosed, recessed porch, under the same roof as the main building and with the same type and quality foundation (includes shape costs).	3/4
Roofed, enclosed, recessed porch with the same type roof and foundation as the main building (includes shape costs).	4/4
Good arbor or pergola with floor.	1/4 to 1/3

Uncovered Concrete Decks, cost per square foot, 4" thick, no reinforcing

	On Grade	1' High	2' High	3' High	4' High
Less than 100 square feet	$3.70	$5.20	$8.30	$11.30	$14.20
100 to 200 square feet	3.30	4.70	6.30	8.40	11.00
200 to 400 square feet	2.85	3.50	5.40	7.30	9.00
Over 400 square feet	2.70	3.30	5.10	6.75	8.00

Uncovered Wood Decks, cost per square foot

1" board floor on 2" x 6" joists 24" o.c.	$11.00 to $12.50
2" board floor on 4" x 6" girders 48" o.c.	12.50 to 14.40

Porch Roofs, cost per square foot

Type	Cost per Square Foot	Alternate Roof Covers		Cost Difference per S.F.
Unceiled shed roof	$3.55 to $4.10	Corrugated aluminum	Deduct	$.20 to $.40
Ceiled shed roof	6.75 to 7.30	Roll asphalt	Deduct	.25 to .35
Unceiled gable roof	3.90 to 4.90	Fiberglass shingles	Deduct	.30 to .40
Ceiled gable roof	6.10 to 8.50	Wood shakes	Add	.50 to .80
(Assumes 3 ply built-up roof & gravel)		Clay or concrete tile	Add	4.00 to 4.50
		Slate	Add	4.80 to 6.90

Residential Basements, cost per square foot

Size	Unfinished Basements	Finished Basements
Less than 400 square feet	$32.30 to $47.70	$36.40 to $53.70
400 - 1,000 square feet	23.20 to 27.60	29.40 to 37.50
Over 1,000 square feet	14.20 to 20.30	21.60 to 28.30

All basement costs assume normal soil conditions, 7'6" headroom, no plumbing, partitions, or windows. Unfinished basements are based on reinforced concrete floors and walls, open ceilings and minimum lighting. Finished basement costs are based on reinforced concrete floors with resilient tile cover, reinforced concrete walls with plywood panel finish, acoustical tile ceiling and lighting similar to average quality residences. Basement costs will be lower in areas where normal footing depths approach 8 feet. These figures assume a normal footing depth of 2 feet below grade.

Additional Costs for Residential Structures

Balconies, Standard Wood Frame, cost per square foot

Supported by 4" x 4" posts, 1" wood floor, open on underside, open 2" x 4" railing.	$9.60 to $10.70
Supported by 4" x 4" posts, 1" wood floor, sealed on underside, solid stucco or wood siding on railing.	10.80 to 11.90
Supported by steel columns, lightweight concrete floors, sealed on underside, solid stucco or open grillwork railing	13.00 to 14.70

Heating and Cooling Equipment

Prices include wiring and minimum duct work.
Use the higher figures for smaller residences and in more extreme climates where greater heating and cooling density is required. Cost per square foot of heated or cooled area.

Type	Perimeter Outlets	Overhead Outlets
Central Ducted Air Systems, Single Family		
Forced air heating	$3.50 to $3.85	$2.85 to $3.50
Forced air heating and cooling	4.40 to 4.70	3.95 to 4.15
Gravity heat	2.70 to 3.60	----
Central Ducted Air Systems, Multi-Family		
Forced air heating	3.20 to 3.50	3.10 to 3.50
Forced air heating and cooling	4.45 to 4.85	3.85 to 4.15
Motel Units		
Forced air heating	2.70 to 2.85	3.60 to 3.90
Forced air heating and cooling	4.50 to 4.85	4.30 to 4.50
Circulating hot and cold water system	8.80 to 10.35	8.80 to 10.35

Floor and Wall Furnaces, cost ea.

Single floor unit	$ 810 to $ 925
Dual floor unit	1,440 to 1,560
Single wall unit	550 to 665
Dual wall unit	615 to 730
Thermostat control, add	130 to 185

Electric Baseboard Units, cost ea.

500 watts, 3'	$240 to $260
1,000 watts, 4'	330 to 360
1,500 watts, 6'	360 to 380
2,000 watts, 8'	415 to 470
2,500 watts, 10'	520 to 550
3,000 watts, 12'	590 to 655

Outside Stairways, cost per square foot of horizontal area

Standard wood frame, wood steps with open risers, open on underside, open 2" x 4" railing, unpainted.	$9.30 to $10.00
Standard wood frame with solid wood risers, sealed on underside, solid stucco or wood siding on railing.	11.30 to 12.80
Precast concrete steps with open risers, steel frame, pipe rail with ornamental grillwork.	23.60 to 26.90

Refrigerated Room Coolers, cost ea.

1/3 ton	$505 to $ 595
1/2	540 to 675
3/4	610 to 930
1	710 to 820
1-1/2	850 to 1,030
2	1,030 to 1,265
Ton = 12,000 BTU	

Electric Wall Heaters, cost ea.

1,000 watts	$300 to $360
2,000	360 to 400
3,000	370 to 470
3,500	380 to 490
4,000	490 to 585
4,500	580 to 730
Add for circulating fan	65 to 85
Add for thermostat	70 to 80

Additional Costs for Residential Structures

Built-In Equipment

Add these costs only when the item is not included in the specification for the applicable quality class. These costs include plumbing, wiring and installation.

Appliances, cost each

Drop-in type, gas or electric		Microwave, built-in	$295 to	$ 585
Range with single oven below	$715 to $1,020	Range hood and fan	270 to	360
Built-in type, gas or electric		Refrigerators, built-in		
Range with single oven below	900 to 1,175	Under counter, 5 cubic feet	390 to	585
Range with oven above and below	1,155 to 1,430	Single door, 12 cubic feet	670 to	810
Surface cooking units, four		Double door, 12 cubic feet	1,005 to	1,140
elements without grill	360 to 500	Double door, 17 cubic feet	1,660 to	1,970
Oven units, single gas or electric	580 to 780	Refrigerator, range, oven and sink	1,760 to	2,020
Oven units, double, gas or electric	960 to 1,140	Dishwashers	470 to	695
Electronic ovens, with range		Garbage disposers	235 to	350
and microwave	1,665 to 2,500	Trash compactors	350 to	520

Fireplaces, cost each

	1 Story	2 Story
Freestanding wood burning heat circulating prefab metal fireplace with interior flue, base and cap.	$ 1,000	$ 1,330
Zero-clearance insulated prefab metal fireplace, brick face	1,420	1,880
5' base, common brick on interior face	1,870	2,050
6' base, common brick, used brick or natural stone on interior face.	2,950	3,050
8' base, used brick or natural stone on interior face, raised hearth.	4,110	4,540

Residential Garages and Carports

Attached and detached garages for single family dwellings are usually classified in the same quality class as the main structure. The attached garage assumes a 20 foot wall in common with the main structure. The costs are per square foot of floor area and include only light excavation for foundations. Multiply the square foot cost below by the correct location factor on page 7 to find the square foot cost for any building site. Costs include no interior finish. Where the interior finish is similar to the main dwelling and the garage walls are in vertical alignment with second floor walls, the square foot cost will be about 80% of the main dwelling square foot cost.

Square Foot Area for Attached Garages for Single Family Dwellings

Quality Class	220	260	280	320	360	400	440	480	540	600	720
1, Best	25.68	24.42	23.94	23.14	22.53	22.06	21.66	21.35	20.95	20.64	20.19
1 & 2	24.22	23.02	22.56	21.81	21.24	20.80	20.42	20.12	19.75	19.46	19.02
2, Good	23.61	21.71	21.28	20.58	20.03	19.59	19.26	18.96	18.62	18.35	17.95
2 & 3	21.40	20.35	19.95	19.30	18.79	18.39	18.06	17.79	17.47	17.21	16.81
3, Average	20.07	19.10	18.72	18.10	17.62	17.23	16.94	16.68	16.38	16.15	15.78
3 & 4	18.15	17.29	16.92	16.36	15.95	15.60	15.32	15.11	14.82	14.61	14.28
4, Low	16.38	15.59	15.27	14.76	14.38	14.07	13.82	13.61	13.38	13.16	12.89

Square Foot Area for Detached Garages for Single Family Dwellings

Quality Class	220	260	280	320	360	400	440	480	540	600	720
1, Best	33.00	30.59	29.65	28.16	27.03	26.11	25.39	24.79	24.06	23.48	22.66
1 & 2	31.06	28.81	27.91	26.49	25.44	24.57	23.88	23.32	22.62	22.11	21.33
2, Good	29.07	26.97	26.15	24.83	23.82	23.00	22.37	21.84	21.20	20.70	19.96
2 & 3	27.43	25.43	24.65	23.42	22.46	21.71	21.09	20.59	19.99	19.52	18.84
3, Average	25.58	23.71	22.98	21.82	20.93	20.22	19.66	19.20	18.66	18.20	17.56
3 & 4	23.03	21.36	20.71	19.66	18.87	18.21	17.72	17.30	16.79	16.38	15.81
4, Low	20.66	19.15	18.56	17.63	16.90	16.35	15.88	15.52	15.06	14.70	14.18

Carports, including asphalt floor, cost between $6.00 and $7.30 per square foot, depending on the type of roof cover.

Additional Costs for Residential Structures

Costs for Multi-Family Residential Bathrooms beyond 1 per unit

	Class 1 Best Quality	Class 2 Good Quality	Class 3 Average Quality	Class 4 Low Quality
2 or 3 units				
2 fixture bath	$3,600	$3,300	$2,800	$2,200
3 fixture bath	5,450	4,900	4,150	3,650
4 fixture bath	6,800	6,550	5,700	4,650
4 to 9 units				
2 fixture bath	3,400	3,100	2,500	2,100
3 fixture bath	5,100	4,550	3,800	3,050
4 fixture bath	6,700	6,050	4,900	4,150
10 or more units				
2 fixture bath	3,300	3,000	2,400	1,925
3 fixture bath	4,750	4,200	3,650	2,750
4 fixture bath	6,600	5,700	4,800	3,750

Half Story Areas

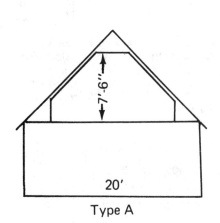

Type A

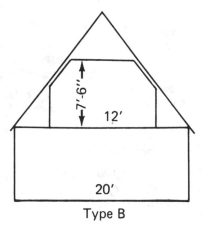

Type B

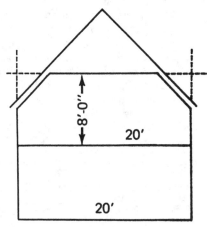

Type C

Use a fraction of the basic square foot cost for figuring the reduced headroom floor area

Type "C" includes typical dormers.

Type	Same Finish As Main Area	Interior Finish
A	1/3	1/4
B	1/2	1/3
C	2/3	1/2

Split Level Construction

All classes, all quality add 3% to base costs

Hydraulic Elevators, per shaft cost for car and machinery, 2 stops

Capacity	100 F.P.M.	125 F.P.M.	150 F.P.M.
1,200 lbs	$29,200	---	
2,000	33,700	$43,700	$55,000
2,500	39,000	49,200	57,400
3,000	41,800	54,100	61,100
3,500	---	57,300	63,800
4,000	---	59,900	67,400

Add for deluxe car, $6,350. Add for each additional stop over 2: baked enamel doors $6,850, stainless steel doors $7,400.

Garages built at ground level under a multi-family or motel unit. The costs below include the following components:
1. A reinforced concrete floor in all areas.
2. Exterior walls, on one long side and two short sides, made up of a wood frame and good quality stucco, wood siding or masonry veneer.
4. A finished ceiling in all areas.
5. The difference between the cost of a standard wood frame floor structure at second floor level and one at ground level.
6. An inexpensive light fixture for each 600 square feet.

Where no exterior walls enclose the two short sides, use $9.50 per square foot.

Garages built as separate structures for multi-family or motel units. The costs below include the following components:
1. Foundations.
2. A reinforced concrete floor in all areas.
3. Exterior walls on one long side and two short sides, made up of a wood frame and good quality stucco, wood siding or masonry veneer.
4. Steel support columns supporting the roof.
5. A wood frame roof structure with composition tar and gravel, wood shingle or light shake cover. No interior ceiling finish.

6. An inexpensive light fixture for each 600 square feet.

Use the location modifiers on page 7 to adjust garage costs to any area.

Basement Garages

Costs are listed below for basement garages built on one level, approximately five feet below grade, directly below 2 to 4 story dwelling structures with perimeter walls in vertical alignment. These costs include:

1. Excavation to 5' below ground line.

2. Full wall enclosure.

3. Typical storage facilities.

4. Minimum lighting.

5. Concrete floors.

Access stairways and driveways ramps outside the perimeter walls should be included when calculating the area of the garage.

Use the location modifiers on page 7 to adjust garage costs to any area.

Ground Level Garages

Area	400	800	1,200	2,000	3,000	5,000	10,000	20,000
Cost	19.20	17.10	14.80	14.20	13.60	13.00	12.20	9.20

Separate Structure Garages

Area	400	800	1,200	2,000	3,000	5,000	10,000	20,000
Cost	21.60	19.90	17.60	17.00	16.90	15.40	14.70	14.40

Basement Garages

Type	5,000	7,500	10,000	15,000	20,000	30,000	40,000	60,000
Reinforced concrete exterior walls and columns. Flat concrete roof slab.	28.00	27.00	25.30	24.00	23.60	23.30	22.80	22.70
Concrete block exterior walls, reinforced concrete columns. Flat concrete roof slab.	27.00	26.30	23.70	22.40	22.30	21.90	21.80	21.60
Concrete block exterior walls, steel posts and beams, light concrete/metal roof fireproofed with spray plaster.	24.60	23.30	21.90	21.40	20.50	20.30	19.90	19.30
Concrete block exterior walls, wood posts and beams, light concrete/metal roof fireproofed with spray plaster.	23.20	22.20	20.50	19.30	19.10	18.40	18.00	17.70

Cabins and Recreational Dwellings

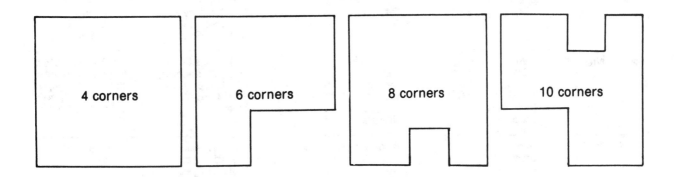

Example of Dwelling Shapes

Cabins and recreational dwellings are designed for single family occupancy, usually on an intermittent basis. These structures are characterized by a more rustic interior and exterior finish and often have construction details which would not meet building requirements in metropolitan areas. Classify these structures into either "conventional type" or "A-frame" construction. Conventional dwellings have an exterior wall which is approximately 8 feet high on all sides. A-frame cabins have a sloping roof which reduces the horizontal area 8 feet above the first floor to between 50% and 75% of the first floor area.

The shape of the outside perimeter is an important consideration in estimating the construction cost. Generally, the more complex the shape, the more expensive the structure per square foot of floor area. The shape classification is based on the outline formed by the outer-most exterior walls, including the garage area, regardless of the varying level. Most cabins and recreational dwellings have 4, 6, 8 or 10 corners as illustrated.

Cabins and recreational dwellings are often built under difficult working conditions and in remote sites. Allowance is made here for transportation of materials to typical recreational locations and for the higher labor cost which is usually associated with greater travel distances and adverse conditions. Individual judgments may be necessary in evaluating the cost impact of the dwelling location. The costs assume construction by skilled professional craftsmen. Where non-professional labor or second quality materials are used, use the next lower quality classification than might otherwise apply. If the structure is assembled from prefabricated components, use costs for the next lower half class.

Quality Classification

	Class 1 Best Quality	Class 2 Good Quality	Class 3 Average Quality	Class 4 Low Quality
Floor Framing	4" x 8" girders 48" o.c. with 2" T&G subfloor, or 2" x 8" to 2" x 10" joists 16" o.c. with 1" plywood subfloor.	4" x 8" girders 48" o.c. with 1-1/4" plywood or 2" T&G subfloor, or 2" x 8" to 10" joists 16" o.c. with 1" plywood subfloor.	4" x 6" girders 48" o.c. with 1-1/4" plywood or 2" T&G subfloor, or 2" x 6" to 8" joists 16" o.c. with 1" plywood subfloor.	4" x 6" girders 48" o.c. with 1-1/4" plywood or 2" x 6" joists 16" o.c. with 1" plywood subfloor.
Roof Framing	Same as floor or heavy glu-lam beams with 2" x 8" or 2" x 10" purlins, 3" T&G flat or low pitch deck and built-up roof.	4" x 8" girders 48" o.c. with 2" or 3" T&G sheathing, or 2" x 6" to 10" rafters 12" to 16" o.c. with 1" sheathing.	4" x 8" girders 48" o.c. with 2" T&G sheathing, or 2" x 6" to 8" rafters 12" to 24" o.c. with 1" sheathing.	4" x 8" girders 48" o.c. with 1-1/4" plywood or 2" T&G sheathing, or 2" x 6" to 8" rafters up to 24" o.c. with 1" sheathing.
Exterior Finish	Good plywood, lap or board and batt siding. Good trim.	Average to good plywood or board siding or good shingles. Some trim.	Average plywood composition or board siding or shingles. Little or no trim.	Low cost plywood or wood siding or low cost shingle or composition siding.
Windows	Larger area of good insulated wood or metal windows.	Adequate number of good wood or metal windows.	Average quality wood or metal windows.	Few low cost windows.
Roofing	Heavy shakes, usually with steep pitch and several roof planes.	Medium shakes, usually with medium to steep pitch.	Wood or composition shingles with medium to steep pitch.	Low cost composition shingles with medium pitch.
Flooring	Good carpet or hardwood. Sheet vinyl in kitchen and baths.	Average carpet with sheet vinyl or tile in kitchen and baths.	Low cost carpet with resilient tile in kitchen and baths.	Composition tile.
Interior Finish	Good quality hardwood veneer paneling, with some decorative exposed structural members.	Good textured gypsum wallboard with some plywood or knotty pine paneling.	Textured gypsum wallboard or average quality plywood or hardboard paneling.	Low cost hardboard paneling or gypsum wallboard.
Bathrooms	Two 3-fixture baths and one 2-fixture bath. Good fixtures.	Two 3-fixture baths. Good fixtures.	Two 3-fixture baths.	One 3-fixture bath.
Kitchen	15' to 18' of good hardwood veneer base and wall cabinet. 15' to 18' of good plastic or ceramic drainboard.	12' to 16' of hardwood veneer base cabinet with matching wall cabinets. 12' to 16' of plastic or ceramic drainboard.	8' to 12' of plywood veneer or painted base cabinet with matching wall cabinets. 8' to 12' of plastic drainboard.	8' to 10' of minimum base cabinets with matching wall cabinets. 8' to 10' of minimum plastic drainboard.
Plumbing	Nine good quality fixtures with one large or two 30 gallon water heaters, copper supply piping.	Seven good quality fixtures and one water heater.	Seven average quality fixtures and one water heater.	Four low cost fixtures and one water heater. Plastic supply piping.
Special Features	Built-in oven, range, dishwasher, disposer, range hood with good insulation, good lighting fixtures, insulated sliding glass door and ornate entry door.	Built-in range, oven and range hood, some insulation, 8' sliding glass door, average electric fixtures.	Drop-range and hood, some insulation, low cost electric fixtures.	Minimum electric fixtures.

Conventional Recreational Dwellings

Four Corners

Estimating Procedure

1. Establish the structure quality class by applying the information on page 30.
2. Multiply the structure floor area by the appropriate cost listed below.
3. Multiply the total from step 2 by the correct location factor listed on page 7.
4. Add, when appropriate, the cost of a deck or porch, paving, fireplace, garage or carport, heating, extra plumbing fixtures, supporting walls, half story areas, construction on hillside lots, and construction in remote areas. See page 39.

Conventional Recreational Dwelling, Class 3

Square Foot Area

Quality Class	400	500	600	700	800	900	1,000	1,100	1,200	1,300	1,400
1, Best	102.73	92.26	85.04	79.73	75.66	72.40	69.74	67.53	65.64	64.02	62.60
1 & 2	94.49	84.85	78.22	73.35	69.59	66.59	64.16	62.11	60.37	58.88	57.59
2, Good	86.13	77.36	71.31	66.87	63.44	60.72	58.48	56.62	55.05	53.67	52.49
2 & 3	81.34	73.04	67.33	63.13	59.89	57.33	55.22	53.46	51.96	50.68	49.55
3, Average	76.47	68.68	63.31	59.36	56.32	53.91	51.92	50.26	48.87	47.66	46.60
3 & 4	70.97	63.74	58.74	55.07	52.27	50.02	48.18	46.64	45.34	44.23	43.24
4, Low	65.28	58.63	54.04	50.66	48.07	46.00	44.32	42.89	41.70	40.68	39.77

Square Foot Area

Quality Class	1,500	1,600	1,700	1,800	2,000	2,200	2,400	2,600	2,800	3,000	3,200
1, Best	60.94	59.88	58.94	58.08	56.62	55.39	54.36	53.48	52.70	52.01	51.40
1 & 2	55.97	55.00	54.15	53.36	52.03	50.89	49.93	49.12	48.41	47.77	47.22
2, Good	51.31	50.41	49.63	48.91	47.68	46.64	45.79	45.01	44.37	43.78	43.27
2 & 3	48.47	47.64	46.87	46.21	45.03	44.08	43.24	42.54	41.92	41.37	40.88
3, Average	45.70	44.91	44.20	43.55	42.46	41.55	40.77	40.11	39.51	39.01	38.56
3 & 4	42.85	42.11	41.44	40.84	39.81	38.96	38.22	37.60	37.05	36.57	36.14
4, Low	39.99	39.28	38.67	38.12	37.16	36.35	35.67	35.09	34.58	34.77	33.74

Note: Add 4% to the square foot cost for floors above the second floor level.

Conventional Recreational Dwellings

Six Corners

Estimating Procedure

1. Establish the structure quality class by applying the information on page 30.
2. Multiply the structure floor area by the appropriate cost listed below.
3. Multiply the total from step 2 by the correct location factor listed on page 7.
4. Add, when appropriate, the cost of a deck or porch, paving, fireplace, garage or carport, heating, extra plumbing fixtures, supporting walls, half story areas, construction on hillside lots, and construction in remote areas. See page 39.

Conventional Recreational Dwelling, Class 4

Square Foot Area

Quality Class	400	500	600	700	800	900	1,000	1,100	1,200	1,300	1,400
1, Best	103.96	93.41	86.17	80.87	76.79	73.56	70.93	68.73	66.87	65.26	63.87
1 & 2	95.67	85.95	79.29	74.40	70.66	67.68	65.26	63.24	61.53	60.06	58.77
2, Good	87.28	78.42	72.33	67.89	64.48	61.75	59.55	57.71	56.13	54.79	53.62
2 & 3	82.48	74.09	68.36	64.14	60.91	58.35	56.26	54.52	53.04	51.78	50.68
3, Average	77.60	69.72	64.31	60.37	57.32	54.93	52.94	51.30	49.92	48.72	47.68
3 & 4	72.06	64.74	59.72	56.04	53.23	50.98	49.15	47.63	46.35	45.24	44.27
4, Low	66.16	59.42	54.81	51.45	48.86	46.81	45.14	43.74	42.56	41.53	40.65

Square Foot Area

Quality Class	1,500	1,600	1,700	1,800	2,000	2,200	2,400	2,600	2,800	3,000	3,200
1, Best	62.33	61.25	60.31	59.43	57.95	56.71	55.69	54.81	54.02	53.35	52.76
1 & 2	57.38	56.40	55.50	54.71	53.35	52.21	51.26	50.46	49.74	49.12	48.56
2, Good	52.40	51.50	50.70	49.96	48.72	47.69	46.81	46.06	45.42	44.86	44.35
2 & 3	49.61	48.75	47.98	47.30	46.13	45.13	44.32	43.61	42.99	42.45	41.98
3, Average	46.77	45.96	45.23	44.59	43.48	42.56	41.78	41.12	40.53	40.02	39.59
3 & 4	43.92	43.16	42.47	41.87	40.83	39.95	39.24	38.61	38.08	37.59	37.17
4, Low	41.02	40.30	39.68	39.10	38.13	37.31	36.64	36.06	35.55	35.10	34.70

Note: Add 4% to the square foot cost for floors above the second floor level.

Conventional Recreational Dwellings

Eight Corners

Estimating Procedure

1. Establish the structure quality class by applying the information on page 30.
2. Multiply the structure floor area by the appropriate cost listed below.
3. Multiply the total from step 2 by the correct location factor listed on page 7.
4. Add, when appropriate, the cost of a deck or porch, paving, fireplace, garage or carport, heating, extra plumbing fixtures, supporting walls, half story areas, construction on hillside lots, and construction in remote areas. See page 39.

Conventional Recreational Dwelling, Class 1 & 2

Square Foot Area

Quality Class	400	500	600	700	800	900	1,000	1,100	1,200	1,300	1,400
1, Best	104.91	94.46	87.28	82.03	77.98	74.76	72.14	69.94	68.08	66.48	65.08
1 & 2	96.61	86.98	80.37	75.53	71.79	68.84	66.43	64.40	62.68	61.21	59.93
2, Good	88.17	79.40	73.37	68.94	65.54	62.84	60.62	58.79	57.23	55.88	54.71
2 & 3	83.37	75.07	69.37	65.18	61.97	59.42	57.33	55.57	54.11	52.84	51.74
3, Average	78.45	70.62	65.28	61.33	58.32	55.89	53.94	52.30	50.90	49.71	48.67
3 & 4	72.60	65.38	60.41	56.77	53.95	51.73	49.91	48.40	47.12	46.01	45.04
4, Low	67.42	60.70	56.10	52.70	50.11	48.04	46.34	44.93	43.75	42.72	41.83

Square Foot Area

Quality Class	1,500	1,600	1,700	1,800	2,000	2,200	2,400	2,600	2,800	3,000	3,200
1, Best	63.39	62.36	61.43	60.59	59.13	57.90	56.86	55.92	55.14	54.43	53.79
1 & 2	58.55	57.61	56.73	55.97	54.63	53.48	52.53	51.68	50.94	50.28	49.70
2, Good	53.53	52.66	51.87	51.17	49.93	48.90	48.00	47.24	46.56	45.96	45.43
2 & 3	50.63	49.78	49.04	48.38	47.22	46.23	45.41	44.67	44.03	43.46	42.95
3, Average	47.78	46.99	46.30	45.67	44.56	43.64	42.85	42.15	41.55	41.01	40.53
3 & 4	44.93	44.19	43.54	42.95	41.92	41.03	40.29	39.66	39.09	38.58	38.13
4, Low	42.03	41.33	40.72	40.16	39.19	38.38	37.68	37.09	36.56	36.08	35.65

Note: Add 4% to the square foot cost for floors above the second floor level.

Conventional Recreational Dwellings
Ten Corners

Estimating Procedure

1. Establish the structure quality class by applying the information on page 30.
2. Multiply the structure floor area by the appropriate cost listed below.
3. Multiply the total from step 2 by the correct location factor listed on page 7.
4. Add, when appropriate, the cost of a deck or porch, paving, fireplace, garage or carport, heating, extra plumbing fixtures, supporting walls, half story areas, construction on hillside lots, and construction in remote areas. See page 39.

Conventional Recreational Dwelling, Class 1

Square Foot Area

Quality Class	400	500	600	700	800	900	1,000	1,100	1,200	1,300	1,400
1, Best	106.10	95.69	88.52	83.23	79.19	75.94	73.30	71.09	69.22	67.61	66.20
1 & 2	97.73	88.14	81.55	76.68	72.94	69.95	67.52	65.50	63.77	62.29	60.98
2, Good	89.26	80.51	74.49	70.03	66.61	63.90	61.67	59.82	58.24	56.88	55.69
2 & 3	84.48	76.19	70.48	66.27	63.03	60.46	58.36	56.60	55.12	53.82	52.70
3, Average	79.62	71.82	66.43	62.49	59.43	56.98	55.02	53.37	51.94	50.74	49.69
3 & 4	74.10	66.83	61.82	58.14	55.31	53.05	51.20	49.66	48.34	47.21	46.24
4, Low	68.47	61.76	57.13	53.73	51.10	49.03	47.32	45.88	44.68	43.64	42.73

Square Foot Area

Quality Class	1,500	1,600	1,700	1,800	2,000	2,200	2,400	2,600	2,800	3,000	3,200
1, Best	64.67	63.60	62.65	61.79	60.32	59.09	58.03	57.16	56.39	55.70	55.11
1 & 2	59.65	58.64	57.77	56.98	55.61	54.48	53.51	52.71	51.99	51.37	50.81
2, Good	54.49	53.58	52.77	52.05	50.80	49.77	48.90	48.15	47.49	46.91	46.42
2 & 3	51.73	50.85	50.10	49.41	48.24	47.24	46.41	45.70	45.08	44.53	44.07
3, Average	48.85	48.03	47.31	46.66	45.55	44.62	43.84	43.17	42.58	42.06	41.60
3 & 4	45.96	45.20	44.52	43.91	42.87	41.99	41.25	40.63	40.08	39.59	39.16
4, Low	43.10	42.36	41.73	41.16	40.19	39.35	38.67	38.08	37.56	37.11	36.71

Note: Add 4% to the square foot cost for floors above the second floor level.

"A-Frame" Cabins
Quality Classification

	Class 1 Best Quality	Class 2 Good Quality	Class 3 Average Quality	Class 4 Low Quality
Floor Framing	4" x 8" girders 48" o.c. with 2" T&G subfloor, or 2" x 6" to 2" x 8" joists 16" o.c. with 1" subfloor.	4" x 8" girders 48" o.c. with 1-1/4" plywood or 2" T&G subfloor, or 2" x 6" to 2" x 8" joists 16" o.c. with 1" subfloor.	4" x 6" girders 48" o.c. with 1-1/4" plywood or 2" T&G subfloor, or 2" x 6" joists 16" o.c. with 1" subfloor.	4" x 6" girders 48" o.c. with 1-1/4" plywood or 2" T&G subfloor, or 2" x 6" joists 16" o.c. with 1"" subfloor.
Roof Framing	4" x 8" at 48" o.c. with 2" or 3" T&G sheathing.	4" x 8" at 48" o.c. with 2" or 3" T&G sheathing.	4" x 8" at 48" o.c. with 2" T&G sheathing.	4" x 8" at 48" o.c. with 1-1/4" plywood or 2" T&G sheathing.
Gable End Finish	Good plywood, lap board or board and batt.	Average to good plywood, or boards.	Average plywood, board or wood shingle.	Low cost plywood, shingle or composition siding.
Windows	Good quality large insulated wood or metal windows.	Average quality insulated wood or metal windows.	Average quality wood or metal windows.	Small glass area of low cost windows.
Roofing	Heavy wood shakes.	Medium wood or aluminum shakes.	Wood or composition shingles.	Low cost composition shingles.
Flooring	Good carpet or hardwood with sheet vinyl in kitchen and baths.	Average to good quality carpet with good tile or sheet vinyl in kitchen and baths.	Average quality carpet with resilient tile in kitchen and baths.	Composition tile.
Interior Finish	Good quality hardwood veneer paneling.	Good textured gypsum wallboard, good plywood or knotty pine paneling.	Textured gypsum wallboard or plywood paneling.	Low cost paneling or wallboard.
Bathrooms	Two 3-fixture baths and one 2-fixture bath, good fixtures.	Two 3-fixture baths, good fixtures.	Two 3-fixture baths, average fixtures.	One 3-fixture bath.
Kitchen	15' to 18' good quality hardwood veneer base cabinet with matching wall cabinets. 15' to 18' of good quality plastic or ceramic tile drain board.	12' to 16' of hardwood veneer base cabinet with matching wall cabinets. 12' to 16' of plastic or ceramic tile drainboard.	8' to 12' of average quality veneer or painted base cabinets with matching wall cabinets. 8' to 12' of plastic drainboard.	6' to 8' of minimum base cabinets with matching wall cabinets. 6' to 8' of minimum plastic drainboard.
Plumbing	Nine good quality fixtures and one larger or two 30 gallon water heaters. Copper supply piping.	Seven good quality fixtures and one water heater.	Seven average quality fixtures and one water heater.	Four low cost fixtures and one water heater. Plastic supply pipe.
Special Features	Built-in oven, range, dishwasher, disposer, range hood with good insulation, good lighting fixtures, insulated sliding glass door and ornate entry door.	Built-in range, oven and range hood, some insulation, 8' sliding glass door, average electric fixtures.	Drop-in range and hood, some insulation, low cost electric fixtures.	Minimum electric fixtures.

"A-Frame" Cabins
Four Corners

Estimating Procedure

1. Establish the structure quality class by applying the information on page 35.
2. Multiply the structure floor area by the appropriate cost listed below.
3. Multiply the total from step 2 by the correct location factor listed on page 7.
4. Add, when appropriate, the cost of a deck or porch, paving, fireplace, garage or carport, heating, extra plumbing fixtures, supporting walls, half story areas, construction on hillside lots, and construction in remote areas. See page 39.

"A-Frame" Cabin, Class 3 & 4

Square Foot Area

Quality Class	400	500	600	700	800	900	1,000	1,100	1,200	1,300	1,400
1, Best	117.52	106.04	98.09	92.20	87.63	84.00	81.01	78.51	76.39	74.54	72.95
1 & 2	107.46	96.97	89.68	84.30	80.14	76.76	74.08	71.79	69.84	68.16	66.69
2, Good	97.19	87.70	81.11	76.24	72.49	69.47	67.01	64.92	63.17	61.65	60.32
2 & 3	91.31	82.39	76.20	71.64	68.10	65.27	62.94	61.00	59.35	57.91	56.66
3, Average	85.24	76.92	71.14	66.88	63.57	60.93	58.77	56.94	55.41	54.08	52.91
3 & 4	77.03	69.50	64.26	60.42	57.43	55.05	53.10	51.45	50.06	48.85	47.80
4, Low	69.09	62.35	57.66	54.21	51.53	49.39	47.64	46.17	44.90	43.82	42.88

Square Foot Area

Quality Class	1,500	1,600	1,700	1,800	2,000	2,200	2,400	2,600	2,800	3,000	3,200
1, Best	70.34	69.20	68.16	67.22	65.60	64.25	63.12	62.11	61.24	60.48	59.79
1 & 2	64.61	63.55	62.59	61.75	60.26	59.01	57.97	57.05	56.25	55.54	54.92
2, Good	58.75	57.77	56.91	56.13	54.79	53.66	52.70	51.87	51.14	50.49	49.93
2 & 3	55.50	54.59	53.78	53.04	51.76	50.70	49.79	49.00	48.31	47.72	47.17
3, Average	51.97	51.10	50.35	49.67	48.48	47.47	46.63	45.89	45.26	44.68	44.18
3 & 4	47.58	46.81	46.10	45.48	44.39	43.47	42.69	42.02	41.44	40.92	40.46
4, Low	43.10	42.39	41.76	41.19	40.20	39.37	38.66	38.05	37.52	37.06	36.63

"A-Frame" Cabins
Six Corners

Estimating Procedure
1. Establish the structure quality class by applying the information on page 35.
2. Multiply the structure floor area by the appropriate cost listed below.
3. Multiply the total from step 2 by the correct location factor listed on page 7.
4. Add, when appropriate, the cost of a deck or porch, paving, fireplace, garage or carport, heating, extra plumbing fixtures, supporting walls, half story areas, construction on hillside lots, and construction in remote areas. See page 39.

"A-Frame" Cabin, Class 2 & 3

Square Foot Area

Quality Class	400	500	600	700	800	900	1,000	1,100	1,200	1,300	1,400
1, Best	119.42	107.76	99.75	93.83	89.30	85.68	82.72	80.25	78.16	76.33	74.76
1 & 2	109.00	98.37	91.05	85.66	81.52	78.21	75.50	73.25	71.33	69.69	68.16
2, Good	98.68	89.05	82.42	77.54	73.78	70.79	68.35	66.31	64.57	63.08	61.79
2 & 3	92.73	83.67	77.45	72.86	69.35	66.54	64.24	62.31	60.67	59.28	58.06
3, Average	86.15	77.75	71.95	67.70	64.41	61.81	59.67	57.89	56.38	55.06	53.93
3 & 4	78.24	70.62	65.36	61.49	58.51	56.13	54.20	52.59	51.21	50.03	48.99
4, Low	70.03	63.21	58.50	55.05	52.38	50.26	48.53	47.08	45.83	44.77	43.85

Square Foot Area

Quality Class	1,500	1,600	1,700	1,800	2,000	2,200	2,400	2,600	2,800	3,000	3,200
1, Best	72.31	71.12	70.06	69.10	67.48	66.11	64.92	63.91	63.03	62.23	61.54
1 & 2	66.25	65.17	64.22	63.33	61.83	60.58	59.50	58.57	57.76	57.05	56.41
2, Good	60.27	59.27	58.39	57.61	56.24	55.10	54.11	53.28	52.54	51.89	51.30
2 & 3	56.91	55.97	55.15	54.39	53.12	52.03	51.10	50.32	49.62	48.99	48.46
3, Average	53.46	52.58	51.82	51.09	49.89	48.88	48.02	47.26	46.61	46.03	45.51
3 & 4	48.84	48.06	47.34	46.69	45.59	44.65	43.87	43.18	42.58	42.05	41.58
4, Low	44.18	43.45	42.82	42.25	41.23	40.39	39.69	39.05	38.52	38.04	37.61

"A-Frame" Cabins
Eight Corners

Estimating Procedure

1. Establish the structure quality class by applying the information on page 35.
2. Multiply the structure floor area by the appropriate cost listed below.
3. Multiply the total from step 2 by the correct location factor listed on page 7.
4. Add, when appropriate, the cost of a deck or porch, paving, fireplace, garage or carport, heating, extra plumbing fixtures, supporting walls, half story areas, construction on hillside lots, and construction in remote areas. See page 39.

"A-Frame" Cabin, Class 2

Square Foot Area

Quality Class	400	500	600	700	800	900	1,000	1,100	1,200	1,300	1,400
1, Best	121.38	109.78	101.76	95.84	91.28	87.61	84.63	82.12	79.99	78.16	76.54
1 & 2	110.77	100.19	92.88	87.48	83.31	79.96	77.24	74.96	73.02	71.33	69.86
2, Good	100.08	90.52	83.93	79.03	75.28	72.26	69.79	67.73	65.97	64.46	63.12
2 & 3	94.01	85.04	78.82	74.25	70.71	67.87	65.55	63.62	61.97	60.54	59.29
3, Average	87.64	79.28	73.48	69.21	65.92	63.28	61.11	59.32	57.77	56.44	55.28
3 & 4	79.26	71.70	66.45	62.59	59.60	57.22	55.28	53.64	52.24	51.05	50.01
4, Low	71.06	64.27	59.57	56.11	53.44	51.28	49.54	48.08	46.83	45.76	44.82

Square Foot Area

Quality Class	1,500	1,600	1,700	1,800	2,000	2,200	2,400	2,600	2,800	3,000	3,200
1, Best	74.14	72.97	71.88	70.93	69.28	67.91	66.72	65.74	64.85	64.08	63.40
1 & 2	66.71	65.63	64.66	63.81	62.32	61.08	60.03	59.13	58.32	57.65	57.03
2, Good	61.77	60.76	59.88	59.09	57.71	56.56	55.59	54.74	54.02	53.37	52.81
2 & 3	58.21	57.27	56.44	55.69	54.38	53.30	52.40	51.59	50.90	50.31	49.76
3, Average	54.64	53.76	52.98	52.27	51.05	50.05	49.19	48.45	47.80	47.23	46.72
3 & 4	49.89	49.09	48.36	47.72	46.61	45.68	44.91	44.22	43.64	43.12	42.65
4, Low	45.20	44.47	43.82	43.24	42.24	41.39	40.67	40.06	39.52	39.05	38.65

Cabins and Recreational Dwellings

Additional Costs

Half-Story Costs

For conventional recreational dwellings, use the suggested fractions found on page 27 in the section "Additional Costs for Residential Structures." For "A-Frame" cabins, use one of the following costs: A simple platform with low cost floor cover, minimum partitions, and minimum lighting costs $20.50 to $28.00 per square foot. Average quality half story area with average quality carpet, average number of partitions finished with gypsum wallboard or plywood veneer and average lighting costs $29.00 to $39.00 per square foot. A good quality half story area with good carpet, decorative rustic partitions, ceiling beams and good lighting costs $46.00 to $58.00 per square foot.

Decks and Porches, per square foot

Small (10 to 50 S.F.) uncovered wood porch with steps and railing	
1' above ground level	$15.20 to $16.70
4' above ground level	16.70 to 19.30
2" wood deck with steps and railing (300 S.F. base)	
1' above ground level	11.10 to 16.20
4' above ground level	14.00 to 17.70

Fireplaces, cost each

Metal hood with concrete slab	$1,000 to $1,330
Simple concrete block	1,790 to 2,320
Simple block with stone facing	2,340 to 2,970
Simple natural stone	2,870 to 3,050
Prefabricated, zero clearance	1,420 to 1,880

Extra Plumbing, cost each

Lavatory	$770 to $990
Water closet	930 to 1,030
Tub	990 to 1,190
Stall shower	710 to 930
Sink	770 to 980

Heating, cost each

Wall furnaces	
35,000 BTU	$770
65,000 BTU	1,020
Central heating, perimeter ducts	$7.00 per S.F.

Garages and Carports, per S.F.

Average carport, no slab (single)	$10.25 to $14.40
Average single garage with slab	22.60 to 25.60
Average double garage with slab	21.20 to 24.50

Flatwork, per square foot

Asphalt paving	$2.30 to $2.50
4" concrete	2.90 to 3.90
6" concrete	3.40 to 4.80

Reinforced concrete walls, per C.F.

Formed one side only	$11.00 to $12.70
Formed both sides	14.00 to 16.00

Supporting Wall Costs

Cabins and recreational dwellings built on sloping lots cost more than if they are built on level lots. The cost of supporting walls of a building that do not enclose any living area should be estimated by using the figures below. These costs include everything above a normal foundation (12" to 18" above ground) up to the bottom of the next floor structure where square foot costs can be applied. In addition to the cost of supporting walls, add the cost of any extra structural members and the higher cost of building on a slope. A good rule of thumb for this is to add $520 for each foot of vertical distance between the highest and the lowest points of intersection of foundation and ground level.

Wood posts, per foot of height

4" x 4"	$1.90 to $3.00
4" x 6"	3.00 to 5.15
6" x 6"	3.90 to 7.30
8" x 8"	8.75 to 13.70
10" x 10"	16.25 to 23.30
12" x 12"	24.60 to 33.60

Brick, per square foot of wall

8" common brick	$23.50 to $29.00
12" common brick	36.00 to 44.80
8" common brick, 1 side face brick	29.80 to 36.70
12" common brick, 1 side face brick	46.50 to 58.00

Reinforced concrete block, per square foot of wall

8" natural	$5.60 to $6.70
8" colored	7.60 to 8.90
8" detailed blocks, natural	6.40 to 8.20
8" detailed blocks, colored	8.60 to 9.80
8" sandblasted	6.70 to 7.80
8" splitface, natural	5.70 to 6.85
8" splitface, colored	8.90 to 10.20
8" slump block, natural	6.40 to 7.60
8" slump block, colored	8.50 to 9.90
12" natural	10.80 to 12.20

Typical Physical Lives in Years
for Residential Structures by Quality Class

	Masonry				Wood Frame			
Quality Class	1	2	3	4	1	2	3	4
1 or 2 family	70	60	60	55	70	60	60	55
3 or more units	60	55	55	50	60	55	55	50
Motels	60	55	55	50	60	55	55	50

Raise half classes to the next higher whole class

Average Life in Years (Percent Good)
for Residential and Agricultural Structures

Age	20 Years Rem. Life	20 Years % Good	25 Years Rem. Life	25 Years % Good	30 Years Rem. Life	30 Years % Good	40 Years Rem. Life	40 Years % Good
0	20	100	25	100	30	100	40	100
1	19	94	24	95	29	96	39	98
2	18	88	23	90	28	93	38	96
3	17	81	22	86	27	89	37	94
4	16	75	21	81	26	86	36	92
5	15	69	20	77	25	82	35	90
6	14	63	19	72	24	79	34	87
7	13	59	18	68	23	75	33	84
8	12	57	17	63	22	71	32	82
9	11	55	16	60	21	67	31	80
10	11	53	16	58	20	64	30	77
11	10	50	15	56	19	60	29	74
12	9	48	14	54	19	59	28	72
13	8	46	13	53	18	57	27	70
14	7	44	12	51	17	56	27	67
15	7	42	11	49	16	54	26	65
16	6	40	11	48	15	53	25	62
17	5	38	10	46	14	52	24	60
18	5	36	9	44	13	50	23	59
19	4	33	8	43	13	49	22	58
20	4	31	7	41	12	47	21	58
21	3	29	7	39	11	46	21	55
22	3	27	6	37	11	44	20	54
23	3	25	6	35	10	43	19	53
24	3	23	5	34	9	42	18	52
25	2	21	5	32	9	40	17	51
26	2	19	4	30	8	39	17	50
27	2	16	4	29	7	37	16	49
28	2	14	4	27	7	36	15	48
29	2	12	3	25	6	34	14	47
30	1	10	3	24	6	33	14	46
31	--	--	3	22	5	31	13	45
32	--	--	3	20	5	30	12	44
33	--	--	2	18	5	29	12	43
34	--	--	2	17	4	27	11	42
35	--	--	2	15	4	26	11	41
36	--	--	2	13	4	24	10	40
38	--	--	1	10	3	21	9	38
40	--	--	--	--	2	19	7	35
42	--	--	--	--	2	16	6	33
46	--	--	--	--	1	10	5	29
50	--	--	--	--	--	--	4	25
55	--	--	--	--	--	--	3	20
60	--	--	--	--	--	--	2	14
64	--	--	--	--	--	--	1	10

Age	45 Years Rem. Life	45 Years % Good	50 Years Rem. Life	50 Years % Good	55 Years Rem. Life	55 Years % Good	60 Years Rem. Life	60 Years % Good	70 Years Rem. Life	70 Years % Good
0	45	100	50	100	55	100	60	100	70	100
2	43	97	48	97	53	98	58	98	68	99
4	41	93	46	94	51	96	56	96	66	98
6	39	89	44	91	49	94	54	94	64	97
8	37	85	42	88	47	91	52	92	62	96
10	35	81	39	85	45	88	50	90	60	95
12	33	77	38	82	43	85	48	88	58	93
14	32	73	36	78	41	82	46	86	56	92
16	30	69	35	74	40	79	45	83	54	90
18	28	65	33	70	38	76	43	80	52	89
20	26	60	31	67	36	73	41	77	50	87
22	24	58	29	63	34	69	39	74	48	86
24	23	56	28	60	32	65	37	71	46	84
26	22	54	26	58	31	62	35	68	44	82
28	20	52	24	56	29	60	34	65	42	80
30	18	50	23	54	27	58	32	63	40	78
32	17	48	21	53	26	56	30	60	38	76
34	15	47	20	51	24	55	29	58	36	73
36	14	45	18	49	23	53	27	57	34	71
38	12	43	17	47	21	51	26	55	32	68
40	11	41	16	45	20	50	24	54	30	65
42	10	39	14	44	19	48	23	52	28	62
44	9	37	13	42	17	46	21	51	26	59
46	8	35	12	40	16	45	20	49	25	56
48	7	33	11	38	15	43	19	47	23	54
50	6	31	10	37	14	41	18	46	21	49
52	5	29	9	35	12	40	16	44	19	45
54	5	28	8	33	11	38	15	43	18	44
56	4	26	7	31	10	36	14	41	16	42
58	4	24	6	30	9	35	13	40	15	38
60	3	22	5	28	8	33	12	38	14	36
62	3	20	4	26	7	31	11	37	12	31
64	3	17	4	24	6	30	10	35	11	30
66	2	16	3	22	5	28	9	33	10	27
68	2	14	3	21	5	27	8	32	9	25
70	2	12	3	19	4	25	7	30	9	24
72	1	10	2	17	4	23	6	29	8	21
74	--	--	2	15	4	21	5	27	7	20
76	--	--	2	14	3	20	5	26	7	19
80	--	--	1	10	2	17	4	23	7	18
82	--	--	--	--	2	15	3	20	6	17
84	--	--	--	--	1	10	2	17	5	16
96	--	--	--	--	--	--	1	10	3	14
98	--	--	--	--	--	--	--	--	2	12
100	--	--	--	--	--	--	--	--	1	10

Commercial Structures Section

Section Contents

Urban Stores

Urban store buildings are designed for retail sales and are usually found in strip or downtown commercial developments. Square foot costs in this section are representative of a building situation where construction activities are restricted to the immediate site. This restriction tends to make the cost slightly higher than suburban type stores where unlimited use of modern machinery and techniques is possible. Do not use these figures for department stores, discount houses or suburban stores. These building types are evaluated in later sections.

Costs are for shell-type buildings without permanent partitions and include all labor, material and equipment costs for the following:
1. Foundations as required for normal soil conditions.
2. Floor, rear wall, side wall and roof structures.
3. A front wall consisting of vertical support columns or pilasters and horizontal beams spanning the area between these members leaving an open space to receive a display front.
4. Interior floor, wall and ceiling finishes.
5. Exterior wall finish on the side and rear walls.
6. Roof cover.
7. Basic lighting and electrical systems.
8. Rough and finish plumbing.
9. Design and engineering fees.
10. Permits and fees.
11. Utility hook-up.
12. Contractor's contingency, overhead and profit.

The in-place costs of the following components should be added to the basic building cost to arrive at the total structure cost. See the section "Additional Costs for Commercial Structures" on page 201.
1. Heating and air conditioning systems.
2. Elevators and escalators.
3. Fire sprinklers and fire escapes.
4. All display front components.
5. Finish materials on the front wall.
6. Canopies, ramps and docks.
7. Interior partitions.
8. Exterior signs.
9. Mezzanines and basements.
10. Communication systems.

For valuation purposes, urban stores are divided into two building types: 1) masonry or concrete frame and, 2) wood or wood and steel frame. Masonry or concrete urban stores vary widely in cost. Consequently, 6 quality classifications are established. Wood or wood and steel frame urban stores are divided into 4 quality classes.

Urban Stores - Masonry or Concrete
Quality Classification

	Class 1 Best Quality	Class 2 Good Quality	Class 3 High Average Quality	Class 4 Low Average Quality	Class 5 Low Quality	Class 6 Minimum Quality
Foundation	Reinforced concrete.	Reinforced concrete.	Reinforced concrete.	Reinforced concrete.	Reinforced concrete.	Reinforced concrete.
Floor Structure	6" reinforced concrete on 6" rock fill or 2" x 12" joists 16" o.c.	4" to 6" reinforced concrete on 6" rock fill or 2" x 10" joists 16" o.c.	4" reinforced concrete on 6" rock fill or 2" x 10" joists 16" o.c.	4" reinforced concrete on 6" rock fill or 2" x 8" joists 16" o.c.	4" reinforced concrete on 6" rock fill or 2" x 6" joists 16" o.c.	4" reinforced concrete on 4" rock fill or 2" x 6" joists 16" o.c.
Wall Structure	Reinforced 8" concrete or 12" common brick or block.	Reinforced 8" concrete or 12" common brick or block.	Reinforced 8" concrete or 12" common brick or block.	Reinforced 8" concrete or 12" common brick or block.	Reinforced 8" concrete block or reinforced 6" concrete.	Reinforced 8" concrete block or reinforced 6" concrete or 8" clay tile or 8" brick.
Roof Covering	5 ply composition roof on 1" sheathing with insulation.	5 ply composition roof on 1" sheathing with insulation.	5 ply composition roof on 1" sheathing with insulation.	5 ply composition roof on 1" sheathing with insulation.	4 ply composition roof on 1" sheathing.	4 ply composition roof on 1" sheathing.
Floor Finish	Combination solid vinyl tile and terrazzo or very good carpet.	Combination solid vinyl tile and terrazzo or very good carpet.	Solid vinyl tile with some terrazzo or good carpet.	Vinyl tile with small areas of terrazzo, carpet or solid vinyl tile.	Resilient tile.	Composition tile.
Interior Wall Finish	Plaster on gypsum or metal lath or 2 layers of 5/8" gypsum wallboard with expensive wall paper or vinyl wall cover.	Plaster on gypsum or metal lath or 2 layers of 5/8" gypsum wallboard with average wall paper or vinyl wall cover.	Plaster with putty coat finish on gypsum or metal lath, or 5/8" gypsum wallboard taped, textured and painted, some vinyl wall covering.	Plaster with putty coat finish or gypsum or metal lath, or 5/8" gypsum wallboard taped, textured and painted or with wall paper.	Lath, 2 coats plaster with putty coat finish or 1/2" gypsum wallboard taped, textured and painted.	Interior plaster on masonry. Colored finish.
Ceiling Finish	Acoustical plaster or suspended anodized acoustical metal panels.	Acoustical plaster or suspended anodized acoustical metal panels.	Plaster with putty coat finish and some acoustical plaster or suspended acoustical tile with gypsum wallboard backing.	Plaster with putty coat finish or suspended acoustical tile.	Gypsum wallboard taped and textured or lath, 2 coats of plaster and putty coat finish.	Ceiling tile or gypsum wallboard and paint.
Exterior Wall Finish	Waterproofed and painted finish with face brick on exposed walls.	Waterproofed and painted finish with face brick on exposed walls.	Waterproofed and painted finish, face brick on exposed walls.	Painted finish, face brick on exposed wall.	Painted finish.	Unfinished.
Lighting	Encased modular units and custom designed chandeliers. Many spotlights.	Encased modular units and stock design chandeliers. Many spotlights.	Encased modular units and stock chandeliers. Many spotlights.	Quad open strip fixtures or triple encased louvered strip fixtures. Average number of spotlights.	Triple open strip fixtures or double encased louvered strip fixtures. Some spotlights.	Double open strip fluorescent fixtures.
Plumbing (Per each 5,000 S.F.)	6 good fixtures, metal or marble toilet partitions.	6 good fixtures, metal or marble toilet partitions.	6 standard fixtures, metal or marble toilet partitions.	6 standard fixtures, metal toilet partitions.	4 standard commercial fixtures, metal toilet partitions.	4 standard commercial fixtures, wood toilet partitions.
Bath Wall Finish	Ceramic tile or marble or custom mosaic tile.	Ceramic tile or marble or custom mosaic tile.	Ceramic tile or marble or plain mosaic tile.	Gypsum wallboard and paint, some ceramic tile or plastic finish wall board.	Gypsum wallboard and paint.	Gypsum wallboard and paint.

Urban Stores - Masonry or Concrete
First Floor, Length Less than Twice Width

Estimating Procedure

1. Establish the structure quality class by applying the information on page 43.
2. Compute the building ground floor area. This should include everything within the exterior walls and all insets outside the walls but under the main roof.
3. Add to or subtract from the cost below the appropriate amount from the Wall Height Adjustment Table on page 48 if the first floor wall height is more or less than 16 feet for large stores or 12 feet for small stores.
4. Multiply the adjusted square foot cost by the building area.
5. Deduct, if appropriate, for common walls, using the figures on page 48.
6. Multiply the total cost by the location factor on page 7.
7. Add the cost of heating and cooling equipment, elevators, escalators, fire escapes, fire sprinklers, display fronts, canopies, ramps, docks, interior partitions, mezzanines, basements, and communication systems from pages 201 to 213.
8. Add the cost of second and higher floors from page 47.

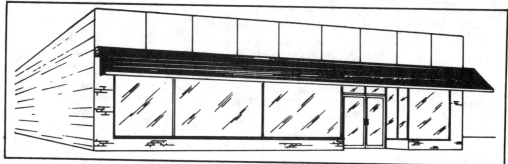

Urban Store, Class 4 & 5

Smaller Stores
Square Foot Area

Quality Class	500	600	700	800	900	1,000	1,200	1,500	1,700	2,000	2,500
4, Low Avg.	80.79	76.77	73.63	71.12	69.03	67.26	64.41	61.25	59.62	57.67	55.22
4 & 5	75.79	72.01	69.08	66.70	64.74	63.09	60.40	57.45	55.93	54.09	51.78
5, Low	70.80	67.26	64.52	62.32	60.47	58.93	56.42	53.66	52.25	50.53	48.38
5 & 6	67.32	63.97	61.36	59.24	57.50	56.03	53.66	51.03	49.68	48.05	46.00
6, Minimum	63.82	60.64	58.17	56.18	54.53	53.12	50.87	48.38	47.10	45.56	43.63

Larger Stores
Square Foot Area

Quality Class	3,000	3,500	4,000	4,500	5,000	6,000	7,500	10,000	15,000	20,000
1, Best	85.39	83.00	81.09	79.55	78.25	76.18	73.95	71.45	68.56	66.86
1 & 2	79.75	77.51	75.74	74.29	73.08	71.15	69.06	66.73	64.02	62.44
2, Good	74.38	72.31	70.66	69.32	68.17	66.37	64.42	62.25	59.73	58.26
2 & 3	70.71	68.74	67.16	65.87	64.79	63.09	61.24	59.16	56.78	55.36
3, Hi. Avg.	65.54	65.12	63.63	62.41	61.39	59.78	58.02	56.05	53.79	52.47
3 & 4	61.84	60.08	58.65	57.48	56.50	54.92	53.18	51.25	48.97	47.62
4, Low Avg.	56.02	54.63	53.34	52.27	51.37	49.95	48.38	46.61	44.55	43.32
4 & 5	53.28	51.75	50.52	49.50	48.66	47.31	45.82	44.14	42.17	41.01
5, Low	50.28	48.82	47.68	46.73	45.94	44.65	43.24	41.67	39.80	38.73
5 & 6	47.58	46.21	45.11	44.20	43.45	42.25	40.90	39.42	37.67	36.64
6, Minimum	44.74	43.46	42.42	41.57	40.88	39.73	38.48	37.08	35.44	34.45

Urban Stores - Masonry or Concrete
First Floor, Length Between 2 and 4 Times Width

Estimating Procedure

1. Establish the structure quality class by applying the information on page 43.
2. Compute the building ground floor area. This should include everything within the exterior walls and all insets outside the walls but under the main roof.
3. Add to or subtract from the cost below the appropriate amount from the Wall Height Adjustment Table on page 48 if the first floor wall height is more or less than 16 feet for large stores or 12 feet for small stores.
4. Multiply the adjusted square foot cost by the building area.
5. Deduct, if appropriate, for common walls, using the figures on page 48.
6. Multiply the total cost by the location factor on page 7.
7. Add the cost of heating and cooling equipment, elevators, escalators, fire escapes, fire sprinklers, display fronts, canopies, ramps, docks, interior partitions, mezzanines, basements, and communication systems from pages 201 to 213.
8. Add the cost of second and higher floors from page 47.

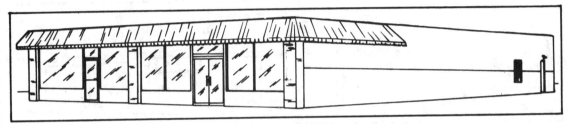

Urban Store, Class 4

Smaller Stores
Square Foot Area

Quality Class	500	600	700	800	900	1,000	1,250	1,500	1,700	2,000	2,500
4, Low Avg.	88.61	84.98	81.26	78.28	75.81	73.74	69.69	66.74	64.87	62.61	59.81
4 & 5	84.42	79.88	76.38	73.57	71.25	69.29	65.51	62.72	60.97	58.83	56.21
5, Low	78.90	74.66	71.37	68.75	66.59	64.75	61.22	58.61	56.97	54.99	52.52
5 & 6	75.27	71.21	68.09	65.59	63.52	61.78	58.39	55.92	54.35	52.47	50.12
6, Minimum	71.67	67.81	64.84	62.46	60.48	58.83	55.61	53.24	51.75	49.95	47.71

Larger Stores
Square Foot Area

Quality Class	3,000	3,500	4,000	4,500	5,000	6,000	7,500	10,000	15,000	20,000
1, Best	90.24	87.59	85.47	83.72	82.26	79.88	77.26	74.34	70.85	68.79
1 & 2	84.52	82.07	80.08	78.43	77.06	74.83	72.41	69.64	66.39	64.44
2, Good	78.72	76.42	74.57	73.04	71.75	69.68	67.42	64.83	61.82	60.01
2 & 3	74.84	72.65	70.90	69.45	68.24	66.26	64.11	61.66	58.78	57.07
3, Hi. Avg.	70.95	68.87	67.21	65.83	64.67	62.81	60.76	58.43	55.70	54.10
3 & 4	65.92	63.83	62.15	60.77	59.61	57.75	55.73	53.43	50.74	49.15
4, Low Avg.	59.88	57.97	56.47	55.19	54.16	52.46	50.60	48.52	46.08	44.65
4 & 5	56.91	55.10	53.67	52.47	51.46	49.85	48.11	46.13	43.79	42.44
5, Low	53.73	52.01	50.65	49.52	48.59	47.07	45.41	43.54	41.35	40.06
5 & 6	50.78	49.15	47.88	46.81	45.91	44.48	42.91	41.15	39.09	37.86
6, Minimum	47.86	46.33	45.12	44.11	43.28	41.92	40.44	38.79	36.83	35.67

Urban Stores - Masonry or Concrete
First Floor, Length More Than 4 Times Width

Estimating Procedure

1. Establish the structure quality class by applying the information on page 43.
2. Compute the building ground floor area. This should include everything within the exterior walls and all insets outside the walls but under the main roof.
3. Add to or subtract from the cost below the appropriate amount from the Wall Height Adjustment Table on page 48 if the first floor wall height is more or less than 16 feet for large stores or 12 feet for small stores.
4. Multiply the adjusted square foot cost by the building area.
5. Deduct, if appropriate, for common walls, using the figures on page 48.
6. Multiply the total cost by the location factor on page 7.
7. Add the cost of heating and cooling equipment, elevators, escalators, fire escapes, fire sprinklers, display fronts, canopies, ramps, docks, interior partitions, mezzanines, basements, and communication systems from pages 201 to 213.
8. Add the cost of second and higher floors from page 47.

Urban Store, Class 3 & 4

Smaller Stores
Square Foot Area

Quality Class	500	600	700	800	900	1,000	1,200	1,500	1,700	2,000	2,500
4, Low Avg.	98.04	94.13	89.63	86.04	83.08	80.58	76.61	72.26	70.03	67.36	64.06
4 & 5	94.01	88.50	84.25	80.86	78.09	75.74	72.01	67.92	65.84	63.32	56.97
5, Low	87.84	82.71	78.73	75.57	72.98	70.80	67.30	63.47	61.53	59.19	56.27
5 & 6	84.12	79.17	75.38	72.35	69.86	67.77	64.43	60.77	58.90	56.64	53.88
6, Minimum	80.33	75.62	72.00	69.11	66.73	64.72	61.52	58.05	56.27	54.11	51.46

Larger Stores
Square Foot Area

Quality Class	3,000	3,500	4,000	4,500	5,000	6,000	7,500	10,000	15,000	20,000
1, Best	97.10	93.86	91.28	89.16	87.35	84.47	81.32	77.77	73.60	71.13
1 & 2	90.99	87.97	85.53	83.55	81.87	79.16	76.20	72.90	68.98	66.68
2, Good	84.74	81.93	79.66	77.80	76.23	73.74	70.97	67.89	64.24	62.09
2 & 3	77.61	75.03	72.95	71.25	69.82	67.52	65.01	62.16	58.83	56.86
3, Hi. Avg.	76.34	73.79	71.77	70.08	68.68	66.42	63.94	61.15	57.86	55.93
3 & 4	70.91	68.43	66.45	64.81	63.43	61.21	58.77	56.02	52.79	50.87
4, Low Avg.	64.73	62.47	60.66	59.16	57.90	55.88	53.66	51.15	48.20	46.44
4 & 5	61.39	59.24	57.50	56.09	54.90	52.97	50.86	48.50	45.70	44.04
5, Low	57.90	55.87	54.25	52.92	51.79	49.97	47.99	45.73	43.10	41.54
5 & 6	54.81	52.89	51.35	50.08	49.02	47.31	45.43	43.30	40.80	39.33
6, Minimum	51.67	49.86	48.42	47.21	46.20	44.59	42.82	40.81	38.47	37.07

Second and Higher Floors
Estimating Procedure

1. Establish the structure quality class. The class for second and higher floors will usually be the same as the first floor.
2. Compute the square foot area of the second floor and each higher floor.
3. Add to or subtract from the square foot cost below the appropriate amount from the Wall Height Adjustment Table on page 48 if the wall height is more or less than 12 feet.
4. Multiply the adjusted square foot cost by the area of each floor.
5. Deduct, if appropriate, for common walls, using the figures on page 48.
6. Add 2% to the cost of each floor above the second floor. For example, the third floor cost would be 102% of the second floor cost and the fourth floor cost would be 104% of the second floor cost.
7. Multiply the total cost for each floor by the location factor on page 7.
8. Add the cost of heating and cooling equipment, escalators, fire escapes, fire sprinklers, and interior partitions from pages 201 to 213.

Length less than twice width
Square Foot Area

Quality Class	2,500	3,000	3,500	4,000	4,500	5,000	6,000	7,500	10,000	15,000	20,000
1, Best	75.16	73.03	71.37	70.04	68.93	68.01	66.51	64.87	62.99	60.78	59.46
1 & 2	69.79	67.81	66.28	65.04	64.00	63.16	61.76	60.24	58.49	56.43	55.23
2, Good	64.39	62.57	61.14	60.00	59.05	58.26	56.98	55.57	53.97	52.07	50.93
2 & 3	60.60	58.89	57.55	56.47	55.59	54.83	53.63	52.30	50.78	48.99	47.95
3, High Avg.	56.95	55.35	54.09	53.06	52.25	51.52	50.40	49.15	47.73	46.06	45.06
3 & 4	52.46	50.83	49.56	48.54	47.70	46.98	45.82	44.55	43.13	41.42	40.41
4, Low Avg.	47.20	45.74	44.61	43.68	42.93	42.27	41.26	40.10	38.81	37.27	36.36
4 & 5	44.13	42.77	41.70	40.83	40.12	39.53	38.54	37.50	36.29	34.84	34.00
5, Low	40.92	39.65	38.66	37.86	37.20	36.65	35.76	34.76	33.64	32.30	31.51
5 & 6	38.56	37.35	36.43	35.67	35.05	34.53	33.68	32.75	31.70	30.44	29.69
6, Minimum	36.10	34.98	34.11	33.40	32.83	32.32	31.54	30.67	29.68	28.51	27.80

Length between 2 and 4 times width
Square Foot Area

Quality Class	2,500	3,000	3,500	4,000	4,500	5,000	6,000	7,500	10,000	15,000	20,000
1, Best	78.08	75.73	73.92	72.44	71.23	70.20	68.54	66.70	64.62	62.15	60.68
1 & 2	72.50	70.33	68.63	67.27	66.13	65.20	63.64	61.92	59.99	57.71	56.34
2, Good	66.84	64.84	63.28	62.02	60.98	60.09	58.67	57.11	55.33	53.22	51.95
2 & 3	63.01	61.13	59.65	58.47	57.49	56.65	55.30	53.83	52.15	50.15	48.95
3, Hi. Avg.	59.18	57.41	56.03	54.93	54.01	53.23	51.95	50.58	48.98	47.11	46.00
3 & 4	54.61	52.79	51.39	50.25	49.31	48.52	47.24	45.83	44.25	42.37	41.24
4, Low Avg.	49.14	47.50	46.24	45.20	44.36	43.66	42.52	41.26	39.80	38.11	37.11
4 & 5	46.00	44.46	43.28	42.32	41.54	40.86	39.78	38.61	37.26	35.67	34.74
5, Low	42.75	41.31	40.20	39.32	38.60	37.97	36.98	35.87	34.63	33.15	32.28
5 & 6	40.27	38.94	37.88	37.05	36.36	35.78	34.84	33.80	32.64	31.24	30.41
6, Minimum	37.77	36.53	35.54	34.77	34.12	33.57	32.68	31.71	30.62	29.30	28.53

Length over 4 times width
Square Foot Area

Quality Class	2,500	3,000	3,500	4,000	4,500	5,000	6,000	7,500	10,000	15,000	20,000
1, Best	81.21	79.09	77.33	75.86	74.60	73.49	71.66	69.54	67.04	63.91	61.96
1 & 2	75.38	73.41	71.79	70.43	69.25	68.23	66.52	64.56	62.24	59.33	57.51
2, Good	69.48	67.67	66.18	64.92	63.84	62.90	61.33	59.51	57.37	54.69	53.02
2 & 3	65.50	63.80	62.39	61.21	60.18	59.29	57.81	56.11	54.08	51.55	49.99
3, Hi. Avg.	61.51	59.91	58.60	57.47	56.51	55.67	54.30	52.70	50.78	48.42	46.95
3 & 4	57.09	55.39	54.01	52.83	51.85	51.00	49.57	47.94	46.03	43.66	42.20
4, Low Avg.	47.75	49.87	48.62	47.57	46.69	45.92	44.64	43.16	41.45	39.32	38.00
4 & 5	48.09	46.65	45.48	44.51	43.68	42.95	41.75	40.40	38.78	36.77	35.54
5, Low	44.75	43.42	42.32	41.41	40.63	39.98	38.86	37.58	36.09	34.23	33.07
5 & 6	42.17	40.92	39.89	39.03	38.29	37.67	36.64	35.42	34.01	32.26	31.16
6, Minimum	39.61	38.45	37.48	36.68	35.99	35.40	34.41	33.29	31.95	30.32	29.28

Wall Height Adjustment

The square foot costs for urban stores are based on the wall heights listed on each page. The main or first floor height is the distance from the bottom of the floor slab or joists to the top of the roof slab or ceiling joists. Second and higher floors are measured from the top of the floor slab or floor joists to the top of the roof slab or ceiling joists. Add or subtract the amount listed to the square foot cost for each foot more or less than the standard wall height in the tables. For second and higher floors use only 75% of the wall height adjustment cost.

Area	500	600	700	800	900	1,000	1,250	1,500	1,750	2,000	2,500
Adjustment	2.33	2.13	1.96	1.86	1.76	1.66	1.50	1.34	1.26	1.18	1.07

Area	3,000	3,500	4,000	4,500	5,000	6,000	7,500	10,000	15,000	20,000
Adjustment	.96	.89	.84	.79	.76	.68	.60	.52	.43	.38

Perimeter Wall Adjustment

A common wall exists when two buildings share one wall. Adjust for common walls by deducting the linear foot costs below from the total structure cost. In some structures one or more walls are not owned at all. In this case, deduct the "No Ownership" cost per linear foot of wall not owned.

Small Urban Stores (to 2,500 S.F.)
Common wall, deduct $74.00 per linear foot
No ownership, deduct $148.00 per linear foot

Large Urban Stores (over 2,500 S.F.)
Common wall, deduct $142.00 per linear foot
No ownership, deduct $284.00 per linear foot

Urban Stores - Wood or Wood and Steel Frame
Quality Classification

	Class 1 Best Quality	Class 2 Good Quality	Class 3 Average Quality	Class 4 Low Quality
Foundation	Reinforced concrete.	Reinforced concrete.	Reinforced concrete.	Reinforced concrete.
Floor Structure	4" reinforced concrete on 6" rock fill or 2" x 10" joists 16" o.c.	4" reinforced concrete on 6" rock fill or 2" x 8" joists 16" o.c.	4" reinforced concrete on 6" rock fill or 2" x 6" joists 16" o.c.	4" reinforced concrete on 4" rock fill or 2" x 6" joists 16" o.c.
Wall Structure	2" x 6" studs 16" o.c.	2" x 4" or 2" x 6" studs 16" o.c.	2" x 4" studs 16" o.c. up to 14' high, 2" x 6" studs 16" o.c. over 14' high.	2" x 4" studs 16" o.c. up to 14' high, 2" x 6" studs 16" o.c. over 14' high.
Roof Covering	5 ply composition roof on 1" x 6" sheathing with insulation.	5 ply composition roof on 1" x 6" sheathing with insulation.	4 ply composition roof on 1" x 6" sheathing.	4 ply composition roof on 1" x 6" sheathing.
Floor Finish	Sheet vinyl with some terrazzo or good carpet.	Resilient tile with small areas of terrazzo, carpet or solid vinyl tile.	Composition tile.	Minimum grade tile.
Interior Wall Finish	Plaster with putty coat finish on gypsum or metal lath, or 5/8" gypsum wallboard taped, textured and painted or some vinyl wall cover.	Plaster with putty coat finish on gypsum or metal lath, or 5/8" gypsum wallboard taped, textured and painted or with wall paper.	Lath, 2 coats plaster with putty coat finish or 1/2" gypsum wallboard taped, textured and painted.	1/2" gypsum wallboard taped, textured and painted.
Ceiling Finish	Plaster with putty coat finish and some acoustical plaster or suspended acoustical tile with gypsum wallboard backing.	Plaster with putty coat finish or suspended acoustical tile with exposed grid.	Gypsum wallboard taped and textured or lath, 2 coats of plaster and putty coat finish.	Ceiling tile or gypsum wallboard and paint.
Exterior Wall Finish	Good wood siding.	Average wood siding.	Stucco or average wood siding.	Stucco or inexpensive wood siding.
Lighting	Encased modular units and stock chandeliers. Many spotlights.	Quad open strip fixtures or triple encased louvered strip fixtures. Average number of spotlights.	Triple open strip fixtures or double encased louvered strip fixtures. Some spotlights.	Double open strip fixtures.
Plumbing *(Per 5,000 S.F.)*	6 standard fixtures, metal or marble toilet partitions.	6 standard fixtures, metal toilet partitions.	4 standard commercial fixtures, metal toilet partitions.	4 standard commercial fixtures, wood toilet partitions.
Bath Wall Finish	Ceramic tile, marble or plain mosaic tile.	Gypsum wallboard and paint, some ceramic tile or plastic finish wallboard	Gypsum wallboard and paint.	Gypsum wallboard and paint.

Urban Stores - Wood or Wood and Steel Frame
First Floor, Length Less Than Twice Width

Estimating Procedure

1. Establish the structure quality class by applying the information on page 49.
2. Compute the building ground floor area. This should include everything within the exterior walls and all insets outside the walls but under the main roof.
3. Add to or subtract from the cost below the appropriate amount from the Wall Height Adjustment Table on page 54 if the first floor wall height is more or less than 16 feet for large stores or 12 feet for small stores.
4. Multiply the adjusted square foot cost by the building area.
5. Deduct, if appropriate, for common walls, using the figures on page 54.
6. Multiply the total cost by the location factor on page 7.
7. Add the cost of heating and cooling equipment, elevators, escalators, fire escapes, fire sprinklers, display fronts, canopies, ramps, docks, interior partitions, mezzanines, basements, and communication systems from pages 201 to 213.
8. Add the cost of second and higher floors from page 53.

Smaller Stores
Square Foot Area

Quality Class	500	600	700	800	900	1,000	1,250	1,500	1,750	2,000	2,500
2, Good	61.80	59.58	57.86	56.44	55.27	54.27	52.32	50.87	49.74	48.83	47.42
2 & 3	57.62	55.56	53.94	52.63	51.54	50.60	48.79	47.42	46.39	45.53	44.21
3, Average	53.32	51.40	49.90	48.68	47.69	46.83	45.14	43.90	42.90	42.12	40.91
3 & 4	48.71	46.95	45.59	44.48	43.55	42.77	41.24	40.09	39.19	38.49	37.37
4, Low	44.05	42.49	41.25	40.25	39.40	38.69	37.31	36.27	35.46	34.81	33.80

Larger Stores
Square Foot Area

Quality Class	3,000	3,500	4,000	4,500	5,000	6,000	7,500	10,000	15,000	20,000
1, Best	59.98	58.77	57.79	56.97	56.28	55.19	53.97	52.59	50.94	49.97
1 & 2	54.03	52.94	52.06	51.35	50.72	49.73	48.62	47.38	45.91	45.02
2, Good	49.05	48.05	47.25	46.59	46.02	45.13	44.13	42.99	41.66	40.86
2 & 3	46.30	45.36	44.60	43.98	43.45	42.61	41.66	40.59	39.33	38.57
3, Average	43.31	42.43	41.72	41.12	40.64	39.85	38.96	37.97	36.78	36.07
3 & 4	40.50	39.69	39.02	38.47	38.02	37.28	36.44	35.51	34.40	33.75
4, Low	37.85	37.10	36.46	35.97	35.54	34.83	34.08	33.19	32.15	31.55

First Floor, Length Between 2 and 4 Times Width

Estimating Procedure

1. Establish the structure quality class by applying the information on page 49.
2. Compute the building ground floor area. This should include everything within the exterior walls and all insets outside the walls but under the main roof.
3. Add to or subtract from the cost below the appropriate amount from the Wall Height Adjustment Table on page 54 if the first floor wall height is more or less than 16 feet for large stores or 12 feet for small stores.
4. Multiply the adjusted square foot cost by the building area.
5. Deduct, if appropriate, for common walls, using the figures on page 54.
6. Multiply the total cost by the location factor on page 7.
7. Add the cost of heating and cooling equipment, elevators, escalators, fire escapes, fire sprinklers, display fronts, canopies, ramps, docks, interior partitions, mezzanines, basements, and communication systems from pages 201 to 213.
8. Add the cost of second and higher floors from page 53.

Urban Store, Class 3

Smaller Stores
Square Foot Area

Quality Class	500	600	700	800	900	1,000	1,250	1,500	1,750	2,000	2,500
2, Good	67.16	64.47	62.40	60.73	59.34	58.17	55.89	54.21	52.91	51.85	50.24
2 & 3	62.26	59.77	57.86	56.31	55.02	53.94	51.81	50.25	49.04	48.07	46.58
3, Average	57.59	55.33	53.53	52.10	50.92	49.92	47.94	46.51	45.39	44.48	43.11
3 & 4	52.65	50.55	48.91	47.60	46.52	45.60	43.81	42.50	41.47	40.65	39.39
4, Low	47.51	45.61	44.14	42.96	41.99	41.15	39.55	38.35	37.43	36.69	35.55

Larger Stores
Square Foot Area

Quality Class	3,000	3,500	4,000	4,500	5,000	6,000	7,500	10,000	15,000	20,000
1, Best	60.68	61.42	60.27	58.52	58.52	57.23	55.80	54.16	52.24	51.08
1 & 2	57.11	55.81	54.77	53.90	53.17	52.00	50.71	49.23	47.47	46.41
2, Good	51.44	50.28	49.34	48.56	47.90	46.84	45.67	44.34	42.75	41.81
2 & 3	48.49	47.39	46.51	45.77	45.15	44.15	43.04	41.79	40.30	39.42
3, Average	45.63	44.59	43.76	43.09	42.49	41.56	40.51	39.32	37.93	37.09
3 & 4	42.63	41.67	40.89	40.26	39.71	38.82	37.85	36.75	35.44	34.67
4, Low	39.60	38.71	37.99	37.39	36.88	36.06	35.16	34.14	32.92	32.20

First Floor, Length More Than 4 Times Width

Estimating Procedure

1. Establish the structure quality class by applying the information on page 49.
2. Compute the building ground floor area. This should include everything within the exterior walls and all insets outside the walls but under the main roof.
3. Add to or subtract from the cost below the appropriate amount from the Wall Height Adjustment Table on page 54 if the first floor wall height is more or less than 16 feet for large stores or 12 feet for small stores.
4. Multiply the adjusted square foot cost by the building area.
5. Deduct, if appropriate, for common walls, using the figures on page 54.
6. Multiply the total cost by the location factor on page 7.
7. Add the cost of heating and cooling equipment, elevators, escalators, fire escapes, fire sprinklers, display fronts, canopies, ramps, docks, interior partitions, mezzanines, basements, and communication systems from pages 201 to 213.
8. Add the cost of second and higher floors from page 53.

Urban Store, Class 3

Smaller Stores
Square Foot Area

Quality Class	500	600	700	800	900	1,000	1,250	1,500	1,750	2,000	2,500
2, Good	73.66	70.03	67.30	65.16	63.42	61.96	59.20	57.21	55.70	54.51	52.71
2 & 3	68.47	65.10	62.58	60.58	58.96	57.60	55.05	53.20	51.78	50.69	49.00
3, Average	63.27	60.18	57.82	55.98	54.49	53.24	50.86	49.16	47.86	46.84	45.30
3 & 4	57.62	54.79	52.66	50.97	49.62	48.48	46.32	44.76	43.58	42.63	41.24
4, Low	52.42	49.84	47.91	46.38	45.14	44.11	42.15	40.73	39.64	38.81	37.52

Larger Stores
Square Foot Area

Quality Class	3,000	3,500	4,000	4,500	5,000	6,000	7,500	10,000	15,000	20,000
1, Best	66.48	64.74	63.35	62.19	61.22	59.67	57.96	56.05	53.79	52.46
1 & 2	60.43	58.86	57.59	56.53	55.65	54.26	52.70	50.95	48.91	47.70
2, Good	54.51	53.08	51.94	51.00	50.19	48.93	47.53	45.96	44.12	43.03
2 & 3	51.42	50.08	48.99	48.11	47.36	46.16	44.85	43.36	41.62	40.59
3, Average	48.32	47.05	46.03	45.20	44.49	43.38	42.13	40.74	39.10	38.15
3 & 4	45.11	43.92	42.98	42.20	41.56	40.49	39.34	38.03	36.51	35.61
4, Low	41.84	40.75	39.88	39.15	38.53	37.56	36.49	35.27	33.86	33.02

Urban Stores - Wood or Wood and Steel Frame
Second and Higher Floors

Estimating Procedure

1. Establish the structure quality class. The class for second and higher floors will usually be the same as the first floor.
2. Compute the square foot area of the second floor and each higher floor.
3. Add to or subtract from the square foot cost below the appropriate amount from the Wall Height Adjustment Table on page 54 if the wall height is more or less than 12 feet.
4. Multiply the adjusted square foot cost by the area of each floor.
5. Deduct, if appropriate, for common walls, using the figures on page 54.
6. Add 2% to the cost for each floor above the second floor. For example, the third floor cost would be 102% of the second floor cost and the fourth floor cost would be 104% of the second floor cost.
7. Multiply the total cost by the location factor on page 7.
8. Add the cost of heating and cooling equipment, escalators, fire escapes, fire sprinklers, and interior partitions from pages 201 to 213.

Length less than twice width
Square Foot Area

Quality Class	2,500	3,000	3,500	4,000	4,500	5,000	6,000	7,500	10,000	15,000	20,000
1, Best	51.12	50.06	49.21	48.52	47.98	47.50	46.74	45.92	44.95	43.82	43.15
1 & 2	46.19	45.22	44.46	43.85	43.36	42.93	42.24	41.47	40.61	39.59	38.98
2, Good	41.30	40.43	39.77	39.21	38.75	38.39	37.75	37.09	36.32	35.39	34.85
2 & 3	38.11	37.32	36.69	36.20	35.77	35.43	34.86	34.22	33.51	32.68	32.17
3, Average	35.02	34.29	33.73	33.26	32.88	32.55	32.04	31.46	30.80	30.02	29.57
3 & 4	32.83	32.15	31.60	31.18	30.83	30.52	30.04	29.49	28.88	28.15	27.70
4, Low	30.60	29.98	29.46	29.06	28.74	28.44	28.00	27.50	26.91	26.24	25.83

Length between 2 and 4 times width
Square Foot Area

Quality Class	2,500	3,000	3,500	4,000	4,500	5,000	6,000	7,500	10,000	15,000	20,000
1, Best	52.60	51.40	50.49	49.74	49.13	48.60	47.75	46.81	45.75	44.50	43.73
1 & 2	47.51	46.43	45.60	44.92	44.37	43.91	43.13	42.29	41.33	40.20	39.51
2, Good	42.47	41.50	40.75	40.17	39.66	39.24	38.56	37.80	36.94	35.93	35.31
2 & 3	39.25	38.36	37.68	37.12	36.66	36.27	35.64	34.95	34.15	33.21	32.64
3, Average	36.06	35.24	34.62	34.12	33.68	33.32	32.75	32.11	31.37	30.51	30.00
3 & 4	33.76	33.01	32.42	31.94	31.53	31.20	30.66	30.07	29.38	28.57	28.08
4, Low	31.50	30.78	30.23	29.79	29.42	29.11	28.60	28.04	27.41	26.64	26.13

Length over 4 times width
Square Foot Area

Quality Class	2,500	3,000	3,500	4,000	4,500	5,000	6,000	7,500	10,000	15,000	20,000
1, Best	54.85	53.45	52.36	51.49	50.77	50.15	49.16	48.06	46.82	45.36	44.49
1 & 2	49.57	48.29	47.33	46.53	45.87	45.31	44.42	43.43	42.32	41.00	40.21
2, Good	44.31	43.18	42.30	41.61	41.02	40.53	39.73	38.84	37.84	36.64	35.96
2 & 3	41.00	39.95	39.13	38.49	37.93	37.48	36.74	35.93	35.00	33.90	33.25
3, Average	37.63	36.68	35.93	35.33	34.83	34.40	33.74	32.98	32.13	31.12	30.52
3 & 4	35.28	34.38	33.68	33.13	32.65	32.26	31.61	30.92	30.13	29.18	28.61
4, Low	32.81	31.98	31.33	30.80	30.38	30.02	29.42	28.76	28.02	27.15	26.62

Wall Height Adjustment

The square foot costs for urban stores are based on the wall heights listed on each page. The main or first floor height is the distance from the bottom of the floor slab or joists to the top of the roof slab or ceiling joists. Second and higher floors are measured from the top of the floor slab or floor joists to the top of the roof slab or ceiling joists. Add or subtract the amount listed to the square foot cost for each foot more or less than the standard wall height in the tables. For second and higher floors use only 75% of the wall height adjustment cost.

Area	500	600	700	800	900	1,000	1,250	1,500	1,750	2,000	2,500
Adjustment	1.00	.90	.84	.78	.74	.70	.64	.58	.54	.51	.46

Area	3,000	3,500	4,000	4,500	5,000	6,000	7,500	10,000	15,000	20,000
Adjustment	.40	.37	.36	.34	.33	.31	.29	.27	.25	.24

Perimeter Wall Adjustment

A common wall exists when two buildings share one wall. Adjust for common walls by deducting the linear foot costs below from the total structure cost. In some structures one or more walls are not owned at all. In this case, deduct the "No Ownership" cost per linear foot of wall not owned.

Small Urban Stores (2,500 S.F. or less)
Common wall, deduct $50.00 per linear foot
No ownership, deduct $100.00 per linear foot

Large Urban Stores (over 2,500 S.F.)
Common wall, deduct $73.00 per linear foot
No ownership, deduct $147.00 per linear foot

Suburban Stores

Suburban stores are usually built as part of shopping centers. They differ from urban stores in that they are built in open areas where modern construction techniques, equipment and more economical designs can be used. They are also subject to greater variations in size and shape than are urban stores. Do not use the figures in this section for department stores, discount houses or urban stores. These building types are evaluated in other sections.

Costs identified "building shell only" do not include permanent partitions, display fronts or finish materials on the front of the building. Costs for "multi-unit buildings" include partitions, display fronts and finish materials on the front of the building. All figures include the following costs:
1. Foundations as required for normal soil conditions.
2. Floor, rear wall, side wall and roof structures.
3. A front wall consisting of vertical support columns or pilasters and horizontal beams spanning the area between these members, leaving an open space to receive a display front.
4. Interior floor, wall and ceiling finishes.
5. Exterior wall finish on the side and rear walls.
6. Roof cover.
7. Basic lighting and electrical systems.
8. Rough and finish plumbing.
9. A usual or normal parapet wall.
10. Design and engineering fees.
11. Permits and fees.
12. Utility hook-ups.
13. Contractor's contingency, overhead and mark-up.

The in-place costs of these extra components should be added to the basic building cost to arrive at total structure cost. See the section "Additional Costs for Commercial and Industrial Structures" on page 201.
1. Heating and air conditioning systems.
2. Fire sprinklers.
3. All display front components (shell-type buildings only).
4. Finish materials on the front wall of the building (shell-type building only).
5. Canopies.
6. Interior partitions (shell-type buildings only).
7. Exterior signs.
8. Mezzanines and basements.
9. Loading docks and ramps.
10. Miscellaneous yard improvements.
11. Communications systems.

For valuation purposes suburban stores are divided into two building types: masonry or concrete frame, or wood or wood and steel frame. Each building type is divided into four shape classes:

1) Buildings in which the depth is greater than the front. 2) Buildings in which the front is between one and two times the depth. 3) Buildings in which the front is between two and four times the depth. 4) Buildings in which the front is greater than four times the depth. Angular buildings should be classed by comparing the sum of the length of all wings to the width of the wings. All areas should be included, but no area should be included as part of two different wings. Note the example at the right.

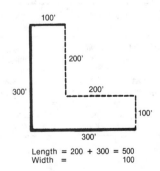

Length = 200 + 300 = 500
Width =   100

Suburban Stores - Masonry or Concrete
Quality Classification

	Class 1 Best Quality	Class 2 Good Quality	Class 3 Average Quality	Class 4 Low Quality
Foundation	Reinforced concrete.	Reinforced concrete.	Reinforced concrete.	Reinforced concrete.
Floor Structure	6" reinforced concrete on 6" rock base.	6" reinforced concrete on 6" rock base.	6" reinforced concrete on 6" rock base.	4" reinforced concrete on 6" rock base.
Wall Structure	8" reinforced decorative concrete block, 6" concrete tilt-up or 8" reinforced brick.	8" reinforced decorative concrete block, 6" concrete tilt-up or 8" reinforced brick.	8" reinforced concrete block, 6" concrete tilt-up or 8" reinforced common brick.	8" reinforced concrete block or 6" concrete tilt-up.
Roof	Glu-lam or steel beams on steel intermediate columns. Panelized roof system, 1/2" plywood sheathing, 5 ply built-up roof with insulation.	Glu-lam or steel beams on steel intermediate columns. Panelized roof system, 1/2" plywood sheathing, 5 ply built-up roof with insulation.	Glu-lam beams or steel intermediate columns. Panelized roof system, 1/2" plywood sheathing, 4 ply built-up roof.	Glu-lam beams on steel intermediate columns. Panelized roof system, 1/2" plywood sheathing, 4 ply built-up roof.
Floor Finish	Terrazzo, sheet vinyl or very good carpet.	Resilient tile with 50% solid vinyl tile, terrazzo, or good carpet.	Composition tile.	Minimum grade tile.
Interior Wall Finish	Inside of exterior walls furred out with gypsum wallboard or lath and plaster cover. Exterior walls and partitions finished with vinyl wall covers and hardwood veneers.	Interior stucco on inside of exterior walls, gypsum wallboard and texture or paper on partitions, some vinyl wall cover and plywood paneling.	Interior stucco on inside of exterior walls, gypsum wallboard and texture and paint on partitions.	Paint on inside of exterior walls, gypsum wallboard with texture and paint on partitions.
Ceiling Finish	Suspended good grade acoustical tile with gypsum wallboard backing.	Suspended acoustical tile with concealed grid system.	Suspended acoustical tile with exposed grid system.	Exposed beams with ceiling tile or paint.
Lighting	Recessed fluorescent lighting in modular plastic panels.	Continuous recessed 3 tube fluorescent strips with egg crate diffusers, 8' o.c.	Continuous 3 tube fluorescent strips with egg crate diffusers, 8' o.c.	Continuous exposed 2 tube fluorescent strips, 8' o.c.
Exterior	Face brick or stone veneer.	Exposed aggregate, some stone veneer.	Paint on exposed areas, some exposed aggregate.	Paint on exposed areas.
Plumbing	6 good fixtures per 5,000 S.F. of floor area, metal toilet partitions.	6 standard fixtures per 5,000 S.F. of floor area, metal toilet partitions.	4 standard fixtures per 5,000 S.F. of floor area, metal toilet partitions.	4 standard fixtures per 5,000 S.F. of floor area wood toilet partitions.

The costs on page 59 and 60, and 66 to 69 include display fronts. The quality of the display front will help establish the quality class of the building as a whole. Display fronts are classified as follows:

	Class 1 Best Quality	Class 2 Good Quality	Class 3 Average Quality	Class 4 Low Quality
Bulkhead (0 to 4' high)	Vitrolite, domestic marble or stainless steel.	Black flagstone, terrazzo or good ceramic tile.	Average ceramic tile, Roman brick or imitation flagstone.	Stucco, wood or common brick.
Window Frame	Bronze or stainless steel.	Heavy aluminum.	Aluminum.	Light aluminum with with stops.
Glass	1/4" plate glass with mitered joints.	1/4" plate glass, some mitered joints.	1/4" plate glass.	Crystal or 1/4" plate glass.
Sign Area (4' high)	Vitrolite, domestic marble or stainless steel.	Black flagstone, terrazzo, or good ceramic tile	Average ceramic tile, Roman brick or imitation flagstone.	Stucco.
Pilasters	Vitrolite, domestic marble.	Black flagstone, terrazzo or good ceramic tile.	Average ceramic tile, Roman brick or imitation flagstone.	Stucco.

Suburban Stores - Masonry or Concrete
Building Shell Only

Estimating Procedure

1. Use these figures to estimate the cost of shell-type buildings without permanent partitions, display fronts or finish material on the front wall of the building.
2. Establish the structure quality class by applying the information on page 56.
3. Compute the building floor area. This should include everything within the exterior walls and all inset areas outside the main walls but under the main roof.
4. Add to or subtract from the cost below the appropriate amount from the Wall height Adjustment Table (at the bottom of this page) if the wall height is more or less than 16 feet.
5. Multiply the adjusted square foot cost by the building area.
6. Deduct, if appropriate, for common walls, using the figures at the bottom of page 58.
7. Multiply the total cost by the location factor on page 7.
8. Add the cost of the appropriate additional components from page 201: heating and cooling equipment, fire sprinklers, display fronts, finish materials on the front wall, canopies, interior partitions, exterior signs, mezzanines and basements, loading docks and ramps, yard improvements, and communication systems.

Depth greater than length of front
Square Foot Area

Quality Class	500	1,000	2,000	2,500	3,000	4,000	5,000	7,500	10,000	12,500	15,000
1, Best	96.40	80.41	69.12	66.26	64.14	61.16	59.13	55.98	54.09	52.82	51.86
1 & 2	88.78	74.05	63.67	61.03	59.08	56.33	54.45	51.56	49.82	48.65	47.77
2, Good	81.13	67.67	58.18	55.75	53.98	51.47	49.77	47.11	45.52	44.43	43.64
2 & 3	76.46	63.78	54.82	52.56	50.87	48.51	46.90	44.40	42.91	41.88	41.14
3, Average	71.71	59.82	51.42	49.28	47.71	45.49	43.99	41.63	40.24	39.30	38.59
3 & 4	67.27	56.12	48.25	46.23	44.75	42.66	41.25	39.06	37.75	36.84	36.19
4, Low	63.76	53.17	45.71	43.81	42.41	40.44	39.10	37.02	35.76	34.92	34.29

Length of front between 1 and 2 times depth
Square Foot Area

Quality Class	500	1,000	2,000	3,000	5,000	7,500	10,000	15,000	20,000	25,000	35,000
1, Best	92.10	77.38	67.01	62.42	57.84	54.95	53.22	51.18	49.98	49.13	48.05
1 & 2	84.69	71.16	61.61	57.40	53.18	50.53	48.94	47.06	45.94	45.18	44.16
2, Good	77.39	65.01	56.30	52.45	48.58	46.16	44.71	43.00	41.97	41.27	40.36
2 & 3	72.87	61.23	53.02	49.39	45.76	43.45	42.10	40.48	39.51	38.87	38.01
3, Average	68.47	57.53	49.83	46.41	43.00	40.86	39.56	38.05	37.14	36.53	35.71
3 & 4	64.58	54.26	46.98	43.76	40.54	38.52	37.32	35.87	35.03	34.45	33.69
4, Low	60.62	50.93	44.10	41.08	38.05	36.16	35.02	33.68	32.88	32.32	31.61

Wall Height Adjustment: Costs above are based on a 16' wall height, measured from the bottom of the floor slab or floor joists to the top of the roof cover. Add or subtract the amount listed to the square foot cost for each foot more or less than 16 feet.

Area	500	1,000	2,000	2,500	3,000	4,000	5,000	7,500	10,000	12,500
Cost	1.60	1.18	.85	.76	.69	.60	.54	.41	.36	.32

Area	15,000	20,000	25,000	30,000	35,000	50,000	70,000	75,000	100,000	150,000
Cost	.31	.30	.28	.26	.24	.20	.16	.15	.11	.08

Suburban Stores - Masonry or Concrete
Building Shell Only

Estimating Procedure

1. Use these figures to estimate the cost of shell-type buildings without permanent partitions, display fronts or finish material on the front wall of the building.
2. Establish the structure quality class by applying the information on page 56.
3. Compute the building floor area. This should include everything within the exterior walls and all inset areas outside the main walls but under the main building roof.
4. Add to or subtract from the cost below the appropriate amount from the Wall Height Adjustment Table (on page 57) if the wall height is more or less than 16 feet.
5. Multiply the adjusted square foot cost by the building area.
6. Deduct, if appropriate, for common walls, using the figures at the bottom of page 58.
7. Multiply the total cost by the location factor on page 7.
8. Add the cost of the appropriate additional components from page 201: heating and cooling equipment, fire sprinklers, display fronts, finish materials on the front wall, canopies, interior partitions, exterior signs, mezzanines and basements, loading docks and ramps, yard improvements, and communication systems.

Suburban Store, Class 1

Length between 2 and 4 times depth
Square Foot Area

Quality Class	1,000	2,000	3,000	5,000	7,500	10,000	15,000	20,000	30,000	50,000	70,000
1, Best	80.24	69.21	64.34	59.45	56.36	54.53	52.36	45,49	49.51	47.97	47.13
1 & 2	73.53	63.41	58.94	54.46	51.65	49.96	47.96	46.78	45.36	43.96	43.20
2, Good	66.81	57.63	53.57	49.49	46.93	45.41	43.59	42.51	41.23	39.94	39.25
2 & 3	62.98	54.32	50.49	46.66	44.24	42.80	41.09	40.08	38.86	37.65	36.99
3, Average	59.06	50.95	47.38	43.75	41.50	40.14	38.53	37.59	36.44	35.30	34.70
3 & 4	55.82	48.16	44.76	41.36	39.21	37.94	36.42	35.51	34.45	33.37	32.79
4, Low	52.45	45.24	42.05	38.86	36.86	35.64	34.23	33.38	32.36	31.36	30.83

Length greater than 4 times depth
Square Foot Area

Quality Class	2,000	3,000	5,000	10,000	15,000	20,000	30,000	50,000	75,000	100,000	150,000
1, Best	73.02	67.51	61.99	56.42	53.98	52.52	50.79	49.05	47.96	47.31	46.55
1 & 2	66.17	61.17	56.16	51.13	48.91	47.59	46.02	44.45	43.45	42.88	42.18
2, Good	60.09	55.55	50.99	46.44	44.41	43.21	41.78	40.36	39.46	38.93	38.28
2 & 3	56.61	52.32	48.04	43.74	41.84	40.71	39.37	38.03	37.18	36.66	36.08
3, Average	53.08	49.09	45.06	41.03	39.23	38.17	36.92	35.66	34.87	34.40	33.83
3 & 4	49.95	46.18	42.40	38.59	36.91	35.92	34.73	33.56	32.80	32.36	31.84
4, Low	47.05	43.50	39.93	36.36	34.78	33.85	32.72	31.61	30.90	30.49	29.98

Perimeter Wall Adjustment: A common wall exists when two buildings share one wall. Adjust for common walls by deducting $140.00 per linear foot of common wall from the total structure cost. In some structures one or more walls are not owned at all. In this case, deduct $280.00 per linear foot of wall not owned.

Suburban Stores - Masonry or Concrete
Multi-Unit Buildings

Estimating Procedure

1. Use these square foot costs to estimate the cost of stores designed for multiple occupancy. These costs include all components of shell buildings plus the cost of display fronts, finish materials on the front of the building and normal interior partitions.
2. Establish the structure quality class by applying the information on page 56. Evaluate the quality of the display front to help establish the correct quality class of the building as a whole. See also pages 207 to 209.
3. Compute the building floor area. This should include everything within the building exterior walls and all inset areas outside the main walls but under the main building roof.
4. Add to or subtract from the cost below the appropriate amount from the Wall Height Adjustment Table (at the bottom of this page) if the wall height is more or less than 16 feet.
5. Multiply the adjusted square foot cost by the building area.
6. Deduct, if appropriate, for common walls, using the figures at the bottom of page 60.
7. Multiply the total cost by the location factor on page 7.
8. Add the cost of the appropriate additional components from page 201: heating and cooling equipment, fire sprinklers, canopies, exterior signs, mezzanines and basements, loading docks and ramps, yard improvements, and communications systems.

Depth greater than length of front
Square Foot Area

Quality Class	500	1,000	2,000	2,500	3,000	4,000	5,000	7,500	10,000	12,500	15,000
1, Best	137.25	112.06	94.26	89.71	86.36	81.68	78.47	73.49	70.52	68.50	67.00
1 & 2	122.37	99.90	84.03	79.99	77.00	72.81	69.95	65.52	62.89	61.08	59.73
2, Good	107.31	87.59	73.69	70.13	67.52	63.86	61.34	57.45	55.13	53.55	52.37
2 & 3	101.03	80.07	67.35	64.12	61.72	58.36	56.08	52.50	50.40	48.94	47.89
3, Average	88.57	72.29	60.82	57.89	55.75	52.70	50.62	47.43	45.50	44.21	43.24
3 & 4	81.74	66.73	56.13	53.44	51.43	48.65	46.73	43.76	41.99	40.80	39.90
4, Low	74.94	61.17	51.46	48.98	47.16	44.58	42.85	40.13	38.50	37.40	36.58

Length of front between 1 and 2 times depth
Square Foot Area

Quality Class	500	1,000	2,000	3,000	5,000	7,500	10,000	15,000	20,000	25,000	35,000
1, Best	151.00	122.27	101.79	92.67	83.49	77.70	74.24	70.13	67.68	66.00	63.81
1 & 2	133.54	108.13	90.00	81.95	73.84	68.71	65.65	62.01	59.85	58.37	56.44
2, Good	115.89	93.83	78.13	71.13	64.07	59.64	56.98	53.83	51.94	50.66	48.96
2 & 3	105.24	85.22	70.95	64.60	58.21	54.16	51.75	48.89	47.18	46.01	44.48
3, Average	97.78	79.18	65.92	60.01	54.08	50.32	48.08	45.41	43.84	42.74	41.32
3 & 4	92.09	74.57	62.07	56.52	50.93	47.39	45.28	42.78	41.28	40.25	38.91
4, Low	83.61	67.70	56.36	51.32	46.24	43.03	41.12	38.84	37.48	36.55	35.34

Wall Height Adjustment: Costs above are based on a 16' wall height, measured from the bottom of the floor slab or floor joists to the top of the roof cover. Add or subtract the amount listed to the square foot cost for each foot more or less than 16 feet.

Square Foot Area

Class	2,000	3,000	5,000	10,000	15,000	25,000	35,000	50,000	75,000	100,000	150,000
1, Best	3.91	3.18	2.45	1.72	1.44	1.10	.94	.81	.72	.67	.61
2, Good	2.81	2.29	1.78	1.23	1.06	.80	.70	.61	.51	.48	.44
3, Average	2.11	1.72	1.32	.94	.79	.62	.52	.44	.39	.37	.34
4, Low	1.63	1.33	1.02	.72	.62	.47	.40	.35	.31	.28	.27

Suburban Stores - Masonry or Concrete
Multi-Unit Buildings

Estimating Procedure

1. Use these square foot costs to estimate the cost of stores designed for multiple occupancy. These costs include all components of shell buildings plus the cost of display fronts, finish materials on the front of the building and normal interior partitions.
2. Establish the structure quality class by applying the information on page 56. Evaluate the quality of the display front to help establish the correct quality class of the building as a whole. See also pages 207 to 209.
3. Compute the building floor area. This should include everything within the building exterior walls and all inset areas outside the main walls but under the main building roof.
4. Add to or subtract from the cost below the appropriate amount from the Wall Height Adjustment Table (at the bottom of this page) if the wall height is more or less than 16 feet.
5. Multiply the adjusted square foot cost by the building area.
6. Deduct, if appropriate, for common walls, using the figures at the bottom of this page.
7. Multiply the total cost by the location factor on page 7.
8. Add the cost of the appropriate additional components from page 201: heating and cooling equipment, fire sprinklers, canopies, exterior signs, mezzanines and basements, loading docks and ramps, yard improvements, and communications systems.

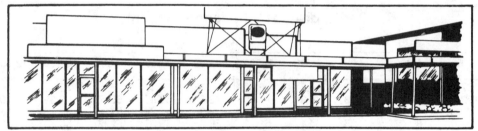

Suburban Store, Class 3 & 4

Length of front between 2 and 4 times depth
Square Foot Area

Quality Class	1,000	2,000	3,000	5,000	7,500	10,000	15,000	20,000	30,000	50,000	70,000
1, Best	137.86	113.10	102.06	90.92	83.88	79.67	74.69	71.70	68.16	64.59	62.70
1 & 2	122.32	100.35	90.55	80.68	74.43	70.69	66.26	63.62	60.47	57.32	55.63
2, Good	105.73	86.73	78.26	69.73	64.32	61.11	57.27	54.99	52.27	49.54	48.08
2 & 3	95.50	78.34	70.71	62.99	58.11	55.20	51.74	49.68	47.22	44.76	43.43
3, Average	87.80	69.86	63.05	56.18	51.81	49.22	46.14	44.29	42.12	39.89	38.72
3 & 4	78.20	64.16	57.88	51.58	47.58	45.21	42.36	40.67	38.66	36.65	35.56
4, Low	71.22	58.44	52.72	46.98	43.34	41.17	38.57	37.05	35.21	33.38	32.40

Length of front greater than 4 times depth
Square Foot Area

Quality Class	2,000	3,000	5,000	10,000	15,000	20,000	30,000	50,000	75,000	100,000	150,000
1, Best	126.43	112.80	99.24	85.70	79.72	76.17	71.98	67.78	65.14	63.56	61.71
1 & 2	111.27	99.28	87.34	75.43	70.18	67.06	63.36	59.66	57.33	55.94	54.31
2, Good	96.02	85.68	75.38	65.09	60.57	57.88	54.68	51.49	49.48	48.29	46.87
2 & 3	86.58	77.25	67.98	58.70	54.61	52.18	49.29	46.42	44.61	43.54	42.25
3, Average	77.35	69.02	60.72	52.43	48.78	46.62	44.04	41.48	39.85	38.88	37.75
3 & 4	71.07	63.41	55.79	48.18	44.83	42.83	40.47	38.11	36.62	35.73	34.69
4, Low	64.55	57.59	50.66	43.75	40.70	38.89	36.74	34.61	33.25	32.45	31.50

Perimeter Wall Adjustment: A common wall exists when two buildings share one wall. Adjust for common walls by deducting $145.00 per linear foot of common wall from the total structure cost. In some structures one or more walls are not owned at all. In this case, deduct $290.00 per linear foot of wall not owned.

Suburban Stores - Wood or Wood and Steel Frame

Quality Classification

	Class 1 Best Quality	Class 2 Good Quality	Class 3 Average Quality	Class 4 Low Quality
Foundation	Reinforced concrete.	Reinforced concrete.	Reinforced concrete.	Reinforced concrete.
Floor Structure	6" reinforced concrete on 6" rock base.	6" reinforced concrete on 6" rock base.	6" reinforced concrete on 6" rock base.	4" reinforced concrete on 6" rock base.
Wall Structure	2" x 6" - 16" o.c.	2" x 6" - 16" o.c.	2" x 6" - 16" o.c.	2" x 4" - 16" o.c.
Roof	Glu-lams or steel beams on steel intermediate columns. Panelized roof system, 1/2" plywood sheathing, 5 ply built-up roof with insulation.	Glu-lams or steel beams on steel intermediate columns. Panelized roof system, 1/2" plywood sheathing, 5 ply built-up roof with insulation.	Glu-lams on steel intermediate columns. Panelized roof system, 1/2" plywood sheathing, 4 ply built-up roof.	Glu-lams on steel intermediate columns. Panelized roof system, 1/2" plywood sheathing 4 ply built-up roof.
Floor Finish	Terrazzo, sheet vinyl, or very good carpet.	Resilient tile with 50% solid vinyl tile, terrazzo, or good carpet.	Composition tile.	Minimum grade tile.
Interior Wall Finish	Gypsum wallboard or lath and plaster on exterior walls and partitions, finished with vinyl wall covers and hardwood veneers.	Gypsum wallboard, texture and paper on exterior walls and partitions, some vinyl wall cover and plywood paneling	Gypsum wallboard, texture and paint on interior walls and partitions.	Gypsum wallboard, texture and paint on interior walls and partitions.
Ceiling Finish	Suspended good grade acoustical tile with gypsum board backing.	Suspended acoustical tile with concealed grid system.	Suspended acoustical tile with exposed grid system.	Exposed beams with ceiling tile or painted.
Lighting	Recessed fluorescent lighting in modular plastic panels.	Continuous recessed 3 tube fluorescent strips with egg crate diffusers, 8' o.c.	Continues 3 tube fluorescent strips with egg crate diffusers, 8' o.c.	Continuous exposed 2 tube fluorescent strips, 8' o.c.
Exterior	Face brick or stone veneer.	Wood siding, some stone veneer.	Stucco on exposed areas, some brick trim.	Stucco on exposed areas.
Plumbing	6 good fixtures per 5,000 S.F. of floor area, metal toilet partitions.	6 standard fixtures per 5,000 S.F. of floor area, metal toilet partitions.	4 standard fixtures per 5,000 S.F. of floor area, metal toilet partitions.	4 standard fixtures per 5,000 S.F. of floor area wood toilet partitions.

Strip and Island Suburban Stores

For estimating purposes, wood frame suburban stores should be divided into strip type units or island type units. Strip type buildings have a front wall made up of display fronts. The side and rear walls, except for delivery or walk-through doors, are made up of solid, continuous wood frame walls. If there are any display areas in the sides or rear of these buildings, the cost of the display front must be added to the building cost and the cost of the wall that it replaces must be deducted from the building costs.

Island type suburban store buildings have display fronts on the major portion of all four sides. Stores may be arranged so that one store fronts on two sides or they may be partitioned in such a way that there are two separate stores fronting on each side of the building.

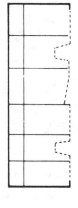

Strip Type

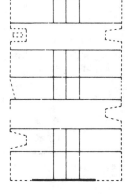

Island Type

Suburban Stores - Wood or Wood and Steel Frame

Building Shell Only, Island Type

Estimating Procedure

1. Use these figures to estimate the cost of shell-type buildings without permanent partitions, display fronts or finish material on the front wall of the building.
2. Establish the structure quality class by applying the information on page 61.
3. Compute the building floor area. This should include everything within the exterior walls and all inset areas outside the main walls but under the main roof.
4. Add to or subtract from the square foot cost below the appropriate amount from the Wall Height Adjustment Table (at the bottom of this page) if the wall height is more or less than 16 feet.
5. Multiply the adjusted square foot cost by the building area.
6. Deduct, if appropriate, for common walls, using the figures at the bottom of this page.
7. Multiply the total cost by the location factor on page 7.
8. Add the cost of the appropriate additional components from page 201: heating and cooling equipment, fire sprinklers, display fronts, finish materials on the front wall, canopies, interior partitions, exterior signs, mezzanines and basements, loading docks and ramps, yard improvements, and communication systems.

Length less than 1-1/2 times depth
Square Foot Area

Quality Class	3,500	5,000	7,500	10,000	12,500	15,000	20,000	30,000	40,000	50,000	75,000
1, Best	54.98	54.07	53.18	52.66	52.31	52.04	51.67	51.22	50.94	50.77	50.49
1 & 2	49.56	48.73	47.93	47.48	47.14	46.91	46.58	46.17	45.92	45.77	45.52
2, Good	44.10	43.37	42.67	42.26	41.96	41.75	41.45	41.09	40.88	40.74	40.50
2 & 3	42.38	40.95	39.61	39.22	38.95	38.76	38.49	38.14	37.94	37.81	37.61
3, Average	37.80	37.17	36.56	36.20	35.96	35.76	35.53	35.23	35.04	34.91	34.71
3 & 4	34.53	33.95	33.39	33.07	32.84	32.68	32.44	32.16	31.99	31.88	31.69
4, Low	31.23	30.70	30.20	29.90	29.70	29.56	29.33	29.10	28.94	28.83	28.67

Length between 1-1/2 and 2 times depth
Square Foot Area

Quality Class	4,500	5,000	7,500	10,000	15,000	20,000	30,000	40,000	50,000	75,000	100,000
1, Best	54.73	54.42	53.38	52.79	52.12	51.74	51.28	51.04	50.87	50.60	50.46
1 & 2	49.31	49.01	48.09	47.55	46.95	46.61	46.21	45.98	45.82	45.59	45.46
2, Good	43.90	43.64	42.81	42.34	41.79	41.50	41.13	40.93	40.80	40.58	40.47
2 & 3	40.70	40.45	39.69	39.25	38.76	38.48	38.14	37.95	37.83	37.63	37.53
3, Average	37.56	37.34	36.62	36.24	35.75	35.50	35.20	35.03	34.90	34.72	34.63
3 & 4	34.31	34.10	33.47	33.10	32.67	32.43	32.17	31.99	31.89	31.73	31.62
4, Low	30.97	30.78	30.20	29.87	29.49	29.28	29.02	28.89	28.78	28.64	28.55

Wall Height Adjustment: Add or subtract the amount listed to the square foot costs above for each foot of wall height more or less than 16 feet.

Area	3,500	5,000	7,500	10,000	12,500	15,000	Over 20,000
Cost	.13	.12	.10	.09	.07	.06	.03

Perimeter Wall Adjustment: For common wall deduct $29.50 per linear foot of common wall. For no wall ownership, deduct $59.00 per linear foot of wall.

Suburban Stores - Wood or Wood and Steel Frame

Building Shell Only, Island Type

Estimating Procedure

1. Use these figures to estimate the cost of shell-type buildings without permanent partitions, display fronts or finish material on the front wall of the building.
2. Establish the structure quality class by applying the information on page 61.
3. Compute the building floor area. This should include everything within the exterior walls and all inset areas outside the main walls but under the main roof.
4. Add to or subtract from the square foot cost below the appropriate amount from the Wall Height Adjustment Table (at the bottom of this page) if the wall height is more or less than 16 feet.
5. Multiply the adjusted square foot cost by the building area.
6. Deduct, if appropriate, for common walls, using the figures at the bottom of this page.
7. Multiply the total cost by the location factor on page 7.
8. Add the cost of the appropriate additional components from page 201: heating and cooling equipment, fire sprinklers, display fronts, finish materials on the front wall, canopies, interior partitions, exterior signs, mezzanines and basements, loading docks and ramps, yard improvements, and communication systems.

Length between 2 and 3 times depth
Square Foot Area

Quality Class	10,000	15,000	20,000	30,000	40,000	50,000	75,000	100,000	150,000	200,000	250,000
1, Best	53.58	52.80	52.33	51.78	51.46	51.24	50.90	50.71	50.48	50.34	50.24
1 & 2	48.26	47.55	47.13	46.64	46.36	46.16	45.86	45.68	45.47	45.33	45.26
2, Good	42.95	42.32	41.94	41.51	41.24	41.07	40.80	40.64	40.46	40.35	40.26
2 & 3	39.84	39.25	38.91	38.50	38.26	38.09	37.85	37.69	37.52	37.41	37.35
3, Average	36.74	36.18	35.87	35.48	35.27	35.11	34.87	34.76	34.61	34.49	34.43
3 & 4	33.52	33.02	32.74	32.39	32.20	32.04	31.85	31.70	31.58	31.47	31.42
4, Low	30.26	29.82	29.56	29.25	29.06	28.94	28.75	28.65	28.49	28.43	28.38

Length greater than 3 times depth
Square Foot Area

Quality Class	7,500	10,000	15,000	20,000	30,000	40,000	50,000	75,000	100,000	150,000	200,000
1, Best	53.45	53.11	52.48	52.11	51.61	51.32	51.09	50.74	50.51	50.23	50.08
1 & 2	49.75	49.43	48.85	46.95	46.52	46.23	46.04	45.71	45.52	45.27	45.13
2, Good	42.82	42.55	42.04	41.74	41.35	41.10	40.92	40.64	40.47	40.25	40.11
2 & 3	39.73	39.47	38.99	38.71	38.36	38.13	37.98	37.70	37.54	37.34	37.21
3, Average	36.66	36.44	35.99	35.72	35.41	35.18	35.04	34.79	34.65	34.45	34.35
3 & 4	33.44	33.23	32.84	32.61	32.30	32.10	31.98	31.75	31.62	31.44	31.34
4, Low	30.23	30.05	29.69	29.48	29.21	29.02	28.91	28.71	28.58	28.42	28.33

Wall Height Adjustment: Add or subtract the amount listed to the square foot costs above for each foot of wall height more or less than 16 feet.

Area	3,500	5,000	7,500	10,000	12,500	15,000	Over 20,000
Cost	.13	.11	.09	.08	.06	.05	.04

Perimeter Wall Adjustment: For common wall, deduct $29.50 per linear foot of common wall. For no wall ownership, deduct $59.00 per linear foot of wall.

Suburban Stores - Wood or Wood and Steel Frame
Building Shell Only, Strip Type

Estimating Procedure

1. Use these figures to estimate the cost of shell-type buildings without permanent partitions, display fronts or finish material on the front wall of the building.
2. Establish the structure quality class by applying the information on page 61.
3. Compute the building floor area. This should include everything within the exterior walls and all inset areas outside the main walls but under the main roof.
4. Add to or subtract from the square foot cost below the appropriate amount from the Wall Height Adjustment Table (at the bottom of this page) if the wall height is more or less than 16 feet.
5. Multiply the adjusted square foot cost by the building area.
6. Deduct, if appropriate, for common walls, using the figures at the bottom of this page.
7. Multiply the total cost by the location factor on page 7.
8. Add the cost of the appropriate additional components from page 201: heating and cooling equipment, fire sprinklers, display fronts, finish materials on the front wall, canopies, interior partitions, exterior signs, mezzanines and basements, loading docks and ramps, yard improvements, and communication systems.

Length less than 1-1/2 times depth
Square Foot Area

Quality Class	500	1,000	2,000	2,500	3,000	4,000	5,000	7,500	10,000	12,500	15,000
1, Best	85.92	74.23	65.96	63.85	62.29	60.12	58.63	56.31	54.92	53.98	53.28
1 & 2	78.12	67.50	59.98	58.06	56.65	54.67	53.31	51.20	49.95	49.10	48.46
2, Good	70.31	60.75	53.99	52.26	50.97	49.21	47.99	46.09	44.95	44.19	43.61
2 & 3	66.39	57.36	50.97	49.35	48.14	46.45	45.31	43.50	42.44	41.72	41.17
3, Average	62.41	53.91	47.91	46.38	45.25	43.66	42.59	40.89	39.89	39.22	38.69
3 & 4	58.18	50.25	44.65	43.23	42.19	40.70	39.69	38.13	37.20	36.56	36.08
4, Low	53.98	46.63	41.44	40.12	39.15	37.78	36.84	35.37	34.51	33.91	33.48

Length between 1-1/2 and 2 times depth
Square Foot Area

Quality Class	500	1,000	2,000	3,000	5,000	6,000	7,500	10,000	15,000	20,000	35,000
1, Best	82.83	71.87	64.07	60.57	57.08	56.02	54.87	53.54	51.98	51.04	49.56
1 & 2	75.34	65.36	58.25	55.08	51.91	50.95	49.91	48.69	47.27	46.41	45.09
2, Good	67.84	58.84	52.45	49.61	46.74	45.87	44.93	43.86	42.57	41.79	40.58
2 & 3	63.95	55.49	49.44	46.75	44.07	43.25	42.35	41.34	40.13	39.41	38.26
3, Average	60.04	52.09	46.42	43.90	41.37	40.60	39.76	38.80	37.67	36.98	35.93
3 & 4	56.15	48.72	43.42	41.06	38.70	37.97	37.20	36.29	35.24	34.60	33.60
4, Low	52.28	45.37	40.44	38.25	36.02	35.36	34.63	33.79	32.81	32.21	31.28

Wall Height Adjustment: Add or subtract the amount listed to the square foot cost above for each foot of wall height more or less than 16 feet.

Area	500	1,000	2,000	2,500	3,000	4,000	5,000	7,500	10,000	12,500
Cost	.94	.67	.48	.44	.39	.35	.32	.28	.25	.24

Area	15,000	20,000	25,000	30,000	35,000	50,000	70,000	75,000	100,000	150,000
Cost	.17	.13	.11	.09	.08	.07	.06	.06	.06	.06

Perimeter Wall Adjustment: For a common wall, deduct $101.80 per linear foot of common wall. For no wall ownership, deduct $203.60 per linear foot of wall.

Suburban Stores - Wood or Wood and Steel Frame

Building Shell Only, Strip Type

Estimating Procedure

1. Use these figures to estimate the cost of shell-type buildings without permanent partitions, display fronts or finish material on the front wall of the building.
2. Establish the structure quality class by applying the information on page 61.
3. Compute the building floor area. This should include everything within the exterior walls and all inset areas outside the main walls but under the main roof.
4. Add to or subtract from the square foot cost below the appropriate amount from the Wall Height Adjustment Table (at the bottom of this page) if the wall height is more or less than 16 feet.
5. Multiply the adjusted square foot cost by the building area.
6. Deduct, if appropriate, for common walls, using the figures at the bottom of this page.
7. Multiply the total cost by the location factor on page 7.
8. Add the cost of the appropriate additional components from page 201: heating and cooling equipment, fire sprinklers, display fronts, finish materials on the front wall, canopies, interior partitions, exterior signs, mezzanines and basements, loading docks and ramps, yard improvements, and communication systems.

Length between 2 and 3 times depth
Square Foot Area

Quality Class	1,000	2,000	3,000	5,000	7,500	10,000	15,000	20,000	30,000	50,000	70,000
1, Best	72.96	65.05	61.52	58.00	55.79	54.47	52.88	51.95	50.85	49.74	49.15
1 & 2	66.35	59.15	55.94	52.75	50.74	49.53	48.11	47.25	46.24	45.23	44.70
2, Good	59.72	53.22	50.35	47.47	45.66	44.57	43.29	42.52	41.62	40.72	40.22
2 & 3	56.42	50.30	47.58	44.85	43.15	42.10	40.89	40.17	39.31	38.48	38.01
3, Average	54.97	48.75	44.78	42.23	40.60	39.65	38.49	37.82	37.01	36.20	35.76
3 & 4	49.68	44.29	41.89	39.49	37.99	37.08	36.02	35.39	34.63	33.87	33.47
4, Low	46.21	41.18	38.96	36.72	35.33	34.48	33.50	32.90	32.19	31.49	31.12

Length greater than 3 times depth
Square Foot Area

Quality Class	2,000	3,000	5,000	10,000	15,000	20,000	30,000	50,000	75,000	100,000	150,000
1, Best	67.18	63.29	59.40	55.50	53.78	52.76	51.54	50.33	49.55	49.11	48.55
1 & 2	60.71	57.20	53.68	50.15	48.60	47.68	46.58	45.48	44.78	44.38	43.87
2, Good	54.66	51.49	48.33	45.15	43.75	42.92	41.93	40.93	40.33	39.95	39.50
2 & 3	51.58	48.58	45.61	42.61	41.29	40.49	39.56	38.62	38.04	37.69	37.29
3, Average	48.51	45.69	42.89	40.07	38.83	38.08	37.21	36.33	35.76	35.45	35.06
3 & 4	45.53	42.90	40.26	37.62	36.45	35.77	34.94	34.11	33.59	33.29	32.91
4, Low	42.50	40.04	37.57	35.11	34.02	33.37	32.60	31.84	31.34	31.06	30.71

Wall Height Adjustment: Add or subtract the amount listed to the square foot cost above for each foot of wall height more or less than 16 feet.

Area	500	1,000	2,000	2,500	3,000	4,000	5,000	7,500	10,000	12,500
Cost	.94	.67	.49	.44	.39	.35	.32	.28	.25	.24

Area	15,000	20,000	25,000	30,000	35,000	50,000	70,000	75,000	100,000	150,000
Cost	.17	.13	.11	.09	.08	.07	.06	.06	.06	.06

Perimeter Wall Adjustment: For a common wall, deduct $101.80 per linear foot of common wall. For no wall ownership, deduct $203.60 per linear foot of wall.

Suburban Stores - Wood or Wood and Steel Frame
Multi-Unit, Island Type

Estimating Procedure

1. Use these figures to estimate the cost of stores designed for multiple occupancy. These costs include all components of shell buildings plus the cost of display fronts, finish materials on the front of the building and normal interior partitions.
2. Establish the structure quality class by applying the information on page 61. Evaluate the quality of the display front to help establish the correct quality class of the building as a whole. The building classes have display fronts as classified on page 56. See also pages 207 to 209.
3. Compute the building floor area. This should include everything within the exterior walls and all inset areas outside the main walls but under the main roof.
4. Add to or subtract from the square foot cost below the appropriate amount from the Wall Height Adjustment Table (at the bottom of this page) if the wall height is more or less than 16 feet.
5. Multiply the adjusted square foot cost by the building area.
6. Deduct, if appropriate, for common walls, using the figures at the bottom of this page.
7. Multiply the total cost by the location factor on page 7.
8. Add the cost of the appropriate additional components from page 201: heating and cooling equipment, fire sprinklers, canopies, exterior signs, mezzanines and basements, loading docks and ramps, yard improvements, and communication systems.

Length less than 1-1/2 times depth
Square Foot Area

Quality Class	3,500	5,000	7,500	10,000	12,500	15,000	20,000	30,000	40,000	50,000	75,000
1, Best	122.34	111.37	101.35	94.39	90.47	87.40	82.40	77.36	74.40	70.18	68.36
1 & 2	106.40	97.02	88.53	82.55	79.13	76.59	72.33	71.74	68.02	63.78	60.89
2, Good	90.09	82.63	75.59	71.37	68.45	66.33	63.32	59.73	57.59	56.12	53.83
2 & 3	79.19	72.65	66.46	62.74	60.19	58.31	55.67	52.51	50.63	49.35	47.33
3, Average	68.25	62.60	57.27	54.07	51.87	50.24	47.96	45.25	43.62	42.53	40.79
3 & 4	60.83	55.80	51.03	48.18	46.22	44.77	42.75	40.31	38.88	37.90	36.36
4, Low	53.54	49.12	44.91	42.41	40.69	39.40	37.62	35.50	34.22	33.35	32.00

Length between 1-1/2 and 2 times depth
Square Foot Area

Quality Class	4,500	5,000	7,500	10,000	15,000	20,000	30,000	40,000	50,000	75,000	100,000
1, Best	116.50	113.20	102.28	96.01	88.78	84.58	79.71	76.86	74.95	72.01	70.29
1 & 2	102.19	99.28	89.71	84.21	77.87	74.19	69.91	67.41	65.74	63.17	61.66
2, Good	87.73	85.25	77.03	72.30	66.85	63.69	60.03	57.90	56.46	54.25	52.99
2 & 3	77.22	75.03	67.79	63.62	58.83	56.05	52.84	50.95	49.68	47.72	46.59
3, Average	66.59	64.70	58.48	54.89	50.76	48.36	45.58	43.93	42.86	41.16	40.18
3 & 4	59.85	58.16	52.55	49.32	45.61	43.45	40.96	39.49	38.50	37.00	36.11
4, Low	53.11	51.59	46.62	43.76	40.47	38.56	36.35	35.04	34.17	32.83	32.03

Wall Height Adjustment: Add or subtract the amount listed to the square foot cost above for each foot of wall height more or less than 16 feet.
Square Foot Area

Quality Class	3,500	5,000	7,500	10,000	12,500	15,000	20,000	30,000	40,000	50,000	75,000	100,000	150,000
1	6.01	5.04	4.15	3.61	3.23	2.96	2.58	2.11	1.85	1.67	1.37	1.21	1.00
2	4.09	3.43	2.81	2.44	2.21	2.01	1.76	1.45	1.26	1.13	.95	.81	.68
3	2.77	2.34	1.91	1.67	1.49	1.37	1.19	.98	.86	.77	.64	.56	.46
4	2.03	1.71	1.40	1.21	1.09	.99	.87	.72	.62	.56	.47	.39	.34

Perimeter Wall Adjustment:
For a common wall, deduct per linear foot: Class 1, $570.00,   Class 2, $370.00, Class 3, $255.00, Class 4, $175.00.
For no wall ownership, deduct per linear foot: Class 1, $1,140.00, Class 2, $740.00, Class 3, $510.00, Class 4, $350.00.

Suburban Stores - Wood or Wood and Steel Frame

Multi-Unit, Island Type

Estimating Procedure

1. Use these figures to estimate the cost of stores designed for multiple occupancy. These costs include all components of shell buildings plus the cost of display fronts, finish materials on the front of the building and normal interior partitions.
2. Establish the structure quality class by applying the information on page 61. Evaluate the quality of the display front to help establish the correct quality class of the building as a whole. The building classes have display fronts as classified on page 56. See also pages 207 to 209.
3. Compute the building floor area. This should include everything within the exterior walls and all inset areas outside the main walls but under the main roof.
4. Add to or subtract from the square foot cost below the appropriate amount from the Wall Height Adjustment Table (at the bottom of this page) if the wall height is more or less than 16 feet.
5. Multiply the adjusted square foot cost by the building area.
6. Deduct, if appropriate, for common walls, using the figures at the bottom of this page.
7. Multiply the total cost by the location factor on page 7.
8. Add the cost of the appropriate additional components from page 201: heating and cooling equipment, fire sprinklers, canopies, exterior signs, mezzanines and basements, loading docks and ramps, yard improvements, and communication systems.

Length between 2 and 3 times depth
Square Foot Area

Quality Class	7,500	10,000	15,000	20,000	30,000	40,000	50,000	75,000	100,000	150,000	200,000
1, Best	103.04	97.07	89.95	85.69	80.60	77.56	75.47	72.22	70.30	67.99	66.61
1 & 2	90.63	85.37	79.08	75.35	70.88	68.21	66.36	63.51	61.81	59.79	58.57
2, Good	78.09	73.57	68.16	64.93	61.08	58.77	57.20	54.74	53.26	51.52	50.48
2 & 3	68.99	64.98	60.20	57.36	53.95	51.93	50.53	48.36	47.05	45.51	44.59
3, Average	59.90	56.42	52.28	49.79	46.84	45.08	43.87	41.98	40.86	39.52	38.72
3 & 4	53.57	50.48	46.77	44.55	41.91	40.32	39.25	37.56	36.56	35.35	34.64
4, Low	47.26	44.53	41.26	39.29	36.96	35.56	34.62	33.11	32.22	31.18	30.56

Length greater than 3 times depth
Square Foot Area

Quality Class	10,000	15,000	20,000	30,000	40,000	50,000	75,000	100,000	150,000	200,000	250,000
1, Best	102.22	93.98	89.11	83.33	79.91	77.57	73.94	71.79	69.23	67.71	66.66
1 & 2	89.94	82.69	78.39	73.32	70.30	68.25	65.06	63.16	60.91	59.56	58.65
2, Good	77.47	71.24	67.55	63.17	60.56	58.79	56.04	54.40	52.46	51.30	50.53
2 & 3	68.55	63.06	59.76	55.91	53.59	52.02	49.60	48.15	46.43	45.42	44.73
3, Average	59.53	54.74	51.89	48.53	46.54	45.17	43.06	41.80	40.31	39.43	38.83
3 & 4	53.29	49.03	46.47	43.46	41.68	40.45	38.55	37.44	36.10	35.30	34.77
4, Low	47.01	43.24	41.00	38.33	36.76	35.68	34.02	33.02	31.85	31.15	30.68

Wall Height Adjustment: Add or subtract the amount listed to the square foot cost above for each foot of wall height more or less than 16 feet.

Square Foot Area

Quality Class	7,500	10,000	12,500	15,000	20,000	30,000	40,000	50,000	75,000	100,000	150,000	200,000	250,000
1	4.19	3.63	3.28	2.97	2.59	2.11	1.85	1.67	1.37	1.20	.99	.87	.77
2	2.86	2.49	2.23	2.02	1.78	1.47	1.26	1.14	.95	.82	.68	.60	.54
3	1.93	1.69	1.51	1.39	1.20	.98	.86	.78	.64	.56	.47	.40	.36
4	1.41	1.23	1.10	1.01	.88	.72	.62	.56	.47	.39	.34	.30	.28

Perimeter Wall Adjustment:
For a common wall, deduct per linear foot: Class 1, $570.00   Class 2, $370.00, Class 3, $255.00, Class 4, $175.00.
For no wall ownership, deduct per linear foot: Class 1, $1,140.00, Class 2, $740.00, Class 3, $510.00, Class 4, $350.00.

Suburban Stores - Wood or Wood and Steel Frame
Multi-Unit, Strip Type

Estimating Procedure

1. Use these figures to estimate the cost of stores designed for multiple occupancy. These costs include all components of shell buildings plus the cost of display fronts, finish materials on the front of the building and normal interior partitions.
2. Establish the structure quality class by applying the information on page 61. Evaluate the quality of the display front to help establish the correct quality class of the building as a whole. The building classes have display fronts as classified on page 56. See also pages 207 to 209.
3. Compute the building floor area. This should include everything within the exterior walls and all inset areas outside the main walls but under the main roof.
4. Add to or subtract from the square foot cost below the appropriate amount from the Wall Height Adjustment Table (at the bottom of this page) if the wall height is more or less than 16 feet.
5. Multiply the adjusted square foot cost by the building area.
6. Deduct, if appropriate, for common walls, using the figures at the bottom of this page.
7. Multiply the total cost by the location factor on page 7.
8. Add the cost of the appropriate additional components from page 201: heating and cooling equipment, fire sprinklers, canopies, exterior signs, mezzanines and basements, loading docks and ramps, yard improvements, and communication systems.

Length less than 1-1/2 times depth
Square Foot Area

Quality Class	500	1,000	2,000	2,500	3,000	4,000	5,000	7,500	10,000	12,500	15,000
1, Best	125.88	104.95	90.21	86.45	83.68	79.79	77.15	73.03	70.57	68.90	67.66
1 & 2	110.96	92.53	79.51	76.20	73.78	70.33	68.00	64.38	62.20	60.72	59.64
2, Good	95.93	80.00	68.74	65.87	63.78	60.81	58.80	55.66	53.79	52.50	51.56
2 & 3	86.92	72.46	62.28	59.69	57.79	55.09	53.26	50.42	48.73	47.58	46.72
3, Average	78.10	65.12	55.97	53.65	51.92	49.52	47.86	45.32	43.79	42.75	41.97
3 & 4	71.59	59.69	51.31	50.99	47.59	45.39	43.89	41.54	40.13	39.18	38.49
4, Low	64.76	53.98	46.40	44.46	43.04	41.04	39.68	37.57	36.30	35.45	34.80

Length between 1-1/2 and 2 times depth
Square Foot Area

Quality Class	500	1,000	2,000	3,000	5,000	7,500	10,000	15,000	20,000	25,000	35,000
1, Best	140.60	115.68	98.07	90.26	82.42	77.48	74.52	71.03	68.95	67.53	65.65
1 & 2	123.44	101.58	86.10	79.25	72.37	68.02	65.43	62.37	60.54	59.27	57.64
2, Good	106.31	87.49	74.14	68.26	62.32	58.57	56.36	53.72	52.14	51.07	49.65
2 & 3	96.04	79.02	66.99	61.65	56.29	52.91	50.90	48.51	47.08	46.12	44.84
3, Average	85.78	70.59	59.84	55.07	50.27	47.26	45.48	38.69	42.06	41.20	40.06
3 & 4	78.10	64.27	54.47	50.14	45.79	43.04	41.41	39.45	38.31	37.52	36.48
4, Low	70.60	58.09	49.23	45.32	41.38	38.89	37.42	35.66	34.62	33.90	32.96

Wall Height Adjustment: Add or subtract the amount listed to the square foot of floor cost for each foot of wall height more or less than 16 feet.

Square Foot Area

Class	500	1,000	2,000	3,000	5,000	7,500	10,000	15,000	20,000	25,000	35,000
1, Best	8.31	5.78	4.07	3.32	2.58	2.10	1.84	1.51	1.32	1.19	1.08
2, Good	5.26	3.68	2.58	2.11	1.62	1.33	1.17	.96	.84	.74	.69
3, Average	3.72	2.60	1.82	1.49	1.16	.95	.83	.67	.59	.54	.49
4, Low	2.66	1.84	1.30	1.07	.83	.68	.58	.49	.41	.37	.35

Perimeter Wall Adjustment: For a common wall, deduct $100.00 per linear foot. For no wall ownership, deduct $200.00 per linear foot.

Suburban Stores - Wood or Wood and Steel Frame
Multi-Unit, Strip Type

Estimating Procedure

1. Use these figures to estimate the cost of stores designed for multiple occupancy. These costs include all components of shell buildings plus the cost of display fronts, finish materials on the front of the building and normal interior partitions.
2. Establish the structure quality class by applying the information on page 61. Evaluate the quality of the display front to help establish the correct quality class of the building as a whole. The building classes have display fronts as classified on page 56. See also pages 207 to 209.
3. Compute the building floor area. This should include everything within the exterior walls and all inset areas outside the main walls but under the main roof.
4. Add to or subtract from the square foot cost below the appropriate amount from the Wall Height Adjustment Table (at the bottom of this page) if the wall height is more or less than 16 feet.
5. Multiply the adjusted square foot cost by the building area.
6. Deduct, if appropriate, for common walls, using the figures at the bottom of this page.
7. Multiply the total cost by the location factor on page 7.
8. Add the cost of the appropriate additional components from page 201: heating and cooling equipment, fire sprinklers, canopies, exterior signs, mezzanines and basements, loading docks and ramps, yard improvements, and communication systems.

Length between 2 and 3 times depth
Square Foot Area

Quality Class	1,000	2,000	3,000	5,000	7,500	10,000	15,000	20,000	30,000	50,000	70,000
1, Best	132.19	109.79	99.87	89.94	83.70	79.97	75.54	72.92	69.80	66.67	65.01
1 & 2	115.95	96.29	87.58	78.91	73.41	70.15	66.27	63.96	61.23	58.49	57.02
2, Good	99.55	82.65	75.20	67.73	63.01	60.21	56.89	54.91	52.56	50.20	48.95
2 & 3	89.70	74.48	67.76	61.04	56.79	54.27	51.27	49.47	47.36	45.24	44.12
3, Average	79.74	66.20	60.23	54.24	50.47	48.24	45.56	43.97	42.10	40.21	39.22
3 & 4	72.60	60.28	54.85	49.37	45.95	43.92	41.48	40.04	38.33	36.62	35.71
4, Low	65.41	54.31	49.43	44.51	41.40	39.57	37.39	36.08	34.55	33.00	32.16

Length greater than 3 times depth
Square Foot Area

Quality Class	2,000	3,000	5,000	10,000	15,000	20,000	30,000	50,000	75,000	100,000	150,000
1, Best	120.65	108.56	96.61	84.77	79.59	76.50	72.86	69.24	66.96	65.61	64.00
1 & 2	105.72	95.14	84.65	74.27	69.74	67.03	63.85	60.67	58.69	57.48	56.09
2, Good	90.67	81.60	72.61	63.70	59.81	57.50	54.76	52.04	50.33	49.31	48.10
2 & 3	78.77	70.88	63.09	55.35	51.97	49.96	47.58	45.21	43.73	42.85	41.78
3, Average	72.42	65.15	57.98	50.87	47.75	45.92	43.73	41.55	40.18	39.37	38.41
3 & 4	66.33	59.68	53.12	46.60	43.74	42.05	40.06	38.05	36.81	36.07	35.20
4, Low	60.11	54.10	48.14	42.24	39.66	38.13	36.31	34.50	33.37	32.70	31.90

Wall Height Adjustment: Add or subtract the amount listed to the square foot cost above for each foot of wall height more or less than 16 feet.

Square Foot Area

Quality Class	2,000	3,000	5,000	10,000	15,000	20,000	30,000	50,000	75,000	100,000	150,000
1, Best	4.07	3.32	2.59	1.85	1.54	1.34	1.10	.91	.74	.67	.58
2, Good	2.55	2.08	1.61	1.17	.96	.84	.70	.57	.47	.41	.35
3, Average	1.83	1.50	1.17	.84	.69	.61	.50	.40	.34	.31	.26
4, Low	1.36	1.10	.86	.60	.51	.46	.36	.30	.25	.23	.19

Perimeter Wall Adjustment: For a common wall, deduct $100.00 per linear foot. For no wall ownership, deduct $200.00 per linear foot.

Supermarkets - Masonry or Concrete
Quality Classification

	Class 1 Best Quality	Class 2 Good Quality	Class 3 Average Quality	Class 4 Low Quality
Foundation	Reinforced concrete.	Reinforced concrete.	Reinforced concrete.	Reinforced concrete.
Floor Structure	4" reinforced concrete on 6" rock fill.	4" reinforced concrete on 6" rock fill.	4" reinforced concrete on 6" rock fill.	4" reinforced concrete on 6" rock fill.
Wall Structure	6" concrete tilt-up or ornamental block or brick.	6" concrete tilt-up, colored concrete block or brick.	6" concrete tilt-up or 8" concrete block.	6" concrete tilt-up or 8" concrete block.
Roof Structure	Glu-lams or steel "I" beams on steel intermediate columns, 2" x 12" purlins 16" o.c., 1/2" plywood sheathing.	Glu-lams or steel "I" beams on steel intermediate columns, 2" x 12" purlins 16" o.c., 1/2" plywood sheathing.	Glu-lams or steel "I" beams on steel intermediate columns, 2" x 12" purlins 16" o.c., 1/2" plywood sheathing.	Glu-lams or steel "I" beams on steel intermediate columns, 3" x 12" purlins 3' o.c., 1/2" plywood sheathing.
Floor Finish	Terrazzo in sales area. Sheet vinyl or carpet in cashiers' area.	Resilient tile in sales area. Terrazzo, solid vinyl tile or carpet in cashiers' area.	Composition tile in sales area.	Minimum grade tile in sales area.
Interior Wall Finish	Inside of exterior walls furred out with gypsum wallboard and paint or interior stucco, interior stucco or gypsum wallboard and vinyl wall cover on partitions.	Paint on inside of exterior walls, gypsum wallboard and paint or vinyl wall cover on partitions.	Paint on inside of exterior walls, wallboard and paint on partitions.	Paint on inside of exterior walls, wallboard and paint on partitions.
Ceiling Finish	Suspended acoustical tile, dropped ceiling over meat and produce departments.	Suspended acoustical tile or gypsum board and acoustical texture, dropped ceiling over meat and produce departments.	Ceiling tile on roof purlins, dropped ceiling over meat department.	Open.
Front	A large amount of plate glass in good aluminum frames (18'-22' high for 3/4 of width), brick or stone veneer on remainder, 1 pair of good automatic doors per 7,000 S.F. of floor area, anodized aluminum sunshade over glass area, 8' canopy across front, 10'-12' raised walk across front.	A large amount of plate glass in good aluminum frames (16'-18' high for 2/3 of width), brick or stone veneer on remainder, 1 pair of good automatic doors per 10,000 S.F. of floor area, 8' canopy across front, 10' raised walk across front.	A moderate amount of plate glass in average quality aluminum frames (12'-16' high for 2/3 of width), exposed aggregate on remainder, 1 pair average automatic doors per 10,000 S.F. of floor area, 6' canopy across front, 8' raised walk across front.	Stucco or exposed aggregate with a small amount of plate glass in an inexpensive aluminum frame (6'-10' high for 1/2 of width), 6' canopy across front, 6' ground level walk across front.
Exterior Wall Finish	Large ornamental rock or brick veneer.	Large ornamental rock or brick veneer.	Paint, some exposed aggregate.	Paint.
Roof Cover	5 ply built-up roofing with large rock.	5 ply built-up roofing with tar and rock.	4 ply built-up roofing.	4 ply built-up roofing.
Plumbing	2 rest rooms with 3 fixtures, floor piping and drains to refrigerated cases, 2 double sinks with drain board.	2 rest rooms with 3 fixtures each, floor piping and drains to refrigerated cases, 2 double sinks with drain board.	2 rest rooms with 2 fixtures each, floor piping and drains to refrigerated cases, 2 double sinks.	1 rest room with 2 fixtures, floor piping and drains to refrigerated cases.
Electrical	Conduit wiring, recessed 4 tube fluorescent fixtures 8' o.c., 30-40 spotlights.	Conduit wiring, 4 tube fluorescent fixtures with diffusers 8' o.c., 30-40 spotlights.	Conduit wiring, 3 tube fluorescent fixtures, 8' o.c., 5 or 10 spotlights	Conduit wiring, double tube fluorescent fixtures, 8' o.c.

Square foot costs include the following components: Foundations as required for normal soil conditions. Floor, wall, and roof structures. Interior floor, wall and ceiling finishes. Exterior wall finish and roof cover. Display fronts. Interior partitions. Entry and delivery doors. A canopy and walk across the front of the building as described in the applicable building specifications. Basic lighting and electrical systems. Rough and finish plumbing. All plumbing, piping and wiring necessary to operate the usual refrigerated cases and vegetable cases. Design and engineering fees. Permits and hook-up fees. Contractor's mark-up.

Supermarkets - Masonry or Concrete

Estimating Procedure

1. Establish the structure quality class by using the information on page 70.
2. Compute the building floor area. This should include everything within the building exterior walls and all insets outside the main walls but under the main building roof.
3. Add to or subtract from the square foot cost below the appropriate amount from the Wall Height Adjustment Column (at the bottom of this page) if the wall height is more or less than 20 feet.
4. Multiply the adjusted square foot cost by the building area.
5. Deduct, if appropriate, for common walls, using the figures at the bottom of this page.
6. Multiply the total cost by the location factor listed on page 7.
7. Add the cost of heating and cooling equipment, fire sprinklers, exterior signs, yard improvements, loading docks, ramps and walk-in boxes if they are an integral part of the building. See pages 201 to 213.

Supermarket, Class 2

Square Foot Area

Quality Class	5,000	7,500	10,000	12,500	15,000	20,000	25,000	30,000	35,000	40,000	50,000
Exceptional	88.56	82.36	78.64	76.08	74.21	71.55	69.72	68.39	67.35	66.49	65.20
1, Best	83.56	77.71	74.20	71.79	70.02	67.52	65.80	64.54	63.55	62.75	61.52
1 & 2	77.09	71.68	68.45	66.22	64.58	62.28	60.69	59.53	58.61	57.88	56.76
2, Good	69.68	64.79	61.86	59.86	58.38	56.30	54.86	53.80	52.99	52.32	51.31
2 & 3	64.69	60.15	57.43	55.58	54.19	52.26	50.93	49.94	49.18	48.56	47.62
3, Average	59.35	55.19	52.70	50.99	49.72	47.96	46.71	45.82	45.12	44.56	43.69
3 & 4	53.56	49.81	47.56	46.02	44.88	43.28	42.18	41.37	40.73	40.23	39.45
4, Low	47.61	44.26	42.26	40.90	39.88	38.47	37.48	36.76	36.20	35.74	35.04
Wall Height Adjustment*	.93	.75	.66	.60	.55	.46	.41	.37	.35	.33	.30

*Wall Height Adjustment: Add or subtract the amount listed in this column to the square foot of floor cost for each foot of wall height more or less than 20 feet.

Perimeter Wall Adjustment: A common wall exists when two buildings share one wall. Adjust for common walls by deducting the linear foot costs below from the total structure cost. In some structures one or more walls are not owned at all. In this case, deduct the "No Ownership" cost per linear foot of wall not owned. For common wall, deduct $160.00 per linear foot. For no wall ownership, deduct $320.00 per linear foot.

Supermarkets - Wood or Wood and Steel Frame

Quality Classification

	Class 1 Best Quality	Class 2 Good Quality	Class 3 Average Quality	Class 4 Low Quality
Foundation	Reinforced concrete.	Reinforced concrete.	Reinforced concrete.	Reinforced concrete.
Floor Structure	4" reinforced concrete on 6" rock fill.	4" reinforced concrete on 6" rock fill.	4 reinforced concrete on 6" rock fill.	4" reinforced concrete on 6" rock fill.
Wall Structure	2" x 6" - 16" o.c.	2" x 6" - 16" o.c.	2" x 4" - 16" o.c.	2" x 4" - 16" o.c.
Roof Structure	Glu-lams or steel "I" beams on steel intermediate columns, 2" x 12" purlins 16" o.c., 1/2" plywood sheathing.	Glu-lams or steel "I" beams on steel intermediate columns, 2" x 12" purlins 16" o.c., 1/2" plywood sheathing.	Glu-lams or steel "I" beams on steel intermediate columns 2" x 12" purlins 16" o.c., 1/2" plywood sheathing.	Glu-lams or steel "I" beams on steel intermediate columns, 3" x 12" purlins 3' o.c., 1/2" plywood sheathing.
Floor Finish	Terrazzo in sales area. Sheet vinyl or carpet in cashiers' area.	Resilient tile in sales area. Terrazzo, solid vinyl tile or carpet in cashiers' area.	Composition tile in sales area.	Minimum tile or inexpensive composition tile in sales area.
Interior Wall Finish	Gypsum wallboard and vinyl wall cover or interior stucco on inside of exterior walls and on partitions.	Gypsum wallboard, texture and paint or vinyl wall cover on inside of exterior walls, and on partitions.	Gypsum wallboard, texture and paint on inside of exterior walls, and on partitions.	Gypsum wallboard and paint on inside of exterior walls, and on partitions.
Ceiling Finish	Suspended acoustical tile, dropped ceiling over meat and produce departments.	Suspended acoustical tile or gypsum board and acoustical texture, dropped ceiling over meat and produce departments.	Ceiling tile on roof purlins, dropped ceiling over meat department.	Open.
Front	A large amount of plate glass in good aluminum frames (18'-22' high for 3/4 of width), brick or stone veneer on remainder, 1 pair good automatic doors per 7,000 S.F. of floor area, anodized aluminum sunshade over glass area, 8' canopy across front, 10'-12' raised walk across front.	A large amount of plate glass in good aluminum frames (16'-18' high for 2/3 of width), brick or stone veneer on remainder, 1 pair good automatic doors per 10,000 S.F. of floor area, 9' canopy across front, 10' raised walk across front.	Moderate amount of plate glass in average quality aluminum frames (12'-16' high for 2/3 of width), wood siding on remainder, 1 pair average automatic doors per 10,000 S.F. of floor area, 6' canopy across front, 8' raised walk across front.	Stucco with small amount of plate glass in an inexpensive aluminum frame (6'-10' high for 1/2 of width), 6' canopy across front, 6' ground level walk across front.
Exterior Wall Finish	Large ornamental rock or brick veneer.	Good wood siding, some masonry veneer.	Stucco or wood siding.	Stucco.
Roof Cover	5 ply built-up roofing with large rock.	5 ply built-up roofing with tar and rock.	4 ply built-up roofing.	4 ply built-up roofing.
Plumbing	2 rest rooms with 3 fixtures each, floor piping and drains to refrigerated cases, 2 double sinks with drain board.	2 rest rooms with 3 fixtures each, floor piping and drains to refrigerated cases, 2 double sinks with drain board.	2 rest rooms with 2 fixtures each, floor piping and drains to refrigerated cases, 2 double sinks.	1 rest room with 2 fixtures, floor piping and drains to refrigerated cases.
Electrical	Conduit wiring, recessed 4 tube fluorescent fixtures, 8' o.c., 30 to 40 spotlights.	Conduit wiring, 4 tube fluorescent fixtures with diffusers, 8' o.c., 30 to 40 spotlights.	Conduit wiring, 3 tube fluorescent fixtures, 8' o.c., 5 or 10 spotlights.	Conduit wiring, double tube fluorescent fixtures, 8' o.c.

Square foot costs include the following components: Foundations as required for normal soil conditions. Floor, wall, and roof structures. Interior floor, wall and ceiling finishes. Exterior wall finish and roof cover. Display fronts. Interior partitions. Entry and delivery doors. A canopy and walk across the front of the building as described in the applicable building specifications. Basic lighting and electrical systems. Rough and finish plumbing. All plumbing, piping and wiring necessary to operate the usual refrigerated cases and vegetable cases. Design and engineering fees. Permits and hook-up fees. Contractor's mark-up.

Supermarkets - Wood or Wood and Steel Frame

Estimating Procedure

1. Establish the structure quality class by using the information on page 72.
2. Compute the building floor area. This should include everything within the building exterior walls and all insets outside the main walls but under the main building roof.
3. Add to or subtract from the square foot cost below the appropriate amount from the Wall Height Adjustment Column (at the bottom of this page) if the wall height is more or less than 20 feet.
4. Multiply the adjusted square foot cost by the building area.
5. Deduct, if appropriate, for common walls, using the figures at the bottom of this page.
6. Multiply the total cost by the location factor listed on page 7.
7. Add the cost of heating and cooling equipment, fire sprinklers, exterior signs, yard improvements, loading docks, ramps and walk-in boxes if they are an integral part of the building. See pages 201 to 213.

Supermarket, Class 3

Square Foot Area

Quality Class	5,000	7,500	10,000	12,500	15,000	20,000	25,000	30,000	35,000	40,000	50,000
Exceptional	83.31	78.50	75.60	73.63	72.16	70.11	68.72	67.66	66.85	66.21	65.19
1, Best	78.70	74.15	71.42	69.56	68.18	66.23	64.90	63.92	63.15	62.53	61.58
1 & 2	71.96	67.80	65.31	63.61	62.33	60.56	59.34	58.45	57.74	57.16	56.30
2, Good	64.42	60.67	58.45	56.92	55.80	54.22	53.12	52.30	51.69	51.18	50.41
2 & 3	59.92	56.46	54.38	52.97	51.92	50.43	49.42	48.67	48.08	47.61	46.90
3, Average	55.06	51.88	49.96	48.65	47.69	46.34	45.39	44.71	44.18	43.73	43.07
3 & 4	50.20	47.29	45.56	44.36	43.48	42.24	41.39	40.78	40.28	39.89	39.28
4, Low	45.00	42.40	40.84	39.76	38.98	37.87	37.10	36.54	36.10	35.74	35.20
Wall Height Adjustment*	.52	.42	.38	.35	.33	.30	.28	.26	.25	.24	.22

*Wall Height Adjustment: Add or subtract the amount listed in this column to the square foot of floor cost for each foot of wall height more or less than 20 feet.

Perimeter Wall Adjustment: A common wall exists when two buildings share one wall. Adjust for common walls by deducting the linear foot costs below from the total structure cost. In some structures one or more walls are not owned at all. In this case, deduct the "No Ownership" cost per linear foot of wall not owned. For common wall, deduct $113.00 per linear foot. For no wall ownership, deduct $226.00 per linear foot.

Small Food Stores - Masonry Construction

Quality Classification

	Class 1 Best Quality	Class 2 Good Quality	Class 3 Average Quality	Class 4 Low Quality
Foundation	Reinforced concrete.	Reinforced concrete.	Reinforced concrete.	Reinforced concrete.
Floor Structure	4" reinforced concrete on 6" rock fill.	4" reinforced concrete on 6" rock fill.	4 reinforced concrete on 6" rock fill.	4" reinforced concrete on 6" rock fill.
Roof Structure	Glu-lams or steel "I" beams on steel intermediate columns, 2" x 12" purlins 16" o.c., 1/2" plywood sheathing.	Glu-lams or steel "I" beams, 2" x 12" purlins 16" o.c., 1/2" plywood sheathing.	Glu-lams, 3" x 12" purlins 3' o.c., 1/2" plywood sheathing.	Glu-lams, 3" x 12" purlins 3' o.c., 1/2" plywood sheathing.
Floor Finish	Resilient tile in sales area.	Composition tile in sales area.	Minimum grade tile in sales area.	Concrete.
Interior Wall Finish	Paint on inside of exterior walls, gypsum wallboard, texture and paint or vinyl wall cover on partitions.	Paint on inside of exterior walls, gypsum wallboard, texture and paint on partitions.	Paint on inside of exterior walls, gypsum wallboard and paint on partitions.	Paint on inside of exterior walls, gypsum wallboard and paint on partitions.
Ceiling Finish	Suspended acoustical tile or gypsum board and acoustical texture.	Ceiling tile on roof purlins.	Open.	Open.
Front	A large amount of plate glass in good aluminum frames (10'-12' high for 2/3 of width), brick or stone veneer on remainder, 1 pair good aluminum and glass doors per 3,000 S.F. of floor area, 8' canopy across front, 10' raised walk across front.	A moderate amount of plate glass in average aluminum frames (8' to 10' high for 2/3 of width), exposed aggregate on remainder, 1 pair average aluminum and glass doors per 3,000 S.F. of floor area, 6' canopy across front, 8' raised walk across front.	Painted concrete block with a small amount of plate glass in an inexpensive aluminum frame (6' to 8' high for 1/2 of width), 6' canopy across front, 6' ground level walk across front.	Painted concrete block with small amount of crystal glass in wood frames, wood and glass doors, small canopy over entrance, 6' ground level walk at entrances.
Exterior Wall Finish	Colored block.	Paint.	Paint.	Paint.
Roof Cover	5 ply built-up roofing with tar and rock.	4 ply built-up roofing	4 ply built-up roofing.	4 ply built-up roofing.
Plumbing	2 rest rooms with 3 fixtures each, floor piping and drains to refrigerated cases.	1 rest room with 3 fixtures, floor piping and drains to refrigerated cases.	1 rest room with 2 fixtures, floor piping and drains to refrigerated cases.	1 rest room with 2 fixtures, floor piping and drains to refrigerated cases.
Electrical	Conduit wiring, 4 tube fluorescent fixtures with diffusers, 8' o.c., 5 spotlights.	Conduit wiring, 3 tube fluorescent fixtures, 8' o.c.	Conduit wiring, double tube fluorescent fixtures, 8' o.c.	Conduit wiring, incandescent fixtures, 10' o.c. or single tube fluorescent fixtures, 8' o.c.

Square foot costs include the following components: Foundations as required for normal soil conditions. Floor, wall, and roof structures. Interior floor, wall and ceiling finishes. Exterior wall finish and roof cover. Display fronts. Interior partitions. Entry and delivery doors. A canopy and walk across the front of the building as described in the applicable building specifications. Basic lighting and electrical systems. Rough and finish plumbing. All plumbing, piping and wiring necessary to operate the usual refrigerated cases and vegetable cases. Design and engineering fees. Permits and hook-up fees. Contractor's mark-up.

Small Food Stores - Masonry Construction

Estimating Procedure

1. Establish the structure quality class by using the information on page 74.
2. Compute the building floor area. This should include everything within the building exterior walls and all insets outside the main walls but under the main building roof.
3. Add to or subtract from the square foot cost below the appropriate amount from the Wall Height Adjustment Column (at the bottom of this page) if the wall height is more or less than 12 feet.
4. Multiply the adjusted square foot cost by the building area.
5. Deduct, if appropriate, for common walls, using the figures at the bottom of this page.
6. Multiply the total cost by the location factor listed on page 7.
7. Add the cost of heating and cooling equipment, fire sprinklers, exterior signs, yard improvements, loading docks, ramps and walk-in boxes if they are an integral part of the building. See pages 201 to 213.

Small Food Store, Class 1 & 2

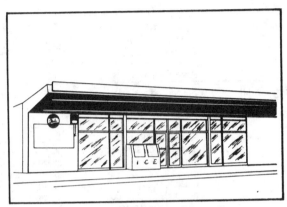

Small Food Store, Class 2

Square Foot Area

Quality Class	500	1,000	1,500	2,000	2,500	3,000	3,500	4,000	4,500	5,000	6,000
1, Best	129.24	97.90	85.60	78.72	74.23	71.00	68.58	66.64	65.07	63.76	61.67
1 & 2	118.47	89.97	78.75	72.48	68.37	65.43	63.21	61.46	60.02	58.82	56.92
2, Good	107.64	81.98	71.89	66.20	62.49	59.83	57.81	56.21	54.90	53.83	52.10
2 & 3	100.20	76.84	67.35	61.94	58.35	55.75	53.77	52.20	50.90	49.82	48.09
3, Average	92.38	70.69	61.96	57.01	53.75	51.40	49.60	48.19	47.02	46.04	44.48
3 & 4	76.15	64.16	57.59	53.34	50.29	47.95	46.07	44.55	43.26	42.16	40.33
4, Low	70.60	59.08	52.95	48.96	46.13	43.97	42.23	40.82	39.62	38.61	36.94
Wall Height Adjustment*	2.93	2.07	1.68	1.45	1.30	1.20	1.10	1.03	.98	.93	.86

Wall Height Adjustment: Add or subtract the amount listed in this column to the square foot of floor cost for each foot of wall height more or less than 12 feet.

Perimeter Wall Adjustment: For common wall, deduct $84.50 per linear foot. For no wall ownership, deduct $169.00 per linear foot.

Small Food Stores - Wood Frame Construction
Quality Classification

	Class 1 Best Quality	Class 2 Good Quality	Class 3 Average Quality	Class 4 Low Quality
Foundation	Reinforced concrete.	Reinforced concrete.	Reinforced concrete.	Reinforced concrete.
Floor Structure	4" reinforced concrete on 6" rock fill.	4" reinforced concrete on 6" rock fill.	4 reinforced concrete on 6" rock fill.	4" reinforced concrete on 6" rock fill.
Wall Structure	2" x 6" - 16" o.c.	2" x 6" - 16" o.c.	2" x 4" - 16" o.c.	2" x 4" - 16" o.c.
Roof Structure	Glu-lams or steel "I" beams on steel intermediate columns, 2" x 12" purlins 16" o.c., 1/2" plywood sheathing.	Glu-lams or steel "I" beams on steel intermediate columns, 2" x 12" purlins 16" o.c., 1/2" plywood sheathing.	Glu-lams, or steel "I" beams, 3" x 12" purlins 3' o.c., 1/2" plywood sheathing.	Glu-lams, 3" x 12" purlins 3' o.c. 1/2" plywood sheathing.
Floor Finish	Resilient tile in sales area.	Resilient tile in sales area.	Inexpensive composition tile in sales area.	Concrete.
Interior Wall Finish	Gypsum wallboard, texture and paint or vinyl wall cover on inside of exterior walls, and on partitions.	Gypsum wallboard, texture and paint on inside of exterior walls, and on partitions.	Gypsum wallboard and paint on inside of exterior walls, and on partitions.	Gypsum wallboard and paint on inside of exterior walls and on partitions.
Ceiling Finish	Suspended acoustical tile or gypsum board and acoustical texture.	Ceiling tile on roof purlins.	Open.	Open.
Front	A large amount of plate glass in good aluminum frames (10'-12' high for 2/3 of width), brick or stone veneer on remainder, 1 pair good aluminum and glass doors per 3,000 S.F. of floor area, 8' canopy across front, 10' raised walk across front.	A moderate amount of plate glass in average quality aluminum frames (8' to 10' high for 2/3 of width), ornamental concrete block on remainder, 1 pair average aluminum and glass doors per 3,000 S.F. of floor area, 6' canopy across front, 8' raised walk across front	Painted stucco with a small amount of plate glass in an inexpensive aluminum frame (6' to 8' high for 1/2 of width) 6' canopy across front. 6' ground level walk across front.	Stucco with small amount of crystal glass in wood frames, wood and glass doors, small canopy over entrance, 6' ground level walk at entrances.
Exterior Wall Finish	Stucco and paint or wood siding.	Stucco and paint.	Stucco.	Stucco.
Roof Cover	5 ply built-up roofing with tar and rock.	4 ply built-up roofing	4 ply built-up roofing.	4 ply built-up roofing.
Plumbing	2 rest rooms with 3 fixtures each, floor piping and drains to refrigerated cases.	1 rest room with 3 fixtures, floor piping and drains to refrigerated cases.	1 rest room with 2 fixtures, floor piping and drains to refrigerated cases.	1 rest room with 2 fixtures, floor piping and drains to refrigerated cases.
Electrical	Conduit wiring, 4 tube fluorescent fixtures with diffusers, 8' o.c., 5 spotlights.	Conduit wiring, 3 tube fluorescent fixtures, 8' o.c.	Conduit wiring, double tube fluorescent fixtures, 8' o.c.	Conduit wiring, incandescent fixtures, 10' o.c. or single tube fluorescent fixtures, 8' o.c.

Square foot costs include the following components: Foundations as required for normal soil conditions. Floor, wall, and roof structures. Interior floor, wall and ceiling finishes. Exterior wall finish and roof cover. Display fronts. Interior partitions. Entry and delivery doors. A canopy and walk across the front of the building as described in the applicable building specifications. Basic lighting and electrical systems. Rough and finish plumbing. All plumbing, piping and wiring necessary to operate the usual refrigerated cases and vegetable cases. Design and engineering fees. Permits and hook-up fees. Contractor's mark-up.

Small Food Stores - Wood Frame Construction

Estimating Procedure

1. Establish the structure quality class by using the information on page 76.
2. Compute the building floor area. This should include everything within the building exterior walls and all insets outside the main walls but under the main building roof.
3. Add to or subtract from the square foot cost below the appropriate amount from the Wall Height Adjustment Column (at the bottom of this page) if the wall height is more or less than 12 feet.
4. Multiply the adjusted square foot cost by the building area.
5. Deduct, if appropriate, for common walls, using the figures at the bottom of this page.
6. Multiply the total cost by the location factor listed on page 7.
7. Add the cost of heating and cooling equipment, fire sprinklers, exterior signs, yard improvements, loading docks, ramps and walk-in boxes if they are an integral part of the building. See pages 201 to 213.

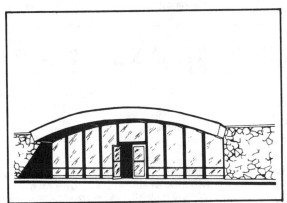

Small Food Store, Class 1

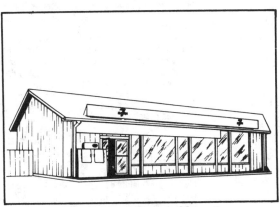

Small Food Store, Class 3

Square Foot Area

Quality Class	500	1,000	1,500	2,000	2,500	3,000	3,500	4,000	4,500	5,000	6,000
1, Best	118.98	90.67	79.97	74.11	70.33	67.67	65.68	64.12	62.83	61.79	60.13
1 & 2	109.45	83.60	73.80	68.43	65.01	62.57	60.73	59.32	58.15	57.18	55.68
2, Good	99.87	76.48	67.61	62.75	59.62	57.41	55.75	54.45	53.39	52.51	51.14
2 & 3	90.55	69.51	61.55	57.20	54.40	52.41	50.92	49.74	48.79	48.01	46.80
3, Average	81.17	62.54	55.48	51.62	49.14	47.39	46.06	45.04	44.20	43.49	42.41
3 & 4	73.24	56.52	50.18	46.74	44.50	42.94	41.75	40.84	40.08	39.47	38.50
4, Low	65.53	50.18	44.56	41.55	39.66	38.33	37.34	36.59	35.98	35.46	34.68
Wall Height Adjustment*	1.84	1.34	1.10	.96	.86	.77	.72	.68	.64	.61	.56

Wall Height Adjustment: Add or subtract the amount listed in this column to the square foot of floor cost for each foot of wall height more or less than 12 feet.

Perimeter Wall Adjustment: For common wall, deduct $57.00 per linear foot. For no wall ownership, deduct $113.00 per linear foot.

Discount Houses - Masonry or Concrete
Quality Classification

	Class 1 Best Quality	Class 2 Good Quality	Class 3 Average Quality	Class 4 Low Quality
Foundation	Reinforced concrete.	Reinforced concrete.	Reinforced concrete.	Reinforced concrete.
Floor Structure	4" reinforced concrete on 6" rock fill.	4" reinforced concrete on 6" rock fill.	4 reinforced concrete on 6" rock fill.	4" reinforced concrete on 6" rock fill.
Wall Structure	6" concrete tilt-up or ornamental block or brick.	6" concrete tilt-up, colored concrete block or brick.	6" concrete tilt-up or 8" concrete block.	6" concrete tilt-up or 8" concrete block.
Roof Structure	Glu-lams or steel "I" beams on steel intermediate columns, plywood sheathing.	Glu-lams or steel "I" beams on steel intermediate columns, plywood sheathing.	Glu-lams, or steel "I" beams on steel intermediate columns, plywood sheathing.	Glu-lams, or steel "I" beams on steel intermediate columns, plywood sheathing.
Floor Finish	Terrazzo, sheet vinyl or good carpet in sales area.	Resilient tile in sales area with some terrazzo, solid vinyl tile, or carpet.	Composition tile in sales area.	Inexpensive composition tile in sales area.
Interior Wall Finish	Inside of exterior walls furred out with gypsum wallboard and paint or interior stucco or gypsum wallboard and vinyl wall cover on partitions.	Paint on inside of exterior walls, gypsum wallboard and paint or vinyl wall cover on partitions.	Paint on inside of exterior walls, gypsum wallboard and paint on partitions.	Paint on inside of exterior walls, gypsum wallboard and paint on partitions.
Ceiling Finish	Suspended acoustical tile, dropped ceilings in some areas.	Suspended acoustical tile or gypsum board and acoustical texture, dropped ceilings in some areas.	Ceiling tile on roof purlins, dropped ceiling over some areas.	Open.
Front	A large amount of plate glass in good aluminum frames (18'-22' high for 3/4 of width), brick or stone veneer on remainder, 1 pair good automatic doors per 7,000 S.F. of floor area, anodized aluminum sunshade over glass area, 8' canopy across front, 10'-12' raised walk across front.	A large amount of plate glass in good aluminum frames (16'-18' high for 2/3 of width), brick or stone veneer on remainder, 1 pair good automatic doors per 10,000 S.F. of floor area, 8' canopy across front, 10' raised walk across front.	A moderate amount of plate glass in average quality aluminum frames (12'-16' high extending 20' on each side of entrances), exposed aggregate on remainder, 1 pair average automatic doors per 20,000 S.F. of floor area, 6' canopy over glass area.	Stucco or exposed aggregate with a small amount of plate glass in inexpensive aluminum frames (6'-10' high for 1/2 of width), 6' canopy at entrances, 6' ground level walk across front.
Exterior Wall Finish	Large ornamental rock or brick veneer on exposed walls.	Exposed aggregate on exposed walls.	Paint, some exposed aggregate.	Paint.
Roof Cover	5 ply built-up roofing with large rock.	5 ply built-up roofing with tar and rock.	4 ply built-up roofing.	4 ply built-up roofing.
Plumbing	2 rest rooms with 8 fixtures each, floor piping and drains to refrigerated cases.	2 rest rooms with 6 fixtures each, floor piping and drains to refrigerated cases.	2 rest rooms with 4 fixtures each, floor piping and drains to refrigerated cases.	2 rest rooms with 2 fixtures each, floor piping and drains to refrigerated cases.
Electrical	Conduit wiring, recessed 4 tube fluorescent fixtures 8' o.c., 60-80 spotlights.	Conduit wiring, 4 tube fluorescent fixtures with diffusers 8' o.c., 60-80 spotlights.	Conduit wiring, 3 tube fluorescent fixtures 8' o.c., 20 to 40 spotlights.	Conduit wiring, double tube fluorescent fixtures, 8' o.c.

Square foot costs include the following components: Foundations as required for normal soil conditions. Floor, wall, and roof structures. Interior floor, wall and ceiling finishes. Exterior wall finish and roof cover. Display fronts. Interior partitions. Entry and delivery doors. A canopy and walk across the front of the building as described in the applicable building specifications. Basic lighting and electrical systems. Rough and finish plumbing. Design and engineering fees. Permits and hook-up fees. Contractor's mark-up.

Discount Houses - Masonry or Concrete

Estimating Procedure

1. Establish the structure quality class by using the information on page 78.
2. Compute the building floor area. This should include everything within the building exterior walls and all insets outside the main walls but under the main building roof.
3. Add to or subtract from the square foot cost below the appropriate amount from the Wall Height Adjustment Column (at the bottom of this page) if the wall height is more or less than 20 feet.
4. Multiply the adjusted square foot cost by the building area.
5. Deduct, if appropriate, for common walls, using the figures at the bottom of this page.
6. Multiply the total cost by the location factor listed on page 7.
7. Add the cost of heating and cooling equipment, fire sprinklers, exterior signs, yard improvements, loading docks, ramps and walk-in boxes if they are an integral part of the building. See pages 201 to 213.

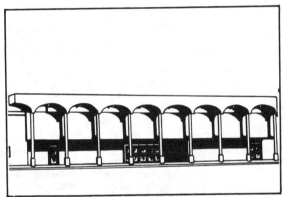

Discount House, Class 3

Discount House, Class 4

Square Foot Area

Quality Class	15,000	20,000	25,000	30,000	35,000	40,000	50,000	75,000	100,000	150,000	200,000
1, Best	57.58	55.61	54.28	53.27	52.49	51.86	50.88	49.37	48.46	47.37	46.70
1 & 2	53.86	52.03	50.79	49.84	49.11	48.52	47.60	46.17	45.32	44.32	43.71
2, Good	50.22	48.52	47.33	46.47	45.79	45.23	44.38	43.07	42.27	41.31	40.74
2 & 3	46.48	44.89	43.82	43.01	42.38	41.86	41.09	39.86	39.11	38.24	37.70
3, Average	42.70	41.27	40.26	39.51	38.95	38.47	37.76	36.62	35.95	35.14	34.65
3 & 4	38.84	37.52	36.61	35.93	35.41	34.98	34.33	33.31	32.68	31.96	31.50
4, Low	34.94	33.76	32.93	32.33	31.86	31.48	30.89	29.97	29.40	28.75	28.34
Wall Height Adjustment*	.58	.51	.44	.39	.37	.35	.32	.27	.25	.22	.21

*Wall Height Adjustment: Add or subtract the amount listed in this column to the square foot of floor cost for each foot of wall height more or less than 20 feet.

Perimeter Wall Adjustment: A common wall exists when two buildings share one wall. Adjust for common walls by deducting the linear foot costs below from the total structure cost. In some structures one or more walls are not owned at all. In this case, deduct the "No Ownership" cost per linear foot of wall not owned. For common wall, deduct $142.00 per linear foot. For no wall ownership, deduct $284.00 per linear foot.

Discount Houses - Wood or Wood and Steel Frame

Quality Classification

	Class 1 Best Quality	Class 2 Good Quality	Class 3 Average Quality	Class 4 Low Quality
Foundation	Reinforced concrete.	Reinforced concrete.	Reinforced concrete.	Reinforced concrete.
Floor Structure	4" reinforced concrete on 6" rock fill.	4" reinforced concrete on 6" rock fill.	4 reinforced concrete on 6" rock fill.	4" reinforced concrete on 6" rock fill.
Wall Structure	2" x 6" - 16" o.c.	2" x 6" - 16" o.c.	2" x 6" - 16" o.c.	2" x 4" - 16" o.c.
Roof Structure	Glu-lams or steel "I" beams on steel intermediate columns, plywood sheathing.	Glu-lams or steel "I" beams on steel intermediate columns, plywood sheathing.	Glu-lams, or steel "I" beams on steel intermediate columns, plywood sheathing.	Glu-lams, or steel "I" beams on steel intermediate columns, plywood sheathing.
Floor Finish	Terrazzo, sheet vinyl or good carpet in sales area.	Resilient tile in sales area with some terrazzo, solid vinyl tile or carpet.	Resilient tile in sales area.	Inexpensive composition tile in sales area.
Interior Wall Finish	Gypsum wallboard and vinyl wall cover or interior stucco on inside of walls and on partitions.	Gypsum wallboard, texture and paint or vinyl wall cover on inside of exterior walls and on partitions.	Gypsum wallboard, texture and paint on inside of exterior walls and on partitions.	Gypsum wallboard and paint on inside of exterior walls and on partitions.
Ceiling Finish	Suspended acoustical tile, dropped ceilings over some areas.	Suspended acoustical tile or gypsum board and acoustical texture, dropped ceilings in some areas.	Ceiling tile on roof purlins, dropped ceiling over some areas.	Open.
Front	A large amount of plate glass in good aluminum frames (18'-22' high for 3/4 of width), brick or stone veneer on remainder, 1 pair good automatic doors 1 per 7,000 S.F. of floor area, anodized aluminum sunshade over glass area, 8' canopy across front, 10'-12' raised walk across front.	A large amount of plate glass in good aluminum frames (16'-18' high for 2/3 of width), brick or stone veneer on remainder, 1 pair good automatic doors per 10,000 S.F. of floor area, 8' canopy across front, 10' raised walk across front.	A moderate amount of plate glass in average quality aluminum frames (12'-16' high extending 20' on each side of entrances), exposed aggregate on remainder, 1 pair average automatic doors per 20,000 S.F. of floor area, 6' canopy over glass areas.	Stucco or exposed aggregate with a small amount of plate glass in inexpensive aluminum frames (6'-10' high for 1/2 of width), 6' canopy at entrances, 6' ground level walk across front.
Exterior Wall Finish	Good wood siding or masonry veneer.	Wood siding, some masonry veneer.	Stucco, some masonry trim.	Stucco.
Roof Cover	5 ply built-up roofing with large rock.	5 ply built-up roofing with tar and rock.	4 ply built-up roofing.	4 ply built-up roofing.
Plumbing	2 rest rooms with 8 fixtures each, floor piping and drains to refrigerated cases.	2 rest rooms with 6 fixtures each, floor piping and drains to refrigerated cases.	2 rest rooms with 4 fixtures each, floor piping and drains to refrigerated cases.	2 rest rooms with 2 fixtures each, floor piping and drains to refrigerated cases.
Electrical	Conduit wiring, recessed 4 tube fluorescent fixtures 8' o.c., 60-80 spotlights.	Conduit wiring, 4 tube fluorescent fixtures with diffusers 8' o.c., 60-80 spotlights.	Conduit wiring, 3 tube fluorescent fixtures 8' o.c., 20 to 40 spotlights.	Conduit wiring, double tube fluorescent fixtures, 8' o.c.

Square foot costs include the following components: Foundations as required for normal soil conditions. Floor, wall, and roof structures. Interior floor, wall and ceiling finishes. Exterior wall finish and roof cover. Display fronts. Interior partitions. Entry and delivery doors. A canopy and walk across the front of the building as described in the applicable building specifications. Basic lighting and electrical systems. Rough and finish plumbing. Design and engineering fees. Permits and hook-up fees. Contractor's mark-up.

Discount Houses - Wood or Wood and Steel Frame

Estimating Procedure

1. Establish the structure quality class by using the information on page 80.
2. Compute the building floor area. This should include everything within the building exterior walls and all insets outside the main walls but under the main building roof.
3. Add to or subtract from the square foot cost below the appropriate amount from the Wall Height Adjustment Column (at the bottom of this page) if the wall height is more or less than 20 feet.
4. Multiply the adjusted square foot cost by the building area.
5. Deduct, if appropriate, for common walls, using the figures at the bottom of this page.
6. Multiply the total cost by the location factor listed on page 7.
7. Add the cost of heating and cooling equipment, fire sprinklers, exterior signs, yard improvements, loading docks, ramps and walk-in boxes if they are an integral part of the building. See pages 201 to 213.

Discount House, Class 2

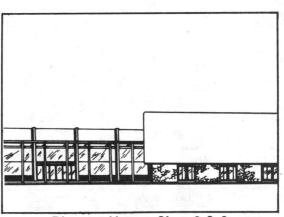

Discount House, Class 2 & 3

Square Foot Area

Quality Class	15,000	20,000	25,000	30,000	35,000	40,000	50,000	75,000	100,000	150,000	200,000
1, Best	54.73	53.39	52.46	51.78	51.24	50.81	50.15	49.13	48.50	47.77	47.34
1 & 2	51.34	50.07	49.21	48.56	48.07	47.65	47.04	46.07	45.51	44.82	44.41
2, Good	47.81	46.62	45.82	45.22	44.75	44.39	43.81	42.90	42.36	41.73	41.35
2 & 3	44.27	43.17	42.43	41.87	41.46	41.10	40.56	39.73	39.23	38.64	38.29
3, Average	40.63	39.65	38.97	38.44	38.04	37.73	37.24	36.46	36.02	35.48	35.15
3 & 4	36.96	36.05	35.43	34.97	31.12	34.32	33.87	33.17	32.76	32.26	31.98
4, Low	33.26	32.45	31.89	31.48	31.15	30.89	30.49	29.86	29.49	29.04	28.77
Wall Height Adjustment*	.32	.30	.28	.27	.25	.24	.23	.20	.18	.14	.13

*Wall Height Adjustment: Add or subtract the amount listed in this column to the square foot of floor cost for each foot of wall height more or less than 20 feet.

Perimeter Wall Adjustment: A common wall exists when two buildings share one wall. Adjust for common walls by deducting the linear foot costs below from the total structure cost. In some structures one or more walls are not owned at all. In this case, deduct the "No Ownership" cost per linear foot of wall not owned. For common wall, deduct $103.00 per linear foot. For no wall ownership, deduct $206.00 per linear foot.

Banks and Savings Offices - Masonry or Concrete
Quality Classification

	Class 1 Best Quality	Class 2 Good Quality	Class 3 High Average Quality	Class 4 Low Average Quality	Class 5 Low Quality
Foundation	Reinforced concrete.	Reinforced concrete.	Reinforced concrete.	Reinforced concrete.	Reinforced concrete.
Floor Structure	Reinforced concrete.	Reinforced concrete.	Reinforced concrete.	Reinforced concrete	Reinforced concrete.
Wall Structure	12" reinforced brick, 8" reinforced concrete or concrete columns with noncombustible filler walls.	12" reinforced brick, 8" reinforced concrete or concrete columns with noncombustible filler walls.	6" concrete tilt-up, 12" reinforced brick, 8" reinforced con. or concrete columns with noncombustible filler walls.	8" decorative concrete block, 6" concrete tilt-up, 8" reinforced brick or concrete columns with noncombustible filler walls.	8" concrete block, 6" reinforced concrete, 6" concrete tilt-up or 8" reinforced brick.
Roof & Cover	5 ply built-up roof, some portions copper or slate.	5 ply built-up roof, some portions heavy shake or mission tile.	4 ply built-up roof, some portions heavy shake or mission tile.	4 ply built-up roof, some portions shake or composition shingles.	4 ply built-up roof.
Exterior Wall Finish	Expensive veneers.	Ornamental brick or stone veneer.	Brick, or ornamental block veneer.	Ornamental rock imbedded in tilt-up panels or stucco with masonry trim.	Exposed aggregate or stucco.
Interior Wall Finish	Ornamental plaster painted with murals or designs. Ornamental moldings, marble and similar expensive finishes.	Ornamental plaster furred out from walls. Ornamental moldings, some hardwood panels or marble finishes.	Hard plaster furred out from walls. Ornamental cove at ceiling. Vinyl wall covering. Some hardwood veneer	Plaster on lath with putty coat finish. Gypsum wallboard, texture and paint. Some vinyl wall cover.	Plaster on lath with putty coat finish or gypsum wallboard, texture and paint.
Glass	Tinted plate glass in customized frames, (50% of exterior wall area.)	Tinted plate glass in bronze frames, (50% of exterior wall area).	Tinted plate glass in heavy anodized aluminum frames or custom wood frames, (50% of exterior wall area).	Moderate amount of plate glass in heavy frames, (25% of exterior wall area).	Small to moderate amount of plate glass in average aluminum frames, (5 to 10% of exterior wall area).
Overhang	4' closed overhang with copper gutters.	4' closed overhang with copper gutters.	4' closed overhang with painted gutters.	3' closed overhang with painted gutters.	None.
Floor Finish *Public Area*	Terrazzo and marble.	Detailed terrazzo.	Detailed terrazzo.	Plain terrazzo.	Sheet vinyl or vinyl asbestos tile.
Floor Finish *Officers Area*	Excellent carpet.	Very good carpet.	Very good carpet.	Good carpet.	Average quality carpet.
Work Area	Good sheet vinyl or carpet.	Average sheet vinyl, tile or carpet.	Resilient tile.	Resilient tile.	Composition tile.
Ceiling Finish	Ornamental plaster with exposed ornamental beams.	Ornamental plaster.	Suspended acoustical tile with gypsum wallboard backing or acoustical plaster.	Suspended acoustical tile with exposed grid system.	Acoustical tile on wood furring.
Lighting	Recessed panelized fluorescent light fixtures, custom light fixtures or chandeliers. Many spotlights.	Recessed panelized fluorescent lighting. Many spotlights.	Recessed panelized fluorescent lighting. Many spotlights.	Recessed panelized fluorescent lighting. Some spotlights.	Nonrecessed panelized lighting.
Plumbing	2 or more rest rooms with special fixtures, good ceramic tile, marble or terrazzo wainscot walls. Hard plaster with putty coat and enamel paint. Marble toilet screens.	2 or more rest rooms with many good fixtures, good ceramic tile, marble or terrazzo wainscot walls. Hard plaster with putty coat and enamel paint. Marble toilet screens.	2 rest rooms with 4 or more fixtures, ceramic tile or terrazzo floors. Ceramic tile or terrazzo wainscot. Hard plaster walls with putty coat and enamel paint. Marble or metal toilet screens.	2 rest rooms with 4 or more fixtures each. Ceramic tile or terrazzo floor. Ceramic tile or terrazzo wainscot plaster walls with putty coat. Enamel paint. Good metal toilet partitions.	2 rest rooms with 4 fixtures each. Ceramic tile or vinyl asbestos tile floors. Plaster walls with enamel paint. Metal toilet screens.

Square foot costs include the following components: Foundations as required for normal soil conditions. Floor, wall, and roof structures. Interior floor, wall and ceiling finishes. Exterior wall finish and roof cover. All glass and glazing. Interior partitions. Roof overhang as described above. Basic electrical systems and lighting fixtures. Rough and finish plumbing. Typical bank vault. Alarm systems. Night depository. Typical record vault and fire doors. Fire exits. Design and engineering fees. Permits and hook-up fees. Contractor's mark-up.

Length Less Than Twice Widths

Estimating Procedure

1. Establish the structure quality class by using the information on page 82.
2. Compute the building floor area. This should include everything within the building exterior walls and all insets outside the main walls but under the main building roof.
3. Add to or subtract from the square foot cost below the appropriate amount from the Wall Height Adjustment Table on page 86 if the wall height is more or less than 16 feet for the first floor or 12 feet for higher floors.
4. Multiply the adjusted square foot cost by the building area.
5. Deduct, if appropriate, for common walls, using the figures on page 86.
6. Multiply the total cost by the location factor listed on page 7.
7. Add the cost of heating and cooling equipment, fire sprinklers, exterior signs, elevators, and yard improvements from pages 201 to 213. Add the cost of mezzanines, bank fixtures, external windows, safe deposit boxes and vault doors from page 92.

Savings Office, Class 3

First Story
Square Foot Area

Quality Class	2,500	3,000	3,500	4,000	4,500	5,000	6,000	7,500	10,000	15,000	20,000
1, Best	207.19	200.83	195.25	190.34	186.01	182.16	175.58	167.75	158.21	145.96	138.18
1 & 2	195.00	189.00	183.75	179.12	175.05	171.43	165.23	157.88	148.90	137.36	130.04
2, Good	182.66	177.05	172.12	167.80	163.99	160.57	154.78	147.89	139.47	128.68	121.82
2 & 3	170.96	165.69	161.09	157.05	153.48	150.29	144.87	138.42	130.55	120.44	114.02
3, Good	159.14	154.25	149.98	146.21	142.87	139.92	134.86	128.93	121.51	112.11	106.13
3 & 4	149.35	144.77	140.75	137.22	134.09	131.32	126.57	120.94	114.05	105.22	99.62
4, Average	139.47	135.18	131.42	128.12	125.21	122.61	118.19	112.92	106.49	98.26	93.01
4 & 5	128.17	124.23	120.78	117.75	115.07	112.67	108.61	103.77	97.87	90.31	85.48
5, Low	116.79	113.20	110.06	107.29	104.86	102.67	98.97	94.56	89.20	82.29	77.90

Second and Higher Stories
Square Foot Area

Quality Class	500	750	1,000	1,500	2,000	3,000	4,000	5,000	7,500	10,000	20,000
1, Best	168.91	151.31	140.91	128.69	121.43	112.89	107.81	104.37	99.02	95.84	89.85
1 & 2	159.72	143.07	133.24	121.66	114.81	106.72	101.94	98.68	93.61	90.61	84.95
2, Good	151.13	135.37	126.06	115.11	108.64	101.00	96.46	93.38	88.59	85.76	80.39
2 & 3	142.36	127.52	118.76	108.45	102.34	95.14	90.86	87.97	83.45	80.77	75.72
2, Good	133.65	119.73	111.49	101.81	96.09	89.31	85.31	82.58	78.35	75.83	71.09
3 & 4	125.44	112.36	104.65	95.56	90.17	83.82	80.07	77.50	73.53	71.17	66.71
4, Average	117.31	105.10	97.87	89.37	84.34	78.40	74.88	72.48	68.77	66.56	62.40
4 & 5	110.15	98.69	91.89	83.93	79.20	73.61	70.31	68.05	64.57	62.50	58.60
5, Low	102.87	92.15	85.82	78.37	73.96	68.74	65.65	63.56	60.30	58.36	54.72

Length Between 2 and 4 Times Width

Estimating Procedure

1. Establish the structure quality class by using the information on page 82.
2. Compute the building floor area. This should include everything within the building exterior walls and all insets outside the main walls but under the main building roof.
3. Add to or subtract from the square foot cost below the appropriate amount from the Wall Height Adjustment Table on page 86 if the wall height is more or less than 16 feet for the first floor or 12 feet for higher floors.
4. Multiply the adjusted square foot cost by the building area.
5. Deduct, if appropriate, for common walls, using the figures on page 86.
6. Multiply the total cost by the location factor listed on page 7.
7. Add the cost of heating and cooling equipment, fire sprinklers, exterior signs, elevators, and yard improvements from pages 201 to 213. Add the cost of mezzanines, bank fixtures, external windows, safe deposit boxes and vault doors from page 92.

Savings Office, Class 4

First Story
Square Foot Area At 16 Foot Wall Height

Quality Class	2,500	3,000	3,500	4,000	4,500	5,000	6,000	7,500	10,000	15,000	20,000
1, Best	220.77	209.37	200.83	194.16	188.78	184.32	177.35	169.92	161.82	152.71	147.54
1 & 2	208.50	197.55	189.38	182.97	177.83	173.54	166.85	159.74	151.98	143.23	138.26
2, Good	196.14	185.65	177.81	171.67	166.72	162.64	156.21	149.41	141.99	133.64	128.91
2 & 3	184.22	174.00	166.41	160.52	155.77	151.88	145.81	139.38	132.49	124.80	120.50
2, Good	171.92	162.51	155.47	149.98	145.53	141.86	136.13	129.99	123.35	115.86	111.62
3 & 4	161.05	152.19	145.59	140.45	136.29	132.87	127.51	121.81	115.65	108.73	104.82
4, Average	150.13	141.83	135.65	130.85	126.98	123.80	118.84	113.57	107.88	101.55	97.98
4 & 5	137.56	129.94	124.29	119.91	116.40	113.50	108.99	104.23	99.12	93.44	90.26
5, Low	124.82	117.96	112.87	108.93	105.77	103.17	99.13	94.90	90.34	85.31	82.49

Second and Higher Stories
Square Foot Area

Quality Class	500	750	1,000	1,500	2,000	3,000	4,000	5,000	7,500	10,000	20,000
1, Best	180.71	161.49	149.98	136.25	128.06	118.31	112.49	108.51	102.31	98.59	91.58
1 & 2	171.19	152.97	142.07	129.08	121.32	112.07	106.55	102.79	96.91	93.40	86.76
2, Very Good	161.91	144.69	134.37	122.10	114.73	105.99	100.79	97.20	91.66	88.34	82.06
2 & 3	152.24	136.06	126.36	114.80	107.90	99.69	94.78	91.41	86.20	83.08	77.16
3, Good	141.92	126.84	117.80	107.02	100.59	92.92	88.35	85.21	80.34	77.44	71.93
3 & 4	133.66	119.43	110.92	100.78	94.72	87.50	83.19	80.25	75.66	72.92	67.73
4, Average	124.98	111.70	103.74	94.25	88.58	81.83	77.80	75.04	70.76	68.21	63.34
4 & 5	117.16	104.70	97.23	88.34	83.03	76.70	72.92	70.34	66.34	63.91	59.38
5, Low	109.33	97.69	90.74	82.44	77.48	71.57	68.05	65.64	61.90	59.65	55.42

Length More Than 4 Times Width

Estimating Procedure

1. Establish the structure quality class by using the information on page 82.
2. Compute the building floor area. This should include everything within the building exterior walls and all insets outside the main walls but under the main building roof.
3. Add to or subtract from the square foot cost below the appropriate amount from the Wall Height Adjustment Table on page 86 if the wall height is more or less than 16 feet for the first floor or 12 feet for higher floors.
4. Multiply the adjusted square foot cost by the building area.
5. Deduct, if appropriate, for common walls, using the figures on page 86.
6. Multiply the total cost by the location factor listed on page 7.
7. Add the cost of heating and cooling equipment, fire sprinklers, exterior signs, elevators, and yard improvements from pages 201 to 213. Add the cost of mezzanines, bank fixtures, external windows, safe deposit boxes and vault doors from page 92.

Bank, Class 5

First Story
Square Foot Area

Quality Class	2,500	3,000	3,500	4,000	4,500	5,000	6,000	7,500	10,000	15,000	20,000
1, Best	235.81	222.11	211.92	204.04	197.70	192.46	184.36	175.79	166.56	156.35	150.60
1 & 2	222.08	209.16	199.59	192.15	186.19	181.28	173.62	165.56	156.87	147.24	141.83
2, Very Good	207.99	195.90	186.91	179.95	174.37	169.77	162.61	155.04	146.91	137.89	132.82
2 & 3	195.07	183.72	175.30	168.78	163.53	159.22	152.50	145.41	137.79	129.33	124.58
2, Good	181.73	171.17	163.31	157.24	152.37	148.34	142.09	135.48	128.36	120.47	116.07
3 & 4	170.05	160.15	152.81	147.13	142.55	138.80	132.95	126.77	120.10	112.74	108.60
4, Average	158.17	148.99	142.16	136.85	132.62	129.11	123.66	117.91	111.73	104.86	101.01
4 & 5	144.86	136.43	130.20	125.35	121.45	118.24	113.25	107.99	102.32	96.04	92.51
5, Low	131.26	123.63	117.97	113.57	110.05	107.15	102.62	97.85	92.73	87.03	83.85

Second and Higher Stories
Square Foot Area

Quality Class	500	750	1,000	1,500	2,000	3,000	4,000	5,000	7,500	10,000	20,000
1, Best	199.99	176.99	163.32	147.13	137.47	126.06	119.25	114.62	107.41	103.12	94.99
1 & 2	189.32	167.55	154.62	139.28	130.14	119.33	112.90	108.50	101.68	97.61	89.93
2, Very Good	178.51	157.98	145.79	131.33	122.73	112.52	106.45	102.31	95.87	92.04	84.79
2 & 3	167.63	148.35	136.89	123.31	115.24	105.65	99.96	96.07	90.02	86.43	79.62
3, Good	156.70	138.68	127.96	115.27	107.73	98.79	93.44	89.81	84.15	80.80	74.42
3 & 4	146.62	129.76	119.74	107.87	100.80	92.42	87.45	84.03	78.75	75.60	69.65
4, Average	136.43	120.74	111.42	100.37	93.78	86.01	81.37	78.19	73.28	70.35	64.81
4 & 5	127.92	113.21	104.46	94.10	87.93	80.62	76.27	73.29	68.69	65.95	60.75
5, Low	119.31	105.59	97.44	87.77	82.02	75.21	71.14	68.38	64.07	61.52	56.68

Wall Adjustments

Wall Height Adjustment

Add or subtract the amount listed in this table to the square foot of floor cost for each foot of wall height more or less than 16 feet, if adjusting for a first floor, and 12 feet if adjusting for upper floors.

Square Foot Area

Quality Class	500	750	1,000	1,500	2,000	2,500	3,000	3,500
1, Best	8.17	6.70	5.83	4.76	4.14	3.71	3.39	3.13
2, Good	7.26	5.96	5.18	4.25	3.68	3.30	3.01	2.79
3, High Avg	6.38	5.24	4.56	3.73	3.23	2.89	2.65	2.44
4, Low Avg	5.21	4.28	3.71	3.04	2.64	2.36	2.16	2.00
5, Low	4.11	3.38	2.93	2.41	2.09	1.87	1.69	1.57

Quality Class	4,000	4,500	5,000	6,000	7,500	10,000	15,000	20,000
1, Best	2.95	2.76	2.63	2.40	2.15	1.85	1.50	1.28
2, Good	2.60	2.46	2.32	2.13	1.89	1.65	1.33	1.14
3, High Avg	2.31	2.15	2.06	1.87	1.67	1.44	1.17	1.01
4, Low Avg	1.87	1.75	1.67	1.52	1.36	1.18	.96	.83
5, Low	1.46	1.37	1.29	1.19	1.05	.91	.73	.64

Perimeter Wall Adjustment
First Story

Class	For a Common Wall, Deduct Per L.F.	For No Wall Ownership, Deduct Per L.F.	For Lack of Exterior Finish, Deduct Per L.F.
1	$485.00	$970.00	$330.00
2	405.00	805.00	260.00
3	330.00	660.00	185.00
4	200.00	405.00	125.00
5	165.00	330.00	33.00

Second and Higher Stories

Class	For a Common Wall, Deduct Per L.F.	For No Wall Ownership, Deduct Per L.F.	For Lack of Exterior Finish, Deduct per L.F.
1	$270.00	$745.00	150.00
2	225.00	455.00	110.00
3	180.00	360.00	72.00
4	150.00	300.00	46.00
5	110.00	215.00	33.00

Note: First floor costs include the cost of overhang as described on page 82. Second floor costs do not include any allowance for overhang.

Banks and Savings Offices - Wood Frame
Quality Classification

	Class 1 Best Quality	Class 2 Good Quality	Class 3 High Average Quality	Class 4 Low Average Quality	Class 5 Low Quality
Foundation	Reinforced concrete.	Reinforced concrete.	Reinforced concrete.	Reinforced concrete.	Reinforced concrete.
Floor Structure	Reinforced concrete.	Reinforced concrete.	Reinforced concrete.	Reinforced concrete	Reinforced concrete.
Wall Structure	2" x 6" - 16" o.c. brick or concrete columns with combustible filler walls.	2" x 6" - 16" o.c. brick or concrete columns with combustible filler walls.	2" x 6" - 16" o.c.	2" x 6" - 16" o.c.	2" x 6" - 16" o.c.
Roof & Cover	Copper or slate.	Mission tile or heavy shakes.	5 ply built-up roof. Some portions heavy shake or mission tile.	4 ply built-up roof. Some portions shingle or composition shingles.	4 ply built-up roof.
Exterior Wall Finish	Expensive veneers.	Expensive ornamental veneer filler walls.	Wood siding combined with brick or stone veneers.	Good wood siding.	Good stucco or wood siding.
Interior Wall Finish	Ornamental plaster painted with murals or designs, ornamental moldings, marbleand similar expensive finishes.	Ornamental plaster ornamental moldings, some hardwood panels or marble finishes.	Hard plaster, ornamental cove at ceiling, vinyl wall covering, some hardwood veneer.	Plaster on lath with putty coat finish. Gypsum wallboard, texture and paint. Some vinyl wall cover.	Plaster on lath with putty coat finish. Gypsum wallboard, texture and paint.
Glass	Tinted plate glass in customized frames (50% of exterior wall area).	Tinted plate glass in bronze frames (50% of exterior wall area).	Tinted plate glass in heavy anodized aluminum frames or custom wood frames (50% of exterior wall area).	Moderate amount of plate glass in heavy frames (25% of exterior wall area).	Small to moderate amount of plate glass in average aluminum frames (5 to 10% of exterior wall area).
Overhang	4' closed overhang with copper gutters.	4' closed overhang with copper gutters.	4' closed overhang with painted gutters.	3' closed overhang with painted gutters.	None.
Floor Finish *Public Area*	Terrazzo and marble.	Detailed terrazzo.	Detailed terrazzo.	Plain terrazzo.	Sheet vinyl or resilient tile.
Floor Finish *Officers Area*	Excellent carpet.	Very good carpet.	Very good carpet.	Good carpet.	Average quality carpet.
Work Area	Good sheet vinyl or carpet.	Average sheet vinyl or carpet.	Resilient tile.	Composition tile.	Composition tile.
Ceiling Finish	Ornamental plaster with exposed ornamental beams.	Ornamental plaster.	Suspended acoustical tile with gypsum wallboard backing or plain acoustical plaster.	Suspended acoustical tile with exposed grid system.	Acoustical tile on wood strips.
Lighting	Recessed panelized fluorescent light fixtures, custom light fixtures or chandeliers. Many spotlights.	Recessed panelized fluorescent lighting. Custom chandeliers. Many spotlights.	Recessed panelized fluorescent lighting. Many spotlights.	Recessed panelized fluorescent lighting. Some spotlights.	Nonrecessed panelized lighting.
Plumbing	2 or more rest rooms with special fixtures. Good ceramic tile, marble or terrazzo wainscot walls. Hard plaster with putty coat and enamel paint. Marble toilet screens. toilet screens.	2 or more rest rooms with many good fixtures. Good ceramic tile, marble or terrazzo wainscot walls. Hard plaster with putty coat and enamel paint. Marble screens.	2 rest rooms with 4 or more fixtures. Ceramic tile or terrazzo wainscot, hard plaster walls with putty coat and enamel paint. Marble or metal toilet screens.	2 rest rooms with 4 or more fixtures each. Ceramic tile or terrazzo floor. Ceramic tile or terrazzo wainscot plaster walls with putty coat. Enamel paint. Good metal toilet partitions.	2 rest rooms with 4 fixtures each. Metal toilet screens. Ceramic tile or vinyl asbestos tile floors. Plaster walls with enamel paint.

Square foot costs include the following components: Foundations as required for normal soil conditions. Floor, wall, and roof structures. Interior floor, wall and ceiling finishes. Exterior wall finish and roof cover. All glass and glazing. Interior partitions. Roof overhang as described above. Basic electrical systems and lighting fixtures. Rough and finish plumbing. Typical bank vault. Alarm systems. Night depository. Typical record vault and fire doors. Fire exits. Design and engineering fees, permits and hook-up fees, contractor's mark-up.

Banks and Savings Offices - Wood Frame
Length Less Than Twice Width

Estimating Procedure

1. Establish the structure quality class by using the information on page 87.
2. Compute the building floor area. This should include everything within the building exterior walls and all insets outside the main walls but under the main building roof.
3. Add to or subtract from the square foot cost below the appropriate amount from the Wall Height Adjustment Table on page 91 if the wall height is more or less than 16 feet for the first floor or 12 feet for higher floors.
4. Multiply the adjusted square foot cost by the building area.
5. Deduct, if appropriate, for common walls, using the figures on page 91.
6. Multiply the total cost by the location factor listed on page 7.
7. Add the cost of heating and cooling equipment, fire sprinklers, exterior signs, elevators, and yard improvements from pages 201 to 213. Add the cost of mezzanines, bank fixtures, external windows, safe deposit boxes and vault doors from page 92.

Savings Office, Class 5

First Story
Square Foot Area

Quality Class	2,500	3,000	3,500	4,000	4,500	5,000	6,000	7,500	10,000	15,000	20,000
1, Best	206.06	194.87	186.75	180.56	175.69	171.75	165.72	159.54	153.15	146.40	142.81
1 & 2	193.51	183.00	175.38	169.58	165.01	161.27	155.64	149.84	143.81	137.48	134.11
2, Good	180.86	171.04	163.91	158.48	154.20	150.73	145.44	140.03	134.41	128.48	125.33
2 & 3	161.23	152.47	146.13	141.30	137.47	134.38	129.67	124.84	119.81	114.55	111.75
3, Good	156.77	148.26	142.09	137.38	133.67	130.67	126.08	121.38	116.50	111.38	108.64
3 & 4	146.07	138.14	132.37	127.99	124.53	121.73	117.48	113.10	108.56	103.77	101.24
4, Average	135.42	128.06	122.72	118.68	115.47	112.89	108.91	104.84	100.64	96.20	93.86
4 & 5	123.46	116.75	111.88	108.17	105.26	102.88	99.29	95.57	91.74	87.70	85.55
5, Low	111.10	105.06	100.69	97.35	94.73	92.60	89.35	86.02	82.57	78.93	76.99

Second and Higher Stories
Square Foot Area

Quality Class	500	750	1,000	1,500	2,000	3,000	4,000	5,000	7,500	10,000	20,000
1, Best	142.14	128.96	121.12	111.81	106.25	99.66	95.74	93.07	88.89	86.41	81.71
1 & 2	131.87	120.11	113.12	104.86	99.95	94.13	90.66	88.30	84.63	82.46	78.31
2, Very Good	122.84	112.29	106.03	98.58	94.15	88.90	85.78	83.66	80.34	78.37	74.64
2 & 3	113.50	104.15	98.59	92.01	88.11	83.48	80.72	78.84	75.92	74.19	70.89
3, Good	104.04	95.98	91.19	85.51	82.11	78.08	75.70	74.08	71.53	70.01	67.15
3 & 4	95.87	88.71	84.47	79.46	76.48	72.96	70.87	69.44	67.24	65.92	63.42
4, Average	87.47	81.40	77.77	73.47	70.92	67.88	66.07	64.84	62.93	61.78	59.63
4 & 5	80.55	75.29	72.15	68.43	66.22	63.61	62.06	61.00	59.34	58.36	56.51
5, Low	75.09	69.73	66.71	63.31	61.36	59.16	57.89	57.04	55.75	54.99	53.62

Banks and Savings Offices - Wood Frame
Length Between 2 and 4 Times Width

Estimating Procedure

1. Establish the structure quality class by using the information on page 87.
2. Compute the building floor area. This should include everything within the building exterior walls and all insets outside the main walls but under the main building roof.
3. Add to or subtract from the square foot cost below the appropriate amount from the Wall Height Adjustment Table on page 91 if the wall height is more or less than 16 feet for the first floor or 12 feet for higher floors.
4. Multiply the adjusted square foot cost by the building area.
5. Deduct, if appropriate, for common walls, using the figures on page 91.
6. Multiply the total cost by the location factor listed on page 7.
7. Add the cost of heating and cooling equipment, fire sprinklers, exterior signs, elevators, and yard improvements from pages 201 to 213. Add the cost of mezzanines, bank fixtures, external windows, safe deposit boxes and vault doors from page 92.

Bank, Class 4

First Story
Square Foot Area

Quality Class	2,500	3,000	3,500	4,000	4,500	5,000	6,000	7,500	10,000	15,000	20,000
1, Best	211.35	200.46	192.38	186.12	181.12	177.00	170.61	163.85	156.62	148.66	144.21
1 & 2	198.56	188.32	180.74	174.86	170.14	166.28	160.27	153.94	147.15	139.66	135.45
2, Good	185.60	176.03	168.93	163.42	159.04	155.43	149.81	143.88	137.54	130.53	126.62
2 & 3	173.16	164.22	157.61	152.49	148.39	144.99	139.76	134.23	128.31	121.78	118.15
3, Good	160.95	152.64	146.49	141.73	137.90	134.77	129.90	124.77	119.27	113.19	109.80
3 & 4	149.30	141.60	135.88	131.47	127.93	125.02	120.49	115.74	110.63	105.00	101.85
4, Average	138.18	131.05	125.78	121.67	118.41	115.71	111.52	107.13	102.40	97.19	94.28
4 & 5	125.58	119.10	114.30	110.58	107.59	105.15	101.36	97.35	93.04	88.31	85.68
5, Low	112.87	107.04	102.74	99.39	96.70	94.51	91.11	87.49	83.63	79.38	76.99

Second and Higher Stories
Square Foot Area

Quality Class	500	750	1,000	1,500	2,000	3,000	4,000	5,000	7,500	10,000	20,000
1, Best	143.56	131.46	124.26	115.72	110.62	104.59	101.01	98.54	94.75	92.47	88.18
1 & 2	135.30	123.90	117.11	109.06	104.27	98.59	95.20	92.89	89.31	87.16	83.11
2, Very Good	126.93	116.24	109.85	102.31	97.81	92.48	89.30	87.13	83.77	81.76	77.95
2 & 3	118.64	108.65	102.67	95.63	91.44	86.43	83.48	81.44	78.31	76.42	72.87
3, Good	110.29	100.99	95.47	88.91	84.99	80.36	77.59	75.71	72.79	71.05	67.75
3 & 4	102.55	93.90	88.75	82.67	79.04	74.74	72.16	70.41	67.67	66.05	62.99
4, Average	94.96	86.97	82.19	76.57	73.19	69.21	66.82	65.20	62.68	61.18	58.34
4 & 5	88.60	81.14	76.69	71.42	68.28	64.57	62.34	60.83	58.47	57.08	54.42
5, Low	82.17	75.27	71.13	66.23	63.34	59.88	57.83	56.41	54.24	52.94	50.48

Banks and Savings Offices - Wood Frame
Length More Than 4 Times Width

Estimating Procedure

1. Establish the structure quality class by using the information on page 87.
2. Compute the building floor area. This should include everything within the building exterior walls and all insets outside the main walls but under the main building roof.
3. Add to or subtract from the square foot cost below the appropriate amount from the Wall Height Adjustment Table on page 91 if the wall height is more or less than 16 feet for the first floor or 12 feet for higher floors.
4. Multiply the adjusted square foot cost by the building area.
5. Deduct, if appropriate, for common walls, using the figures on page 91.
6. Multiply the total cost by the location factor listed on page 7.
7. Add the cost of heating and cooling equipment, fire sprinklers, exterior signs, elevators, and yard improvements from pages 201 to 213. Add the cost of mezzanines, bank fixtures, external windows, safe deposit boxes and vault doors from page 92.

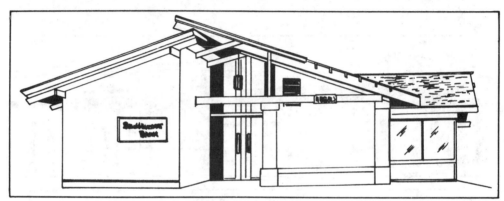

Bank, Class 4 & 5

First Story
Square Foot Area

Quality Class	2,500	3,000	3,500	4,000	4,500	5,000	6,000	7,500	10,000	15,000	20,000
1, Best	222.07	209.87	200.86	193.90	188.34	183.78	176.73	169.30	161.40	152.73	147.92
1 & 2	208.65	197.18	188.71	182.17	176.96	172.67	166.03	159.06	151.64	143.49	138.97
2, Very Good	194.89	184.20	176.27	170.17	165.28	161.29	155.09	148.58	141.65	134.02	129.80
2 & 3	182.23	172.20	164.82	159.10	154.55	150.81	145.00	138.93	132.45	125.32	121.37
3, Good	169.24	159.93	153.07	147.76	143.52	140.06	134.68	129.02	122.99	116.38	112.72
3 & 4	156.87	148.24	141.88	136.97	133.04	129.81	124.83	119.59	114.01	107.88	104.47
4, Average	144.40	136.46	130.61	126.08	122.46	119.50	114.91	110.09	104.95	99.30	96.18
4 & 5	130.73	123.54	118.23	114.14	110.87	108.18	104.02	99.66	95.01	89.90	87.06
5, Low	116.93	110.51	105.77	102.10	99.16	96.76	93.05	89.14	84.98	80.41	77.88

Second and Higher Stories
Square Foot Area

Quality Class	500	750	1,000	1,500	2,000	3,000	4,000	5,000	7,500	10,000	20,000
1, Best	164.95	147.51	137.12	124.83	117.52	108.87	103.70	100.18	94.73	91.47	85.31
1 & 2	153.08	137.36	128.03	116.96	110.38	102.58	97.94	94.78	89.85	86.94	81.40
2, Very Good	141.06	127.10	118.82	109.02	103.20	96.27	92.16	89.35	85.01	82.42	77.52
2 & 3	129.57	117.26	109.94	101.28	96.13	90.04	86.40	83.93	80.09	77.78	73.46
2, Good	117.93	107.30	100.98	93.50	89.02	83.74	80.60	78.45	75.11	73.12	69.35
3 & 4	108.03	98.67	93.10	86.50	82.60	77.94	75.18	73.30	70.37	68.63	65.34
4, Average	98.01	89.92	85.12	79.46	76.07	72.10	69.72	68.10	65.59	64.11	61.30
4 & 5	89.59	82.63	78.50	73.61	70.71	67.27	65.23	63.83	61.67	60.38	57.94
5, Low	81.20	75.35	71.87	67.76	65.31	62.45	60.72	59.56	57.75	56.66	54.63

Banks and Savings Offices - Wood Frame

Wall Adjustments

Wall Height Adjustment

Add or subtract the amount listed in this table to the square foot of floor cost for each foot of wall height more or less than 16 feet, if adjusting for first floor, and 12 feet if adjusting for upper floors.

Square Foot Area

Quality Class	500	750	1,000	1,500	2,000	2,500	3,000	3,500
1, Best	6.88	5.64	4.91	4.03	3.49	3.12	2.87	2.64
2, Good	6.12	5.04	4.37	3.58	3.12	2.79	2.54	2.37
3, High Avg	5.34	4.39	3.81	3.11	2.70	2.42	2.23	2.05
4, Low Avg	3.59	2.95	2.56	2.11	1.83	1.64	1.50	1.38
5, Low	2.34	1.91	1.68	1.37	1.20	1.06	.97	.90

Square Foot Area

Quality Class	4,000	4,500	5,000	6,000	7,500	10,000	15,000	20,000
1, Best	2.61	2.41	2.25	2.01	1.78	1.53	1.29	1.15
2, Good	2.31	2.12	2.00	1.77	1.56	1.35	1.12	1.02
3, High Avg	2.00	1.86	1.73	1.55	1.36	1.19	.99	.88
4, Low Avg	1.34	1.25	1.18	1.05	.92	.81	.67	.61
5, Low	.90	.84	.80	.70	.60	.53	.44	.39

Perimeter Wall Adjustment
First Story

Class	For a Common Wall, Deduct Per L.F.	For No Wall Ownership, Deduct Per L.F.	For Lack of Exterior Finish, Deduct Per L.F.
1	$330.00	$660.00	$320.00
2	265.00	525.00	260.00
3	215.00	435.00	185.00
4	155.00	310.00	125.00
5	73.00	145.00	42.00

Second and Higher Stories

Class	For a Common Wall, Deduct Per L.F.	For No Wall Ownership, Deduct Per L.F.	For Lack of Exterior Finish, Deduct Per L.F.
1	$200.00	$400.00	$185.00
2	160.00	320.00	125.00
3	130.00	260.00	83.00
4	83.00	165.00	57.00
5	57.00	114.00	26.00

Note: First floor costs include the cost of overhang as described on page 87. Second floor costs do not include any allowance for overhang.

Banks and Savings Offices
Additional Costs

Mezzanines Without Partitions

Quality Class	1, Best	2, Good	3, High Avg.	4, Low Avg.	5, Low
S.F. Cost	$50.23	$48.80	$42.19	$38.95	$35.16

With Partitions

Quality Class	1, Best	2, Good	3, High Avg.	4, Low Avg.	5, Low
S.F. Cost	$66.22	$60.47	$53.57	$50.38	$43.95

Mezzanine costs include: Floor system, floor finish, typical stairway, lighting and structural support costs.

Fixtures, cost per square foot of floor

Shell-type counters, plastic finish, no counter screens or drawers.	$15.20 to $17.00
Counters with drawers, good hardwood, plain counter screens.	18.10 to 24.60
Counters with drawers, good hardwood, plastic countertops, average counter screens.	24.60 to 36.20
Counters with drawers, terrazzo or marble finish, marble counter tops, fancy counter screens.	36.20 to 60.80

Costs include counters, screens, and partitions. The square-foot cost should be applied only to the first floor area. Office areas used for purposes other than conducting bank business related to the immediate site should be excluded.

External Access Facilities, cost each unit

Drive-up teller, flush window	$8,350 to $11,500
Drive-up teller, projected window	10,150 to 15,200
Walk-up teller, flush window	5,750 to 7,200
Automatic teller, cash dispensing, with phone	50,200 to 57,750
Night deposit vault whole chest	9,750 to 13,400

Safe Deposit Boxes, cost per box

Box Sizes	Modular Unit	Custom Built
3" x 5"	$72.00 to $82.00	$106.00 to $116.00
5" x 5"	82.00 to 93.00	116.00 to 143.00
3" x 10"	77.00 to 98.00	127.00 to 148.00
5" x 10"	103.00 to 118.00	160.00 to 185.00
10" x 10"	144.00 to 165.00	212.00 to 265.00

Bank fixtures and safe deposit boxes are part of the structure cost only if they are fixed to the building.

Vault Doors Cash and Safe Deposit (rectangular)

Steel thickness	1-1/2"	3-1/2"	7"
In Place Cost	$17,500 to $23,500	$23,500 to $29,000	$29,000 to $39,000

Vault Doors Record Storage

Description	3' x 7', 2 hour	3' x 7', 4 hour	4' x 7', 2 hour	4' x 7', 4 hour
In Place Cost	$3,000	$3,400	$3,900	$4,300

Prices include frames, time locks, and architraves.

Department Stores - Reinforced Concrete
Quality Classification

	Class 1 Best Quality	Class 2 Good Quality	Class 3 Average Quality	Class 4 Low Quality
Foundation	Reinforced concrete.	Reinforced concrete.	Reinforced concrete.	Reinforced concrete.
Ground Floor Structure	6" reinforced concrete on 6" rock base.	6" reinforced concrete on 6" rock base.	6" reinforced concrete on 6" rock base.	4" reinforced concrete on 6" rock base.
Wall Structure	Reinforced concrete.	Reinforced concrete.	Reinforced concrete.	Reinforced concrete.
Upper Floor Structure	Reinforced concrete.	Reinforced concrete.	Reinforced concrete.	Reinforced concrete.
Roof and Cover	Reinforced concrete with 5 ply built-up roofing and insulation.	Reinforced concrete with 5 ply built-up roofing and insulation.	Reinforced concrete with 4 ply built-up roofing.	Reinforced concrete with 4 ply built-up roofing
Floor Finish	Terrazzo and very good carpet.	Resilient tile with 50% vinyl tile, terrazzo or good carpet.	Composition tile.	Minimum grade tile.
Interior Wall Finish	Gypsum wallboard or lath and plaster finished with good paper or vinyl wall cover on hardwood veneer paneling.	Gypsum wallboard and texture or paper. Some vinyl wall cover or - hardwood veneer paneling.	Interior stucco or gypsum wallboard, texture and paint.	Gypsum wallboard, texture and paint.
Ceiling Finish	Suspended good grade acoustical tile with gypsum wallboard backing.	Suspended acoustical tile with concealed grid system.	Suspended acoustical tile with exposed grid system	Painted.
Lighting	Recessed fluorescent lighting in modular plastic panels. Many spotlights.	Continuous recessed 3 tube fluorescent strips with egg crate diffusers 8' o.c. Average number of spotlights.	Continuous 3 tube fluorescent strips with egg crate diffusers, 8' o.c. Some spotlights.	Continuous exposed 2 tube fluorescent strips, 8' o.c.
Display Fronts	Very good front as described on page 56. 15 to 25% of the first floor exterior wall is made up of display fronts.	Good quality front as described on page 56. 15 to 25% of the first floor exterior wall is made up of display fronts.	Average quality front as described on page 56. 10 to 20% of the first floor exterior wall is made up of display fronts.	Average quality flat type as described on page 56. 10 to 20% of the first floor exterior wall is made up of display fronts.
Exterior Wall Finish	Ornamental block or brick, some marble veneer.	Decoative block, some stone veneer.	Paint.	Paint.
Plumbing	6 good fixtures per 20,000 square feet of floor area. Metal toilet partitions.	6 standard fixtures per 20,000 square feet of floor area. Metal toilet partitions.	4 standard fixtures per 20,000 square feet of floor area. Metal toilet partitions.	4 standard fixtures per 20,000 square feet of floor area. Wood toilet partitions.

Wall Height Adjustment: Add or subtract the amount listed to the square foot of floor cost for each foot of first and upper story wall height more or less than 20 feet.

Square Foot Area

Quality Class	20,000	25,000	30,000	35,000	40,000	45,000	50,000	60,000	70,000	80,000	100,000
1, Best	1.32	1.10	.99	.91	.83	.78	.75	.70	.68	.66	.64
2, Good	.90	.74	.67	.61	.57	.55	.53	.48	.46	.43	.42
3, Average	.72	.61	.54	.50	.45	.43	.41	.37	.36	.35	.34
4, Low	.58	.51	.43	.39	.36	.35	.33	.31	.30	.29	.28

Square foot costs include the following components: Foundations as required for normal soil conditions. Floor, wall and roof structures. Interior ceiling, wall and floor finishes (including carpet). Exterior wall finish and roof cover. Display fronts. Interior partitions (including perimeter wall partitions). Entry and delivery doors. Basic lighting and electrical systems. Rough and finish plumbing. Design and engineering fees. Permits and hook-up fees. Contractor's mark-up.

Department Stores - Reinforced Concrete

First Floor

Estimating Procedure

1. Use these square foot costs to estimate the cost of retail stores designed to sell a wider variety of goods. These buildings differ from discount houses in that they have more interior partitions and more elaborate interior and exterior finishes.
2. Establish the structure quality class by using the information on page 93.
3. Compute the building floor area. This should include everything within the building exterior walls and all inset areas outside the main walls but under the main building roof.
4. Add to or subtract from the square foot cost below the appropriate amount from the Wall Height Adjustment Table (near the bottom of page 93) if the wall height is more or less than 20 feet.
5. Multiply the adjusted square foot cost by the building area.
6. Deduct, if appropriate, for common walls, using the figures at the bottom of this page.
7. Multiply the total cost by the location factor listed on page 7.
8. Add the cost of appropriate additional components from page 201: heating and air conditioning systems, elevators and escalators, fire sprinklers, exterior signs, canopies and walks, paving and curbing, loading docks and ramps, miscellaneous yard improvements, and mezzanines.
9. Add the cost of second and higher floors and basements from page 95.

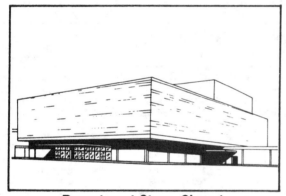

Department Store, Class 1

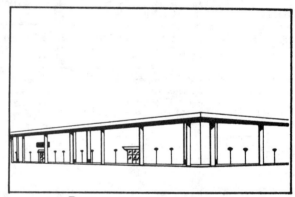

Department Store, Class 2

First Floor
Square Foot Area

Quality Class	20,000	25,000	30,000	35,000	40,000	45,000	50,000	60,000	70,000	80,000	100,000
1, Best	128.14	123.12	119.05	115.67	112.80	110.32	108.19	104.58	101.68	99.27	95.48
1 & 2	118.61	114.21	110.62	107.64	105.12	102.96	101.06	97.89	95.35	93.22	89.88
2, Good	109.01	105.19	102.09	99.52	97.35	95.49	93.85	91.14	88.93	87.13	84.26
2 & 3	100.34	97.08	94.44	92.26	90.41	88.82	87.46	85.14	83.29	81.75	79.32
3, Average	91.61	88.89	86.70	84.92	83.42	82.12	81.01	79.16	77.66	76.43	74.50
3 & 4	82.46	80.29	78.56	77.14	75.96	74.94	74.06	72.58	71.41	70.45	68.91
4, Low	74.79	73.24	71.98	70.96	70.06	69.29	68.62	67.50	66.60	65.83	64.66

Perimeter Wall Adjustment
First Floor

Class	For a Common Wall, Deduct Per L.F.	For No Wall Ownership, Deduct Per L.F.	For Lack of Exterior Finish, Deduct Per L.F.
1	$430.00	$855.00	$280.00
2	320.00	640.00	165.00
3	230.00	455.00	73.00
4	155.00	310.00	43.00

Department Stores - Reinforced Concrete

Estimating Procedure

1. Establish the basement and upper floor quality class. The quality class will usually be the same as the first floor of the building. Square foot costs for unfinished basements will be nearly the same regardless of the structure quality class.
2. Compute the floor area.
3. Add to or subtract from the square foot cost below the appropriate amount from the Wall Height Adjustment Table on page 93 for second and higher floors and from the bottom of this page for basements for each foot of wall height more or less than 20 feet for second and higher floors and 16 feet for basements.
4. Multiply the adjusted square foot cost from one of the 3 tables below by the floor area.
5. Deduct, if appropriate, for common or unfinished upper floor walls, using the costs in the table below titled "Second and Higher Floor Perimeter Wall Adjustments."
6. Multiply the total cost by the location factor listed on page 7.
7. Add the cost of appropriate additional components from page 201: heating and air conditioning systems, elevators and escalators, fire sprinklers, and mezzanines.
8. Add the total from this page to the total from page 94 to find the building cost.

Second and Higher Stories
Square Foot Area

Quality Class	20,000	25,000	30,000	35,000	40,000	45,000	50,000	60,000	70,000	80,000	100,000
1, Best	114.87	110.16	106.30	103.09	100.36	98.00	95.94	92.50	89.71	87.38	83.71
1 & 2	106.61	102.39	98.98	96.12	93.71	91.64	89.80	86.75	84.28	82.25	79.02
2, Good	98.20	94.50	91.51	89.05	86.95	85.16	83.58	80.97	78.85	77.10	74.33
2 & 3	90.67	87.53	84.99	82.88	81.12	79.58	78.26	76.02	74.23	72.73	70.40
3, Average	83.13	80.49	78.38	76.64	75.18	73.91	72.80	71.00	69.52	68.30	66.39
3 & 4	75.53	73.38	71.66	70.26	69.09	68.09	67.22	65.76	64.62	63.65	62.15
4, Low	67.82	66.16	64.85	63.81	62.93	62.20	61.56	60.50	59.66	58.96	57.89

Second And Higher Floor Perimeter Wall Adjustment

Class	For a Common Wall, Deduct Per L.F.	For No Wall Ownership, Deduct Per L.F.	For Lack of Exterior Finish, Deduct Per L.F.
1	$330.00	$660.00	$220.00
2	240.00	475.00	130.00
3	170.00	340.00	62.00
4	145.00	290.00	42.00

Finished Basements
Square Foot Area

Quality Class	20,000	25,000	30,000	35,000	40,000	45,000	50,000	60,000	70,000	80,000	100,000
1, Best	97.88	94.78	92.20	90.02	88.15	86.54	85.09	82.70	80.76	79.12	76.54
1 & 2	91.21	88.46	86.18	84.26	82.63	81.22	79.97	77.90	76.21	74.81	72.56
2, Good	85.93	83.42	81.36	79.64	78.16	76.88	75.76	73.88	72.37	71.11	69.09
2 & 3	79.80	77.67	75.91	74.45	73.23	72.16	71.23	69.67	68.42	67.36	65.71
3, Average	74.83	73.22	71.82	70.60	69.45	68.63	67.81	66.43	65.28	64.32	62.78
3 & 4	68.82	67.31	66.09	65.10	64.27	63.57	62.95	61.92	61.10	60.40	59.34
4, Low	62.58	61.34	60.43	59.69	59.09	58.59	58.15	57.43	56.87	56.42	55.73

Unfinished Basements

Area	20,000	25,000	30,000	35,000	40,000	45,000	50,000	60,000	70,000	80,000	100,000
Cost	43.35	42.57	41.91	41.47	41.02	40.80	40.47	40.03	39.70	39.47	39.02

Basement Wall Height Adjustment: Add or subtract the amount listed to the square foot of floor cost for each foot of basement wall height more or less than 16 feet.

Area	20,000	25,000	30,000	35,000	40,000	45,000	50,000	60,000	70,000	80,000	100,000
Finished	.59	.52	.46	.41	.39	.37	.36	.35	.34	.33	.33
Unfinished	.53	.46	.39	.37	.35	.33	.32	.31	.30	.30	.29

Department Stores - Masonry or Concrete
Quality Classification

	Class 1 Best Quality	Class 2 Good Quality	Class 3 Average Quality	Class 4 Low Quality
Foundation	Reinforced concrete.	Reinforced concrete.	Reinforced concrete.	Reinforced concrete.
Floor Structure	6" reinforced concrete on 6" rock base.	6" reinforced concrete on 6" rock base.	6" reinforced concrete on 6" rock base.	4" reinforced concrete on 6" rock base.
Wall Structure	8" reinforced decorative concrete block, 6" concrete tilt-up or 8" reinforced brick.	8" reinforced decorative concrete block, 6" concrete tilt-up or 8" reinforced brick.	8" reinforced concrete. block, 6" concrete tilt-up or 8" reinforced common brick.	8" reinforced concrete block or 6" concrete tilt-up.
Roof and Cover	Glu-lams or steel beams on steel intermediate columns. Panelized roof system, 1/2" plywood sheathing, 5 ply built-up roof with insulation.	Glu-lams or steel beams on intermediate columns. Panelized roof system, 1/2" plywood sheathing, 5 ply built-up roof with insulation.	Glu-lams or steel intermediate columns. Panelized roof system, 1/2" plywood sheathing, 4 ply built-up roof.	Glu-lams on steel intermediate columns. Panelized roof system, 1/2" plywood sheathing, 4 ply built-up roof.
Floor Finish	Terrazzo and very good carpet.	Resilient tile with 50% sheet vinyl, terrazzo or good carpet.	Composition tile.	Minimum grade tile.
Interior Wall Finish	Gypsum wallboard or lath and plaster, finished with good paper or vinyl wall covers, or hardwood veneer paneling.	Gypsum wallboard and texture or paper, some vinyl wall cover or hardwood veneer paneling.	Interior stucco or gypsum wallboard, texture and paint.	Gypsum wallboard, texture and paint.
Ceiling Finish	Suspended good grade acoustical tile with gypsum wallboard backing.	Suspended acoustical tile with concealed grid system.	Suspended acoustical tile with exposed grid system.	Painted.
Lighting	Recessed fluorescent lighting in modular plastic panels. Many spotlights.	Continuous recessed 3 tube fluorescent strips with egg crate diffusers, 8' o.c. Average number of spotlights.	Continuous 3 tube fluorescent strips with egg crate diffusers, 8' o.c. Some spotlights.	Continuous exposed 2 tube fluorescent strips, 8' o.c.
Display Fronts	Very good front as described on page 56. 15 to 25% of the first floor exterior wall is made up of display fronts.	Good quality front as described on page 56. 15 to 25% of the first floor exterior wall is made up of display fronts.	Average quality front as described on page 56. 10 to 20% of the first floor exterior wall is made up of display fronts.	Average quality flat type as described on page 56. 10 to 20% of the first floor exterior wall is made up of display fronts.
Exterior Wall Finish	Ornamental block or large rock imbedded in tilt-up panels, some stone veneer.	Exposed aggregate or decorative block. Some stone veneer.	Paint, some exposed aggregate.	Paint.
Plumbing	6 good fixtures per 20,000 square feet of floor area. Metal toilet partitions.	6 standard fixtures per 20,000 square feet of floor area. Metal toilet partitions.	4 standard fixtures per 20,000 square feet of floor area. Metal toilet partitions.	4 standard fixtures per 20,000 square feet of floor area. Wood toilet partitions.

Square foot costs include the following components: Foundations as required for normal soil conditions. Floor, wall and roof structures. Interior ceiling, wall and floor finishes (including carpet). Exterior wall finish and roof cover. Display fronts. Interior partitions (including perimeter wall partitions). Entry and delivery doors. Basic lighting and electrical systems. Rough and finish plumbing. Design and engineering fees. Permits and hook-up fees. Contractor's mark-up.

The in-place cost of these extra components should be added to the basic building cost to arrive at the total structure cost. See the section "Additional Costs for Commercial Structures" on page 201. Heating and air conditioning systems. Elevators and escalators. Fire sprinklers. Exterior signs. Canopies and walks. Paving and curbing. Loading docks or ramps. Miscellaneous yard improvements. Mezzanines.

Department Stores - Masonry or Concrete
First Floor

Estimating Procedure

1. Use these square foot costs to estimate the cost of retail stores designed to sell a wider variety of goods. These buildings differ from discount houses in that they have more interior partitions and more elaborate interior and exterior finishes.
2. Establish the structure quality class by using the information on page 96.
3. Compute the building floor area. This should include everything within the building exterior walls and all inset areas outside the main walls but under the main building roof.
4. Add to or subtract from the square foot cost below the appropriate amount from the Wall Height Adjustment Table (near the bottom of this page) if the wall height is more or less than 16 feet.
5. Multiply the adjusted square foot cost by the building area.
6. Deduct, if appropriate, for common or unfinished walls, using the figures at the bottom of this page.
7. Multiply the total cost by the location factor listed on page 7.
8. Add the cost of appropriate additional components from page 201: heating and air conditioning systems, elevators and escalators, fire sprinklers, exterior signs, canopies and walks, paving and curbing, loading docks and ramps, miscellaneous yard improvements, and mezzanines.
9. Add the cost of second and higher floors and basements from page 98.

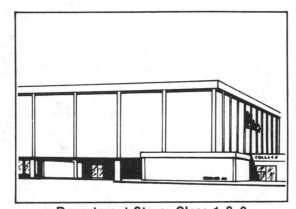

Department Store, Class 1 & 2

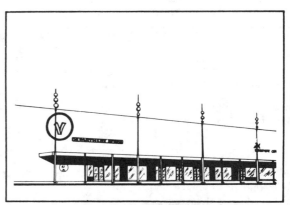

Department Store, Class 2 & 3

First Floor
Square Foot Area

Quality Class	20,000	25,000	30,000	35,000	40,000	45,000	50,000	60,000	70,000	80,000	100,000
1, Best	97.10	92.81	89.39	86.61	84.26	82.27	80.53	77.64	75.31	73.40	70.40
1 & 2	86.66	82.83	79.79	77.29	75.20	73.40	71.86	69.28	67.21	65.52	62.81
2, Good	76.12	72.76	70.10	67.90	66.06	64.49	63.12	60.86	59.04	57.54	55.19
2 & 3	67.75	64.77	62.38	60.42	58.79	57.39	56.19	54.16	52.56	51.22	49.12
3, Average	59.30	56.69	54.60	52.91	51.46	50.25	49.18	47.42	46.01	44.83	43.00
3 & 4	51.50	49.25	47.43	45.96	44.70	43.65	42.73	41.19	39.95	38.95	37.36
4, Low	43.94	42.01	40.46	39.20	38.14	37.23	36.44	35.14	34.09	33.22	31.86
Wall Height Adjustment*	.53	.44	.39	.36	.34	.33	.32	.31	.30	.30	.30

*Wall Height Adjustment: Add or subtract the amount listed in this column to the square foot of floor cost for each foot of wall height more or less than 16 feet.

Perimeter Wall Adjustment: For common wall, deduct $135.00 per linear foot. For no wall ownership, deduct $270.00 per linear foot.

Upper Floors and Basements

Estimating Procedure

1. Establish the basement and upper floor quality class. The quality class will usually be the same as the first floor of the building. Square foot costs for unfinished basements will be nearly the same regardless of the structure quality class.
2. Compute the floor area.
3. Add to or subtract from the square foot cost below the appropriate amount from the Wall Height Adjustment Column for each foot of wall height more or less than 12 feet for second and higher floors and 12 feet for basements.
4. Multiply the adjusted square foot cost by the floor area.
5. Deduct, if appropriate, for common or unfinished upper floor walls, using the Perimeter Wall Adjustment costs listed below.
6. Multiply the total cost by the location factor listed on page 7.
7. Add the cost of the appropriate additional components from page 201: heating and air conditioning systems, elevators and escalators, fire sprinklers, and mezzanines.
8. Add the total from this page to the total from page 97 to find the building cost.

Second and Higher Floors
Square Foot Area

Quality Class	20,000	25,000	30,000	35,000	40,000	45,000	50,000	60,000	70,000	80,000	100,000
1, Best	85.01	81.88	79.00	76.45	74.18	72.17	70.37	67.27	64.71	62.54	59.04
1 & 2	74.63	71.87	69.37	67.12	65.12	63.36	61.78	59.08	56.81	54.91	51.81
2, Good	64.36	61.98	59.81	57.88	56.16	54.64	53.27	50.93	48.98	47.34	44.69
2 & 3	56.86	54.75	52.85	51.14	49.62	48.27	47.06	45.00	43.28	41.82	39.49
3, Average	48.92	47.11	45.47	43.99	42.70	41.51	40.48	38.70	37.23	35.99	33.97
3 & 4	40.10	38.63	37.27	36.06	35.01	34.05	33.19	31.73	30.53	29.51	27.85
4, Low	32.39	31.32	30.49	29.75	29.16	28.63	28.19	27.42	26.82	26.31	25.50
Wall Height Adjustment*	.38	.33	.30	.28	.27	.26	.25	.24	.24	.23	.23

*Wall Height Adjustment: Add or subtract the amount listed in this column to the square foot of floor cost for each foot of wall height more or less than 12 feet.

Perimeter Wall Adjustment: For common wall, deduct $78.00 per linear foot. For no wall ownership, deduct $155.00 per linear foot.

Finished Basements
Square Foot Area

Quality Class	20,000	25,000	30,000	35,000	40,000	45,000	50,000	60,000	70,000	80,000	100,000
1, Best	75.53	72.49	70.05	68.03	66.33	64.88	63.60	61.48	59.78	58.36	56.15
2, Good	58.76	56.41	54.50	52.93	51.62	50.47	49.49	47.84	46.51	45.43	43.68
3, Average	46.35	44.48	42.98	41.76	40.71	39.80	39.03	37.73	36.68	35.82	34.46
4, Low	33.03	31.71	30.63	29.74	29.00	28.38	27.81	26.89	26.15	25.53	24.55

Unfinished Basements

Area	20,000	25,000	30,000	35,000	40,000	45,000	50,000	60,000	70,000	80,000	100,000
Cost	19.51	18.60	17.93	17.45	17.03	16.70	16.41	15.98	15.62	15.35	14.93

Wall Height Adjustment: Add or subtract the amount listed in this table to the square foot of floor cost for each foot of basement wall height more or less than 12 feet.

Area	20,000	25,000	30,000	35,000	40,000	45,000	50,000	60,000	70,000	80,000	100,000
Finished	.41	.38	.37	.35	.34	.34	.33	.32	.32	.31	.30
Unfinished	.39	.36	.35	.34	.33	.32	.32	.31	.30	.30	.29

Department Stores - Wood Frame
Quality Classification

	Class 1 Best Quality	Class 2 Good Quality	Class 3 Average Quality	Class 4 Low Quality
Foundation	Reinforced concrete.	Reinforced concrete.	Reinforced concrete.	Reinforced concrete.
Floor Structure	6" reinforced concrete on 6" rock base.	6" reinforced concrete on 6" rock base.	6" reinforced concrete on 6" rock base.	4" reinforced concrete on 6" rock base.
Wall Structure	2" x 6", 16" o.c.	2" x 6", 16" o.c.	2" x 6", 16" o.c.	2" x 4", 16" o.c.
Roof & Cover	Glu-lams or steel beams on steel intermediate columns. Panelized roof system, 1/2" plywood sheathing, 5 ply built-up roof with insulation.	Glu-lams or steel beams on steel intermediate columns. Panelized roof system, 1/2" plywood sheathing, 5 ply built-up roof with insulation.	Glu-lams or steel intermediate columns. Panelized roof system, 1/2" plywood sheathing, 4 ply built-up roof.	Glu-lams on steel intermediate columns. Panelized roof system, 1/2" plywood sheathing, 4 ply built-up roof.
Floor Finish	Terrazzo and very good carpet.	Resilient tile with 50% sheet vinyl, terrazzo or good carpet.	Composition tile.	Minimum grade tile.
Interior Wall Finish	Gypsum wallboard or lath and plaster finished with good paper or vinyl wall covers, or hardwood veneer paneling.	Gypsum wallboard, texture or paper, some vinyl wall cover or hardwood veneer paneling.	Interior stucco or gypsum wallboard. Texture and paint.	Gypsum wallboard. Texture and paint.
Ceiling Finish	Suspended good grade acoustical tile with gypsum wallboard backing.	Suspended acoustical tile with concealed grid system.	Suspended acoustical tile with exposed grid system.	Painted.
Lighting	Recessed fluorescent lighting in modular plastic panels. Many spotlights.	Continuous recessed 3 tube fluorescent strips with egg crate diffusers, 8' o.c. Average number of spotlights.	Continuous 3 tube fluorescent strips with egg crate diffusers, 8' o.c. Some spotlights.	Continuous exposed 2 tube fluorescent strips, 8' o.c.
Display Fronts	Very good fronts as described on page 56. 15 to 25% of the first floor exterior wall is made up of display fronts.	Good quality front as described on page 56. 15 to 25% of the first floor exterior wall is made up of display fronts.	Average quality front as described on page 56. 10 to 20% of the first floor exterior wall is made up of display fronts.	Average quality flat type as described on page 56. 10 to 20% of the first floor exterior wall is made up of display fronts.
Exterior Wall Finish	Good wood siding. Extensive stone veneer.	Wood siding. Some brick or stone veneer.	Stucco. Some brick trim.	Stucco.
Plumbing	6 good fixtures per 20,000 square feet of floor area. Metal toilet partitions.	6 standard fixtures per 20,000 square feet of floor area. Metal toilet partitions.	4 standard fixtures per 20,000 square feet of floor area. Metal toilet partitions.	4 standard fixtures per 20,000 square feet of floor area. Wood toilet partitions.

Square foot costs include the following components: Foundations as required for normal soil conditions. Floor, wall and roof structures. Interior ceiling, wall and floor finishes (including carpet). Exterior wall finish and roof cover. Display fronts. Interior partitions (including perimeter wall partitions). Entry and delivery doors. Basic lighting and electrical systems. Rough and finish plumbing. Design and engineering fees. Permits and hook-up fees. Contractor's mark-up.

The in-place cost of these extra components should be added to the basic building cost to arrive at the total structure cost. See the section "Additional Costs for Commercial Structures" on page 201. Heating and air conditioning systems. Elevators and escalators. Fire sprinklers. Exterior signs. Canopies and walks. Paving and curbing. Loading docks or ramps. Miscellaneous yard improvements. Mezzanines.

Department Stores - Wood Frame
First Floor

Estimating Procedure

1. Use these square foot costs to estimate the cost of retail stores designed to sell a wider variety of goods. These buildings differ from discount houses in that they have more interior partitions and more elaborate interior and exterior finishes.
2. Establish the structure quality class by applying the information on page 99.
3. Compute the floor area. This should include everything within the building exterior walls and all inset areas outside the main walls but under the main building roof.
4. Add to or subtract from the square foot cost below the appropriate amount from the Wall Height Adjustment Column (near the bottom of this page) if the wall height is more or less than 16 feet.
5. Multiply the adjusted square foot cost by the building area.
6. Deduct, if appropriate, for common walls or no wall ownership, using the figures at the bottom of this page.
7. Multiply the total cost by the location factor on page 7.
8. Add the cost of appropriate additional components from page 201: heating and air conditioning systems, elevators and escalators, fire sprinklers, exterior signs, canopies and walks, paving and curbing, loading docks and ramps, miscellaneous yard improvements, and mezzanines.
9. Add the cost of second and higher floors and basements from page 101.

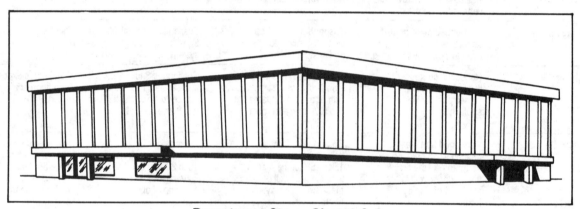

Department Store, Class 1 & 2

Square Foot Area

Quality Class	20,000	25,000	30,000	35,000	40,000	45,000	50,000	60,000	70,000	80,000	100,000
1, Best	94.09	90.27	87.19	84.62	82.47	80.60	78.97	76.27	74.07	72.26	69.39
1 & 2	84.13	80.73	77.96	75.68	73.75	72.08	70.63	68.20	66.24	64.63	62.05
2, Good	73.48	70.50	68.09	66.10	64.42	62.94	61.68	59.58	57.86	56.43	54.20
2 & 3	65.28	62.65	60.49	58.72	57.22	55.93	54.78	52.92	51.40	50.13	48.15
3, Average	57.23	54.91	53.03	51.46	50.16	49.02	48.04	46.38	45.05	43.95	42.21
3 & 4	49.47	47.46	45.83	44.50	43.35	42.38	41.51	40.09	38.95	37.99	36.48
4, Low	41.94	40.24	38.87	37.73	36.76	35.93	35.20	34.00	33.02	32.20	30.93
Wall Height Adjustment*	.25	.24	.23	.22	.21	.21	.20	.19	.18	.17	.12

*Wall Height Adjustment: Add or subtract the amount listed in this table to the square foot of floor cost for each foot of wall height more or less than 16 feet.

Perimeter Wall Adjustment: For common wall, deduct $29.00 per linear foot. For no wall ownership, deduct $58.00 per linear foot.

Department Stores - Wood Frame
Upper Floors and Basements
Estimating Procedure

1. Establish the basement and upper floor quality class. The quality class will usually be the same as the first floor of the building. Square foot costs for unfinished basements will be nearly the same regardless of the structure quality class.
2. Compute the floor area.
3. Add to or subtract from the square foot cost below the appropriate amount from the Wall Height Adjustment Column for each foot of wall height more or less than 12 feet.
4. Multiply the adjusted square foot cost by the floor area.
5. Deduct, if appropriate, for common upper floor walls and walls not owned, using the Perimeter Wall Adjustment costs.
6. Multiply the total cost by the location factor on page 7.
7. Add the cost of appropriate additional components from page 201: heating and air conditioning systems, elevators and escalators, fire sprinklers, and mezzanines.
8. Add the total from this page to the page 100 to find the building cost.

Second and Higher Floors
Square Foot Area

Quality Class	20,000	25,000	30,000	35,000	40,000	45,000	50,000	60,000	70,000	80,000	100,000
1, Best	77.98	75.23	72.91	70.91	69.20	67.70	66.37	64.13	62.32	60.79	58.38
1 & 2	68.74	66.31	64.26	62.49	60.99	59.66	58.50	56.55	54.93	53.59	51.47
2, Good	59.54	57.45	55.66	54.15	52.83	51.70	50.68	48.98	47.60	46.43	44.57
2 & 3	52.65	50.80	49.21	47.88	46.71	45.70	44.81	43.32	42.07	41.04	39.41
3, Average	45.64	44.02	42.66	41.48	40.50	39.61	38.84	37.54	36.48	35.58	34.16
3 & 4	38.52	37.15	35.99	35.01	34.15	33.44	32.78	31.68	30.79	30.02	28.83
4, Low	31.51	30.36	29.44	28.66	27.96	27.35	26.83	25.91	25.18	24.56	23.57
Wall Height Adjustment*	.23	.21	.19	.17	.16	.13	.12	.11	.10	.09	.07

*Wall Height Adjustment: Add or subtract the amount listed in this table to the square foot of floor cost for each foot of basement wall height more or less than 12 feet.

Perimeter Wall Adjustment: For common wall, deduct $76.00 per linear foot. For no wall ownership, deduct $153.00 per linear foot.

Finished Basements
Square Foot Area

Quality Class	20,000	25,000	30,000	35,000	40,000	45,000	50,000	60,000	70,000	80,000	100,000
1, Best	76.26	73.82	71.51	69.39	67.49	65.80	64.27	61.61	59.40	57.51	54.45
2, Good	59.37	57.48	55.67	54.03	52.56	51.22	50.04	47.97	46.24	44.77	42.40
3, Average	47.17	45.67	44.24	42.94	41.77	40.70	39.76	38.13	36.74	35.57	33.69
4, Low	32.48	31.45	30.46	29.57	28.75	28.03	27.37	26.24	25.30	24.49	23.20

Unfinished Basements

Area	20,000	25,000	30,000	35,000	40,000	45,000	50,000	60,000	70,000	80,000	100,000
Cost	18.70	17.97	17.38	16.89	16.47	16.11	15.83	15.31	14.90	14.56	14.02

Wall Height Adjustment: Add or subtract the amount listed in this table to the square foot of floor cost for each foot of basement wall height more or less than 12 feet.

Area	20,000	25,000	30,000	35,000	40,000	45,000	50,000	60,000	70,000	80,000	100,000
Finished	.41	.39	.37	.36	.35	.34	.33	.32	.31	.31	.30
Unfinished	.39	.37	.35	.34	.33	.32	.32	.31	.30	.29	.28

Quality Classification

	Class 1 Best Quality	Class 2 Good Quality	Class 3 Average Quality	Class 4 Low Quality
Foundation	Reinforced concrete.	Reinforced concrete.	Reinforced concrete.	Reinforced concrete
First Floor Structure	Reinforced concrete slab on grade or standard wood frame.	Reinforced concrete slab on grade or standard wood frame.	Reinforced concrete slab on grade or 4" x 6" girders with plywood sheathing.	Reinforced concrete slab on grade.
Upper Floor Structures	Standard wood frame, plywood and 1-1/2" light-weight concrete sub floor.	Standard wood frame, plywood and 1-1/2" light-weight concrete sub floor.	Standard wood frame, 5/8" plywood sub floor.	Standard wood frame, 5/8" plywood sub floor.
Walls	8" decorative concrete block or 6" concrete tilt-up.	8" decorative concrete block or 6" concrete tilt-up.	8" reinforced concrete block or 8" reinforced brick or 8" clay tile.	8" reinforced concrete block or 8" clay tile.
Roof Structure	Standard wood frame, flat or low pitch.	Standard wood frame, flat or low pitch.	Standard wood frame, flat or low pitch.	Standard wood frame, flat or low pitch.
Exterior Finish: **Walls**	Decorative block or large rock imbedded in tilt-up panels with 10 - 20% brick or stone veneer	Decorative block or exposed aggregate and 10 - 20% brick or stone veneer.	Stucco or colored concrete block.	Painted concrete block or tile.
Windows	Average number in good aluminum frame. Fixed plate glass in good frame on front side.	Average number in good aluminum frame. Some fixed plate glass in front.	Average number of average aluminum sliding type.	Average number of low cost aluminum sliding type.
Roof Cover	5 ply built-up roofing on flat roofs. Heavy shake or tile on sloping roofs.	5 ply built-up roofing on flat roofs. Average shake or composition, tar and large rock on sloping roofs.	4 ply built-up roofing on flat roofs. Wood shingle or composition, tar and pea gravel on sloping roofs.	3 ply built-up roofing on flat roofs. Composition shingle on sloping roofs.
Overhang	3' closed overhang. fully guttered.	2' closed overhang, fully guttered.	None on flat roofs. 18" open on sloping roofs. Fully guttered.	None on flat roofs. 12" to 16" open on sloping roofs. Gutters over entrances.
Floor Finish **Offices**	Very good carpet.	Good carpet.	Average grade carpet.	Minimum tile.
Corridors	Solid vinyl tile or carpet.	Resilient tile.	Composition tile.	Minimum tile.
Bathrooms	Sheet vinyl or ceramic tile.	Sheet vinyl or ceramic tile.	Vinyl asbestos tile.	Minimum tile.
Interior Wall Finish **Offices**	Good hardwood veneer.	Hardwood veneer paneling or vinyl wall cover.	Gypsum wallboard, texture and paint.	Gypsum wallboard, texture and paint.
Corridors	Good hardwood veneer.	Gypsum wallboard and vinyl wall cover.	Gypsum wallboard, texture and paint.	Gypsum wallboard, texture and paint.
Bathrooms	Gypsum wallboard and enamel with ceramic tile wainscot.	Gypsum wallboard and enamel or vinyl wall covering.	Gypsum wallboard and enamel.	Gypsum wallboard, texture and paint.
Ceiling Finish	Suspended "T" bar and acoustical tile.	Gypsum wallboard and acoustical tile.	Gypsum wallboard and acoustical texture.	Gypsum walboard and paint.
Utilities **Plumbing**	Copper tubing and top quality fixtures.	Copper tubing and good fixtures.	Copper tubing and standard fixtures.	Copper tubing and economy fixtures.
Lighting	Conduit wiring, good fixtures.	Conduit wiring, good fixtures.	Romex or conduit wiring, average fixtures.	Romex wiring, economy fixtures.

Square foot costs include the following components: Foundations as required for normal soil conditions. Floor, wall and roof structures. Interior ceiling, wall and floor finishes. Exterior wall finish and roof cover. Interior partitions. Cabinets, doors and windows. Basic electrical systems and lighting fixtures. Rough plumbing and fixtures. Permits and fees. Contractor's mark-up.

Exterior Suite Entrances, Length Less Than Twice Width

Estimating Procedure

1. Use these figures to estimate general office buildings in which access to each suite is through an exterior entrance. Medical and dental offices have smaller rooms and more plumbing fixtures than general offices and should be estimated with figures from the Medical and Dental Buildings Section. See page 118.
2. Establish the building quality class by applying the information on page 102.
3. Compute the first floor area. This should include everything within the exterior walls and all insets outside the main walls but under the main roof.
4. If the first floor wall height is more or less than 10 feet, add to or subtract from the first floor square foot cost below the appropriate amount from the Wall Height Adjustment Table on page 109.
5. Multiply the adjusted square foot cost by the first floor area.
6. Deduct, if appropriate, for common walls or no wall finish, Use the figures on page 109.
7. If there are second or higher floors, compute the square foot area on each floor. Locate the appropriate square foot cost from the table at the bottom of this page. Adjust this figure for a wall height more or less than 9 feet, using the figures on page 109. Multiply the adjusted cost by the square foot area on each floor. Use the figures on page 109 to deduct for common walls or no wall finish. Add the result to the cost from step 6 above.
8. Multiply the total cost by the location factor on page 7.
9. Add the cost of heating and air conditioning systems, elevators, fire sprinklers, exterior signs, paving and curbing, miscellaneous yard improvements, covered porches and garages. See pages 201 to 213.

First Story
Square Foot Area

Quality Class	1,000	1,500	2,000	2,500	3,000	4,000	5,000	7,500	10,000	15,000	20,000
Exceptional	133.60	122.62	116.31	112.09	109.04	104.81	101.98	97.65	95.13	92.17	90.44
1, Best	122.07	112.03	106.26	102.40	99.61	95.75	93.17	89.23	86.91	84.20	82.62
1 & 2	110.28	101.22	96.01	92.52	90.01	86.52	84.18	80.62	78.54	76.08	74.66
2, Good	99.53	91.35	86.65	83.49	81.22	78.09	75.97	72.75	70.87	68.67	67.37
2 & 3	90.85	83.38	79.08	76.23	74.14	71.27	69.36	66.41	64.69	62.68	61.49
3, Average	82.01	75.26	71.37	68.80	66.92	64.32	62.58	59.93	58.38	56.56	55.50
3 & 4	73.83	67.75	64.26	61.94	60.24	57.91	56.34	53.96	52.56	50.91	49.98
4, Low	65.37	60.00	56.91	54.84	53.35	51.28	49.91	47.79	46.55	45.09	44.24

Second and Higher Stories
Square Foot Area

Quality Class	1,000	1,500	2,000	2,500	3,000	4,000	5,000	7,500	10,000	15,000	20,000
Exceptional	123.52	114.06	108.54	104.82	102.09	98.32	95.76	91.81	89.48	86.76	85.14
1, Best	112.72	104.09	99.05	95.65	93.17	89.70	87.38	83.78	81.65	79.16	77.67
1 & 2	102.49	94.63	90.06	86.96	84.70	81.56	79.43	76.16	74.23	71.96	70.62
2, Good	93.15	86.02	81.85	79.04	76.99	74.13	72.22	69.22	67.48	65.41	64.19
2 & 3	84.77	78.29	74.50	71.94	70.06	67.47	65.71	63.01	61.40	59.53	58.42
3, Average	75.85	70.06	65.92	64.36	62.70	60.38	58.79	56.38	54.94	53.28	52.28
3 & 4	67.59	62.42	59.41	57.36	55.87	53.79	52.41	50.25	48.97	47.48	46.58
4, Low	60.22	55.61	52.92	51.09	49.76	47.91	46.67	44.75	43.61	42.27	41.49

General Office Buildings - Masonry or Concrete
Exterior Suite Entrances, Length Between 2 and 4 Times Width

Estimating Procedure

1. Use these figures to estimate general office buildings in which access to each suite is through an exterior entrance. Medical and dental offices have smaller rooms and more plumbing fixtures than general offices and should be estimated with figures from the Medical and Dental Buildings Section. See page 118.
2. Establish the building quality class by applying the information on page 102.
3. Compute the first floor area. This should include everything within the exterior walls and all insets outside the main walls but under the main roof.
4. If the first floor wall height is more or less than 10 feet, add to or subtract from the first floor square foot cost below the appropriate amount from the Wall Height Adjustment Table on page 109.
5. Multiply the adjusted square foot cost by the first floor area.
6. Deduct, if appropriate, for common walls or no wall finish. Use the figures on page 109.
7. If there are second or higher floors, compute the square foot area on each floor. Locate the appropriate square foot cost from the table at the bottom of this page. Adjust this figure for a wall height more or less than 9 feet, using the figures on page 109. Multiply the adjusted cost by the square foot area on each floor. Use the figures on page 109 to deduct for common walls or no wall finish. Add the result to the cost from step 6 above.
8. Multiply the total cost by the location factor on page 7.
9. Add the cost of heating and air conditioning systems, elevators, fire sprinklers, exterior signs, paving and curbing, miscellaneous yard improvements, covered porches and garages. See pages 201 to 213.

First Story
Square Foot Area

Quality Class	1,000	1,500	2,000	2,500	3,000	4,000	5,000	7,500	10,000	15,000	20,000
Exceptional	145.13	131.04	123.07	117.84	114.06	108.92	105.51	100.35	97.36	93.91	91.89
1, Best	132.58	119.72	112.45	107.66	104.21	99.51	96.39	91.68	88.95	85.80	83.96
1 & 2	119.97	108.31	101.74	97.39	94.28	90.04	87.22	82.95	80.47	77.63	75.96
2, Good	108.21	97.71	91.78	87.86	85.06	81.21	78.66	74.83	72.60	70.02	68.53
2 & 3	98.88	89.27	83.85	80.28	77.72	74.21	71.88	68.37	66.34	63.98	62.61
3, Average	89.15	80.58	75.67	72.45	70.14	66.98	64.87	61.70	59.85	57.73	56.51
3 & 4	80.57	72.76	68.33	65.42	63.34	60.47	58.58	55.73	54.06	52.14	51.03
4, Low	71.57	64.62	60.69	58.11	56.25	53.71	52.03	49.48	48.02	46.31	45.32

Second and Higher Stories
Square Foot Area

Quality Class	1,000	1,500	2,000	2,500	3,000	4,000	5,000	7,500	10,000	15,000	20,000
Exceptional	134.05	121.34	114.17	109.46	106.07	101.44	98.37	93.74	91.07	87.98	86.15
1, Best	122.35	110.75	104.20	99.89	96.81	92.58	89.78	85.55	83.11	80.30	78.63
1 & 2	111.57	101.00	95.02	91.09	88.27	84.42	81.86	78.02	75.78	73.21	71.72
2, Good	101.24	91.63	86.24	82.66	80.10	76.62	74.29	70.79	68.77	66.44	65.07
2 & 3	92.07	83.33	78.41	75.18	72.84	69.65	67.55	64.37	62.53	60.41	59.18
3, Average	82.88	75.03	70.59	67.68	65.58	62.72	60.82	57.96	56.30	54.40	53.28
3 & 4	74.22	67.20	63.23	60.60	58.73	56.17	54.46	51.91	50.42	48.70	47.71
4, Low	65.70	59.45	54.95	53.65	51.97	49.70	48.20	45.93	44.64	43.11	42.21

General Office Buildings - Masonry or Concrete

Exterior Suite Entrances, Length More Than 4 Times Width

Estimating Procedure

1. Use these figures to estimate general office buildings in which access to each suite is through an exterior entrance. Medical and dental offices have smaller rooms and more plumbing fixtures than general offices and should be estimated with figures from the Medical and Dental Buildings Section. See page 118.
2. Establish the building quality class by applying the information on page 102.
3. Compute the first floor area. This should include everything within the exterior walls and all insets outside the main walls but under the main roof.
4. If the first floor wall height is more or less than 10 feet, add to or subtract from the first floor square foot cost below the appropriate amount from the Wall Height Adjustment Table on page 109.
5. Multiply the adjusted square foot cost by the first floor area.
6. Deduct, if appropriate, for common walls or no wall finish. Use the figures on page 109.
7. If there are second or higher floors, compute the square foot area on each floor. Locate the appropriate square foot cost from the table at the bottom of this page. Adjust this figure for a wall height more or less than 9 feet, using the figures on page 109. Multiply the adjusted cost by the square foot area on each floor. Use the figures on page 109 to deduct for common walls or no wall finish. Add the result to the cost from step 6 above.
8. Multiply the total cost by the location factor on page 7.
9. Add the cost of heating and air conditioning systems, elevators, fire sprinklers, exterior signs, paving and curbing, miscellaneous yard improvements, covered porches and garages. See pages 201 to 213.

First Story
Square Foot Area

Quality Class	1,000	1,500	2,000	2,500	3,000	4,000	5,000	7,500	10,000	15,000	20,000
Exceptional	158.58	141.46	131.83	125.50	120.96	114.79	110.71	104.53	100.99	96.88	94.49
1, Best	144.48	128.87	120.10	114.34	110.21	104.58	100.85	95.24	92.00	88.26	86.10
1 & 2	131.05	116.90	108.94	103.72	99.97	94.86	91.47	86.38	83.45	80.06	78.09
2, Good	117.89	105.15	98.01	93.30	89.94	85.33	82.30	77.72	75.06	72.02	70.24
2 & 3	107.91	96.25	89.70	85.40	82.32	78.11	75.31	71.12	68.72	65.92	64.30
3, Average	97.88	87.32	81.38	77.48	74.68	70.87	68.33	64.53	62.34	59.79	58.33
3 & 4	88.16	78.64	73.28	69.77	67.25	63.82	61.55	58.12	56.15	53.86	52.53
4, Low	78.44	69.97	65.20	62.07	59.84	56.78	54.76	51.72	49.96	47.91	46.74

Second and Higher Stories
Square Foot Area

Quality Class	1,000	1,500	2,000	2,500	3,000	4,000	5,000	7,500	10,000	15,000	20,000
Exceptional	145.61	130.04	121.30	115.57	111.46	105.87	102.16	96.61	93.40	89.69	87.53
1, Best	133.01	118.78	110.81	105.55	101.81	96.70	93.31	88.23	85.31	81.92	79.95
1 & 2	121.30	108.33	101.06	96.28	92.85	88.19	85.11	80.47	77.80	74.72	72.92
2, Good	110.48	98.68	92.03	87.69	84.57	80.33	77.51	73.30	70.86	68.04	64.48
2 & 3	100.91	90.11	84.06	80.09	77.24	73.36	70.80	66.94	64.72	62.15	60.65
3, Average	90.92	81.22	75.76	72.18	69.59	66.11	63.80	60.33	58.32	56.00	54.66
3 & 4	82.09	73.31	68.39	65.16	62.84	59.68	57.59	54.45	52.65	50.57	49.34
4, Low	73.25	65.40	61.00	58.12	56.05	53.23	51.39	48.58	46.97	45.11	44.02

Interior Suite Entrances, Length Less Than Twice Width

Estimating Procedure

1. Use these figures to estimate general office buildings in which access to each suite is through an interior corridor. Medical and dental offices have smaller rooms and more plumbing fixtures than general offices and should be estimated with figures from the Medical and Dental Buildings Section. See page 118.
2. Establish the building quality class by applying the information on page 102.
3. Compute the first floor area. This should include everything within the exterior walls and all insets outside the main walls but under the main roof.
4. If the first floor wall height is more or less than 10 feet, add to or subtract from the first floor square foot cost below the appropriate amount from the Wall Height Adjustment Table on page 109.
5. Multiply the adjusted square foot cost by the first floor area.
6. Deduct, if appropriate, for common walls or no wall finish. Use the figures on page 109.
7. If there are second or higher floors, compute the square foot area on each floor. Locate the appropriate square foot cost from the table at the bottom of this page. Adjust this figure for a wall height more or less than 9 feet, using the figures on page 109. Multiply the adjusted cost by the square foot area on each floor. Use the figures on page 109 to deduct for common walls or no wall finish. Add the result to the cost from step 6 above.
8. Multiply the total cost by the location factor on page 7.
9. Add the cost of heating and air conditioning systems, elevators, fire sprinklers, exterior signs, paving and curbing, miscellaneous yard improvements, covered porches and garages. See pages 201 to 213.

First Story
Square Foot Area

Quality Class	2,000	2,500	3,000	4,000	5,000	7,500	10,000	15,000	20,000	30,000	40,000
Exceptional	112.81	108.67	105.65	101.51	98.73	94.50	92.03	89.13	87.44	85.45	84.28
1, Best	103.02	99.24	96.48	92.70	90.16	86.29	84.04	81.41	79.86	78.03	76.97
1 & 2	94.02	90.57	88.06	84.61	82.29	78.76	76.69	74.30	72.87	71.22	70.25
2, Good	85.72	82.58	80.29	77.14	75.02	71.80	69.93	67.74	66.45	64.94	64.04
2 & 3	78.33	75.45	73.36	70.48	68.55	65.61	63.89	61.90	60.72	59.34	58.52
3, Average	70.67	68.08	66.19	63.59	61.85	59.21	57.65	55.85	54.79	53.53	52.79
3 & 4	63.52	61.20	59.50	57.17	55.59	53.21	51.82	50.20	49.25	48.12	47.46
4, Low	56.15	54.08	52.60	50.52	49.15	47.03	45.81	44.37	43.52	42.54	41.94

Second and Higher Stories
Square Foot Area

Quality Class	2,000	2,500	3,000	4,000	5,000	7,500	10,000	15,000	20,000	30,000	40,000
Exceptional	105.34	101.62	98.91	95.16	92.65	88.80	86.53	83.89	82.33	80.49	79.42
1, Best	96.24	92.83	90.37	86.93	84.64	81.11	79.05	76.64	75.21	73.53	72.54
1 & 2	88.12	85.01	82.74	79.61	77.49	74.28	72.39	70.17	68.86	67.33	66.43
2, Good	80.44	77.09	75.54	72.67	70.74	67.81	66.07	64.06	62.87	61.47	66.43
2 & 3	73.30	70.72	68.81	66.21	64.46	61.78	60.20	58.36	57.29	56.01	55.51
3, Average	65.72	63.41	61.72	59.37	57.80	55.40	53.99	52.34	51.36	50.22	49.55
3 & 4	58.89	56.81	55.29	53.19	51.79	49.64	48.36	46.90	46.02	45.00	44.39
4, Low	51.90	50.06	48.72	46.88	45.64	43.74	42.63	41.33	40.56	39.66	39.13

General Office Buildings - Masonry or Concrete

Interior Suite Entrances, Length Between 2 and 4 Times Width

Estimating Procedure

1. Use these figures to estimate general office buildings in which access to each suite is through an interior corridor. Medical and dental offices have smaller rooms and more plumbing fixtures than general offices and should be estimated with figures from the Medical and Dental Buildings Section. See page 118.
2. Establish the building quality class by applying the information on page 102.
3. Compute the first floor area. This should include everything within the exterior walls and all insets outside the main walls but under the main roof.
4. If the first floor wall height is more or less than 10 feet, add to or subtract from the first floor square foot cost below the appropriate amount from the Wall Height Adjustment Table on page 109.
5. Multiply the adjusted square foot cost by the first floor area.
6. Deduct, if appropriate, for common walls or no wall finish. Use the figures on page 109.
7. If there are second or higher floors, compute the square foot area on each floor. Locate the appropriate square foot cost from the table at the bottom of this page. Adjust this figure for a wall height more or less than 9 feet, using the figures on page 109. Multiply the adjusted cost by the square foot area on each floor. Use the figures on page 109 to deduct for common walls or no wall finish. Add the result to the cost from step 6 above.
8. Multiply the total cost by the location factor on page 7.
9. Add the cost of heating and air conditioning systems, elevators, fire sprinklers, exterior signs, paving and curbing, miscellaneous yard improvements, covered porches and garages. See pages 201 to 213.

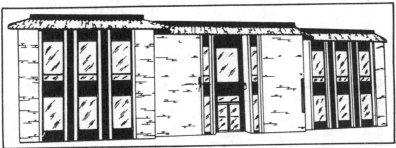

General Offices, Class 1 & 2

First Story
Square Foot Area

Quality Class	2,000	2,500	3,000	4,000	5,000	7,500	10,000	15,000	20,000	30,000	40,000
Exceptional	119.57	114.45	110.77	105.73	102.37	97.27	94.32	90.88	88.87	86.51	85.14
1, Best	109.21	104.56	101.20	96.59	93.52	88.87	86.16	83.02	81.17	79.03	77.78
1 & 2	99.56	95.41	92.34	88.14	85.34	81.08	78.62	75.74	74.08	72.12	70.96
2, Good	90.77	86.99	84.18	80.36	77.80	73.92	71.68	69.07	67.53	65.75	64.70
2 & 3	83.08	79.54	76.98	73.47	71.13	67.59	65.54	63.14	61.76	60.12	59.16
3, Average	75.01	71.81	69.50	66.34	64.23	61.04	59.17	57.01	55.76	54.28	53.41
3 & 4	67.54	64.66	62.59	59.73	57.85	54.97	53.28	51.35	50.20	48.88	48.10
4, Low	59.77	57.21	55.36	52.86	51.18	48.62	47.15	45.43	44.43	43.25	42.56

Second and Higher Stories
Square Foot Area

Quality Class	2,000	2,500	3,000	4,000	5,000	7,500	10,000	15,000	20,000	30,000	40,000
Exceptional	111.23	106.63	103.29	98.74	95.70	91.10	88.42	85.31	83.48	81.35	80.09
1, Best	101.52	97.31	94.29	90.13	87.37	83.15	80.71	77.88	76.21	74.26	73.12
1 & 2	93.06	89.21	86.41	82.60	80.07	76.22	73.98	71.37	69.85	68.06	67.03
2, Good	85.24	81.70	79.15	75.66	73.34	69.81	67.76	65.37	63.97	62.35	61.38
2 & 3	77.59	74.37	72.06	68.88	66.77	63.55	61.68	59.50	58.24	56.75	55.89
3, Average	69.63	66.80	64.72	61.87	59.98	57.09	55.41	53.46	52.32	50.99	50.21
3 & 4	62.65	60.05	58.17	55.60	53.89	51.31	49.79	48.03	47.02	45.81	45.11
4, Low	55.36	53.06	51.42	49.15	47.65	45.34	44.01	42.46	41.55	40.50	39.88

Interior Suite Entrances, Length More Than 4 Times Width

Estimating Procedure

1. Use these figures to estimate general office buildings in which access to each suite is through an interior corridor. Medical and dental offices have smaller rooms and more plumbing fixtures than general offices and should be estimated with figures from the Medical and Dental Buildings Section. See page 118.
2. Establish the building quality class by applying the information on page 102.
3. Compute the first floor area. This should include everything within the exterior walls and all insets outside the main walls but under the main roof.
4. If the first floor wall height is more or less than 10 feet, add to or subtract from the first floor square foot cost below the appropriate amount from the Wall Height Adjustment Table on page 109.
5. Multiply the adjusted square foot cost by the first floor area.
6. Deduct, if appropriate, for common walls or no wall finish. Use the figures on page 109.
7. If there are second or higher floors, compute the square foot area on each floor. Locate the appropriate square foot cost from the table at the bottom of this page. Adjust this figure for a wall height more or less than 9 feet, using the figures on page 109. Multiply the adjusted cost by the square foot area on each floor. Use the figures on page 109 to deduct for common walls or no wall finish. Add the result to the cost from step 6 above.
8. Multiply the total cost by the location factor on page 7.
9. Add the cost of heating and air conditioning systems, elevators, fire sprinklers, exterior signs, paving and curbing, miscellaneous yard improvements, covered porches and garages. See pages 201 to 213.

First Story
Square Foot Area

Quality Class	2,000	2,500	3,000	4,000	5,000	7,500	10,000	15,000	20,000	30,000	40,000
Exceptional	128.09	121.74	117.20	111.03	106.93	100.80	97.28	93.22	90.87	88.13	86.53
1, Best	117.03	111.22	107.07	101.42	97.70	92.09	88.88	85.15	83.01	80.52	79.06
1 & 2	106.77	101.47	97.68	92.54	89.13	84.02	81.07	77.71	75.73	73.45	72.13
2, Good	97.32	92.50	89.02	84.35	81.23	76.59	73.90	70.82	69.04	66.95	65.74
2 & 3	88.95	84.52	81.36	77.07	74.24	69.98	67.54	64.72	63.09	61.19	60.09
3, Average	80.51	76.51	73.66	69.78	67.20	63.36	61.14	58.60	57.11	55.40	54.40
3 & 4	72.61	69.02	66.43	62.94	60.62	57.15	55.15	52.85	51.51	49.97	49.05
4, Low	64.48	61.29	59.00	55.88	53.84	50.74	48.97	46.93	45.75	44.37	43.57

Second and Higher Stories
Square Foot Area

Quality Class	2,000	2,500	3,000	4,000	5,000	7,500	10,000	15,000	20,000	30,000	40,000
Exceptional	117.60	111.84	107.80	102.41	98.94	93.89	91.07	87.90	86.11	84.07	82.91
1, Best	106.55	101.32	97.64	92.80	89.66	85.07	82.52	79.64	78.02	76.19	75.14
1 & 2	98.26	93.45	90.07	85.57	82.68	78.46	76.10	73.45	71.95	70.26	69.28
2, Good	90.28	85.85	82.74	78.61	75.96	72.07	69.92	73.45	71.95	70.26	69.28
2 & 3	82.18	78.15	75.30	71.55	69.14	65.60	63.64	61.43	60.17	58.75	57.95
3, Average	74.18	70.55	68.00	64.61	62.43	59.23	57.46	55.46	54.33	53.05	52.31
3 & 4	74.19	70.55	68.01	64.61	62.42	59.24	57.46	55.45	54.33	53.04	52.31
4, Low	59.27	56.36	54.33	51.62	49.86	47.33	45.90	44.31	43.40	42.38	41.80

General Office Buildings - Masonry or Concrete

Wall Height Adjustments

The square foot costs for general offices are based on the wall heights of 10 feet for first floors and 9 feet for higher floors. The main or first floor height is the distance from the bottom of the floor slab or joists to the top of the roof slab or ceiling joists. Second and higher floors are measured from the top of the floor slab or floor joists to the top of the roof slab or ceiling joists. Add or subtract the amount listed in this table to the square foot of floor cost for each foot of wall height more or less than 10 feet, if adjusting for a first floor, and 9 feet, if adjusting for upper floors.

Square Foot Area

Quality Class	1,000	1,500	2,000	3,000	4,000	5,000	7,500	10,000	15,000	20,000	40,000
1, Best	3.56	2.78	2.36	1.88	1.59	1.42	1.13	.98	.79	.68	.48
2, Good	3.01	2.34	1.98	1.57	1.35	1.19	.97	.82	.67	.58	.41
3, Average	2.56	2.02	1.70	1.36	1.15	1.02	.81	.70	.58	.48	.35
4, Low	2.27	1.77	1.51	1.20	1.02	.90	.72	.63	.51	.42	.31

Perimeter Wall Adjustment

A common wall exists when two buildings share one wall. Adjust for common walls by deducting the linear foot cost below from the total structure cost. In some structures one or more walls are not owned at all. In this case, deduct the "No Ownership" cost per linear foot of wall not owned. Where a perimeter wall remains unfinished, deduct the "lack of exterior finish" cost.

First Story

Class	For a Common Wall, Deduct Per L.F.	For No Wall Ownership, Deduct Per L.F.	For Lack of Exterior Finish, Deduct Per L.F.
1	$180.00	$360.00	$135.00
2	145.00	290.00	88.00
3	108.00	215.00	52.00
4	83.00	165.00	31.00

Second and Higher Stories

Class	For a Common Wall, Deduct Per L.F.	For No Wall Ownership, Deduct Per L.F.	For Lack of Exterior Finish, Deduct Per L.F.
1	$170.00	$340.00	$135.00
2	145.00	290.00	88.00
3	108.00	215.00	52.00
4	103.00	205.00	31.00

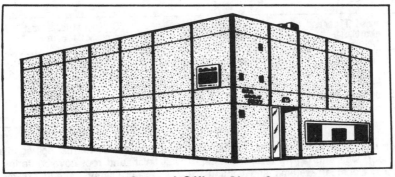

General Office, Class 3

General Office Buildings - Wood Frame

Quality Classification

	Class 1 Best Quality	Class 2 Good Quality	Class 3 Average Quality	Class 4 Low Quality
Foundation	Reinforced concrete.	Reinforced concrete.	Reinforced concrete.	Reinforced concrete.
First Floor Structure	Reinforced concrete slab on grade or standard wood frame.	Reinforced concrete slab on grade or standard wood frame.	Reinforced concrete slab on grade or 4" x 6" girders with 2" T&G subfloor.	Reinforced concrete slab on grade.
Upper Floor Structure	Standard wood frame. Plywood and 1-1/2" light-weight concrete subfloor.	Standard wood frame. Plywood and 1-1/2" light-weight concrete subfloor.	Standard wood frame. 5/8" plywood subfloor.	Standard wood frame. 5/8" plywood subfloor.
Walls	Standard wood frame.	Standard wood frame.	Standard wood frame.	Standard wood frame.
Roof Structure	Standard wood frame, flat or low pitch.	Standard wood frame, flat or low pitch.	Standard wood frame, flat or low pitch.	Standard wood frame, flat or low pitch.
Exterior Finish Walls	Good wood siding with 10 to 20% brick or stone veneer.	Average wood siding or stucco and 10 to 20% brick or stone veneer.	Stucco with some wood trim or cheap wood siding.	Stucco.
Windows	Average number in good aluminum frame. Fixed plate glass in good frame on front side.	Average number in good aluminum frame. Some fixed plate glass in front.	Average number of average aluminum sliding type.	Average number of low cost aluminum sliding type
Roof Cover	5 ply built-up roofing on flat roofs. Heavy shake or tile on sloping roofs.	5 ply built-up roofing on flat roofs. Average shake or composition, tar and large rock on sloping roofs.	4 ply built-up roofing on flat roofs. Wood shingle or composition, tar and pea gravel on sloping roofs.	3 ply built-up roofing on flat roofs. Composition shingle on sloping roofs.
Overhang	3' closed overhang, fully guttered.	2' closed overhang, fully guttered.	None on flat roofs. 18" on sloping roofs, fully guttered.	None on flat roofs. 12" to 16" open on sloping roofs, gutters over entrances.
Floor Finish Offices	Very good carpet.	Good carpet.	Average grade carpet	Minimum tile.
Corridors	Solid vinyl tile or carpet.	Resilient tile.	Composition tile.	Minimum tile.
Bathrooms	Sheet vinyl or ceramic tile.	Sheet vinyl or ceramic tile.	Composition tile.	Minimum tile.
Interior Wall Finish Offices	Good hardwood veneer paneling.	Hardwood veneer paneling or vinyl wall cover.	Gypsum wallboard, texture and paint.	Gypsum wallboard, texture and paint.
Corridors	Good hardwood veneer paneling.	Gypsum wallboard and vinyl wall cover.	Gypsum wallboard, texture and paint.	Gypsum wallboard, texture and paint.
Bathrooms	Gypsum wallboard and enamel with ceramic tile wainscot.	Gypsum wallboard and enamel or vinyl wall covering.	Gypsum wallboard and enamel.	Gypsum wallboard, texture and paint.
Ceiling Finish	Suspended "T" bar and acoustical tile.	Gypsum wallboard and acoustical tile.	Gypsum wallboard and acoustical texture.	Gypsum wallboard and paint.
Utilities Plumbing	Copper tubing, top quality fixtures.	Copper tubing, good fixtures.	Copper tubing, standard fixtures.	Copper tubing, economy fixtures.
Lighting	Conduit wiring, good fixtures.	Conduit wiring, good fixtures.	Romex or conduit wiring, average fixtures.	Romex wiring, economy fixtures.

Square foot costs include the following components: Foundations as required for normal soil conditions. Floor, wall and roof structures. Interior floor, wall and ceiling finishes. Exterior wall finish and roof cover. Interior partitions. Cabinets, doors and windows. Basic electrical systems and lighting fixtures. Rough plumbing and fixtures. Permits and fees. Contractors' mark-up.

General Office Buildings - Wood Frame

Exterior Suite Entrances, Length Less Than Twice Width

Estimating Procedure

1. Use these figures to estimate general office buildings in which access to each suite is through an exterior entrance. Medical and dental offices have smaller rooms and more plumbing fixtures than general offices and should be estimated with figures from the Medical and Dental Buildings Section. See page 118.
2. Establish the building quality class by applying the information on page 110.
3. Compute the first floor area. This should include everything within the exterior walls and all insets outside the main walls but under the main roof.
4. If the first floor wall height is more or less than 10 feet, add to or subtract from the first floor square foot cost below the appropriate amount from the Wall Height Adjustment Table on page 117.
5. Multiply the adjusted square foot cost by the first floor area.
6. Deduct, if appropriate, for common walls or no wall finish. Use the figures on page 117.
7. If there are second or higher floors, compute the square foot area on each floor. Locate the appropriate square foot cost from the table at the bottom of this page. Adjust this figure for a wall height more or less than 9 feet, using the figures on page 117. Multiply the adjusted cost by the square foot area on each floor. Use the figures on page 117 to deduct for common walls or no wall finish. Add the result to the cost from step 6 above.
8. Multiply the total cost by the location factor on page 7.
9. Add the cost of heating and air conditioning systems, elevators, fire sprinklers, exterior signs, paving and curbing, miscellaneous yard improvements, covered porches and garages. See pages 201 to 213.

General Offices, Class 1 & 2

First Story
Square Foot Area

Quality Class	1,000	1,500	2,000	2,500	3,000	4,000	5,000	7,500	10,000	15,000	20,000
Exceptional	112.30	106.19	102.68	100.35	98.65	96.32	94.76	92.39	91.01	89.38	88.43
1, Best	102.62	97.01	93.82	91.68	90.14	88.01	86.58	84.43	83.14	81.66	80.79
1 & 2	92.49	87.44	84.56	82.64	81.23	79.33	78.03	76.09	74.93	73.61	72.83
2, Good	83.04	78.53	75.93	74.20	72.97	71.23	70.08	68.32	67.29	66.10	65.40
2 & 3	75.78	71.66	69.30	67.72	66.57	65.01	63.94	62.36	61.42	60.32	59.67
3, Average	68.33	64.61	62.47	61.05	60.03	58.58	57.66	56.21	55.36	54.37	53.80
3 & 4	60.77	57.47	55.56	54.30	53.38	52.13	51.28	49.99	49.24	48.36	47.85
4, Low	53.19	50.31	48.65	47.52	46.74	45.62	44.89	43.76	43.10	42.34	41.89

Second and Higher Stories
Square Foot Area

Quality Class	1,000	1,500	2,000	2,500	3,000	4,000	5,000	7,500	10,000	15,000	20,000
Exceptional	98.11	94.18	91.94	90.44	89.36	87.87	86.87	85.36	84.47	83.44	82.83
1, Best	89.51	85.90	83.86	82.52	81.52	80.16	79.24	77.86	77.06	76.11	75.57
1 & 2	81.32	78.07	76.20	74.96	74.06	72.83	72.00	70.74	70.01	69.16	68.64
2, Good	73.18	70.27	68.60	67.48	66.66	65.55	64.81	63.69	63.02	62.24	61.80
2 & 3	66.02	63.38	61.88	60.87	60.13	59.14	58.48	57.44	56.85	56.15	55.75
3, Average	58.50	56.15	54.83	53.93	53.28	52.41	51.81	50.89	50.38	49.76	49.40
3 & 4	51.76	49.68	48.51	47.72	47.14	46.36	45.83	45.03	44.56	44.03	43.71
4, Low	44.83	43.04	42.01	41.32	40.82	40.15	39.70	39.00	38.59	38.13	37.85

General Office Buildings - Wood Frame
Exterior Suite Entrances, Length Between 2 and 4 Times Width

Estimating Procedure

1. Use these figures to estimate general office buildings in which access to each suite is through an exterior entrance. Medical and dental offices have smaller rooms and more plumbing fixtures than general offices and should be estimated with figures from the Medical and Dental Buildings Section. See page 118.
2. Establish the building quality class by applying the information on page 110.
3. Compute the first floor area. This should include everything within the exterior walls and all insets outside the main walls but under the main roof.
4. If the first floor wall height is more or less than 10 feet, add to or subtract from the first floor square foot cost below the appropriate amount from the Wall Height Adjustment Table on page 117.
5. Multiply the adjusted square foot cost by the first floor area.
6. Deduct, if appropriate, for common walls or no wall finish. Use the figures on page 117.
7. If there are second or higher floors, compute the square foot area on each floor. Locate the appropriate square foot cost from the table at the bottom of this page. Adjust this figure for a wall height more or less than 9 feet, using the figures on page 117. Multiply the adjusted cost by the square foot area on each floor. Use the figures on page 117 to deduct for common walls or no wall finish. Add the result to the cost from step 6 above.
8. Multiply the total cost by the location factor on page 7.
9. Add the cost of heating and air conditioning systems, elevators, fire sprinklers, exterior signs, paving and curbing, miscellaneous yard improvements, covered porches and garages. See pages 201 to 213.

General Offices, Class 3

First Story
Square Foot Area

Quality Class	1,000	1,500	2,000	2,500	3,000	4,000	5,000	7,500	10,000	15,000	20,000
Exceptional	118.64	110.87	106.47	103.58	101.49	98.65	96.78	93.94	92.28	90.39	89.26
1, Best	108.37	101.27	97.26	94.62	92.72	90.12	88.40	85.80	84.29	82.57	81.54
1 & 2	97.60	91.21	87.59	85.21	83.51	81.17	79.62	77.29	75.92	74.35	73.44
2, Good	87.67	81.93	78.67	76.53	75.01	72.91	71.52	69.41	68.20	66.80	65.97
2 & 3	80.09	74.85	71.88	69.93	68.53	66.62	65.34	63.44	62.31	61.03	60.28
3, Average	72.23	67.48	64.81	63.05	61.79	60.06	58.91	57.17	56.17	55.02	54.34
3 & 4	64.35	60.12	57.74	56.18	55.04	53.51	52.50	50.94	50.06	49.01	48.42
4, Low	56.42	52.70	50.61	49.25	48.26	46.90	46.01	44.66	43.88	42.97	42.44

Second and Higher Stories
Square Foot Area

Quality Class	1,000	1,500	2,000	2,500	3,000	4,000	5,000	7,500	10,000	15,000	20,000
Exceptional	102.14	96.97	94.11	92.24	90.92	89.12	87.94	86.18	85.18	84.01	83.34
1, Best	93.32	88.58	85.99	84.26	83.05	81.41	80.33	78.73	77.81	76.76	76.14
1 & 2	84.42	80.24	77.86	76.32	75.23	73.74	72.77	71.31	70.47	69.52	68.95
2, Good	76.28	72.40	70.27	68.87	67.89	66.55	65.65	64.35	63.60	62.73	62.23
2 & 3	68.88	65.40	63.47	62.21	61.31	60.10	59.30	58.12	57.43	56.67	56.21
3, Average	61.15	58.05	56.34	55.22	54.41	53.34	52.65	51.58	50.99	50.32	49.90
3 & 4	53.70	50.99	49.49	48.51	47.81	46.87	46.24	45.32	44.79	44.17	43.83
4, Low	47.02	44.64	43.31	42.46	41.84	41.02	40.47	39.67	39.21	38.67	38.37

General Office Buildings - Wood Frame

Exterior Suite Entrances, Length More Than 4 Times Width

Estimating Procedure

1. Use these figures to estimate general office buildings in which access to each suite is through an exterior entrance. Medical and dental offices have smaller rooms and more plumbing fixtures than general offices and should be estimated with figures from the Medical and Dental Buildings Section. See page 118.
2. Establish the building quality class by applying the information on page 110.
3. Compute the first floor area. This should include everything within the exterior walls and all insets outside the main walls but under the main roof.
4. If the first floor wall height is more or less than 10 feet, add to or subtract from the first floor square foot cost below the appropriate amount from the Wall Height Adjustment Table on page 117.
5. Multiply the adjusted square foot cost by the first floor area.
6. Deduct, if appropriate, for common walls or no wall finish. Use the figures on page 117.
7. If there are second or higher floors, compute the square foot area on each floor. Locate the appropriate square foot cost from the table at the bottom of this page. Adjust this figure for a wall height more or less than 9 feet, using the figures on page 117. Multiply the adjusted cost by the square foot area on each floor. Use the figures on page 117 to deduct for common walls or no wall finish. Add the result to the cost from step 6 above.
8. Multiply the total cost by the location factor on page 7.
9. Add the cost of heating and air conditioning systems, elevators, fire sprinklers, exterior signs, paving and curbing, miscellaneous yard improvements, covered porches and garages. See pages 201 to 213.

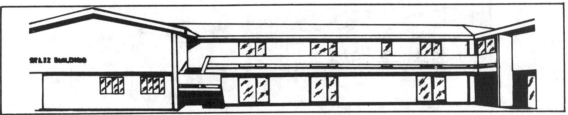

General Offices, Class 2

First Story
Square Foot Area

Quality Class	1,000	1,500	2,000	2,500	3,000	4,000	5,000	7,500	10,000	15,000	20,000
Exceptional	126.21	116.56	111.17	107.62	105.07	101.62	99.34	95.92	93.94	91.67	90.35
1, Best	115.30	106.49	101.53	98.30	95.97	92.82	90.74	87.61	85.80	83.73	82.53
1 & 2	103.94	95.98	91.53	88.62	86.52	83.67	81.80	78.98	77.36	75.48	74.38
2, Good	93.23	86.11	82.10	79.47	77.60	75.05	73.38	70.84	69.38	67.70	66.72
2 & 3	85.26	78.73	75.08	72.69	70.97	68.64	67.10	64.78	63.46	61.91	61.02
3, Average	77.06	71.17	67.85	65.70	64.16	62.04	60.63	58.54	57.35	55.96	55.16
3 & 4	68.66	63.41	60.46	58.53	57.16	55.29	54.04	52.18	51.10	49.87	49.15
4, Low	60.26	55.65	53.08	51.38	50.16	48.52	47.42	45.79	44.84	43.76	43.12

Second and Higher Stories
Square Foot Area

Quality Class	1,000	1,500	2,000	2,500	3,000	4,000	5,000	7,500	10,000	15,000	20,000
Exceptional	106.56	100.44	96.99	94.74	93.13	90.93	89.47	87.28	86.03	84.57	83.71
1, Best	97.34	91.76	88.61	86.56	85.08	83.07	81.74	79.74	78.58	77.27	76.48
1 & 2	88.26	83.19	80.35	78.47	77.13	75.32	74.11	72.29	71.24	70.04	69.34
2, Good	79.65	75.06	72.49	70.81	69.60	67.96	66.87	65.24	64.29	63.20	62.56
2 & 3	72.03	67.88	65.57	64.04	62.95	61.46	60.47	58.99	58.14	57.15	56.58
3, Average	64.07	60.40	58.34	56.98	56.00	54.68	53.81	52.49	51.74	50.86	50.36
3 & 4	56.84	53.58	51.75	50.55	49.67	48.52	47.74	46.57	45.89	45.12	44.65
4, Low	49.45	46.60	45.02	43.96	43.23	42.19	41.52	40.51	39.91	39.24	38.85

Interior Suite Entrances, Length Less Than Twice Width

Estimating Procedure

1. Use these figures to estimate general office buildings in which access to each suite is through an interior corridor. Medical and dental offices have smaller rooms and more plumbing fixtures than general offices and should be estimated with figures from the Medical and Dental Buildings Section. See page 118.
2. Establish the building quality class by applying the information on page 110.
3. Compute the first floor area. This should include everything within the exterior walls and all insets outside the main walls but under the main roof.
4. If the first floor wall height is more or less than 10 feet, add to or subtract from the first floor square foot cost below the appropriate amount from the Wall Height Adjustment Table on page 117.
5. Multiply the adjusted square foot cost by the first floor area.
6. Deduct, if appropriate, for common walls or no wall finish. Use the figures on page 117.
7. If there are second or higher floors, compute the square foot area on each floor. Locate the appropriate square foot cost from the table at the bottom of this page. Adjust this figure for a wall height more or less than 9 feet, using the figures on page 117. Multiply the adjusted cost by the square foot area on each floor. Use the figures on page 117 to deduct for common walls or no wall finish. Add the result to the cost from step 6 above.
8. Multiply the total cost by the location factor on page 7.
9. Add the cost of heating and air conditioning systems, elevators, fire sprinklers, exterior signs, paving and curbing, miscellaneous yard improvements, covered porches and garages. See page 201.

General Offices, Class 3 & 4

First Story
Square Foot Area

Quality Class	2,000	2,500	3,000	4,000	5,000	7,500	10,000	15,000	20,000	30,000	40,000
Exceptional	99.08	96.82	95.19	92.92	91.39	89.06	87.69	86.08	85.15	84.02	83.37
1, Best	90.40	88.35	86.85	84.78	83.39	81.25	80.01	78.54	77.68	76.67	76.07
1 & 2	86.38	80.48	79.12	77.23	75.97	74.04	72.90	71.57	70.77	69.86	69.30
2, Good	74.94	73.22	71.99	70.28	69.11	67.35	66.32	65.11	64.40	63.56	63.06
2 & 3	68.40	66.84	65.71	64.16	63.09	61.48	60.53	59.43	58.78	58.01	57.56
3, Average	61.66	60.26	59.24	57.82	56.87	55.42	54.58	53.58	52.98	52.30	51.89
3 & 4	55.01	53.76	52.85	51.58	50.74	49.45	48.68	47.80	47.27	46.64	46.27
4, Low	48.14	47.03	46.24	45.13	44.39	43.26	42.60	41.82	41.35	40.81	40.51

Second and Higher Stories
Square Foot Area

Quality Class	2,000	2,500	3,000	4,000	5,000	7,500	10,000	15,000	20,000	30,000	40,000
Exceptional	88.41	86.97	85.92	84.46	83.48	82.00	81.13	80.09	79.51	78.79	78.38
1, Best	80.77	79.46	78.48	77.16	76.28	74.92	74.11	73.19	72.62	71.98	71.60
1 & 2	74.02	72.81	71.92	70.71	69.90	68.64	67.92	67.06	66.56	65.96	65.62
2, Good	67.16	66.05	65.25	64.15	63.41	62.26	61.62	60.84	60.38	59.85	59.53
2 & 3	62.50	61.53	58.95	57.97	57.30	56.26	55.67	54.97	54.57	54.08	53.79
3, Average	53.97	53.09	52.45	51.55	50.95	50.06	49.51	48.88	48.52	48.09	47.84
3 & 4	47.79	47.00	46.44	45.66	45.13	44.32	43.85	43.28	42.97	42.58	42.36
4, Low	41.49	40.80	40.31	39.64	39.19	38.48	38.06	37.59	37.31	36.98	36.78

General Office Buildings - Wood Frame

Interior Suite Entrances, Length Between 2 and 4 Times Width

Estimating Procedure

1. Use these figures to estimate general office buildings in which access to each suite is through an interior corridor. Medical and dental offices have smaller rooms and more plumbing fixtures than general offices and should be estimated with figures from the Medical and Dental Buildings Section. See page 118.
2. Establish the building quality class by applying the information on page 110.
3. Compute the first floor area. This should include everything within the exterior walls and all insets outside the main walls but under the main roof.
4. If the first floor wall height is more or less than 10 feet, add to or subtract from the first floor square foot cost below the appropriate amount from the Wall Height Adjustment Table on page 117.
5. Multiply the adjusted square foot cost by the first floor area.
6. Deduct, if appropriate, for common walls or no wall finish. Use the figures on page 117.
7. If there are second or higher floors, compute the square foot area on each floor. Locate the appropriate square foot cost from the table at the bottom of this page. Adjust this figure for a wall height more or less than 9 feet, using the figures on page 117. Multiply the adjusted cost by the square foot area on each floor. Use the figures on page 117 to deduct for common walls or no wall finish. Add the result to the cost from step 6 above.
8. Multiply the total cost by the location factor on page 7.
9. Add the cost of heating and air conditioning systems, elevators, fire sprinklers, exterior signs, paving and curbing, miscellaneous yard improvements, covered porches and garages. See page 201.

First Story
Square Foot Area

Quality Class	2,000	2,500	3,000	4,000	5,000	7,500	10,000	15,000	20,000	30,000	40,000
Exceptional	102.57	99.93	98.00	95.31	93.48	90.68	89.03	87.08	85.92	84.56	83.73
1, Best	94.35	91.93	90.15	87.67	86.02	83.43	81.90	80.11	79.05	77.79	77.04
1 & 2	85.26	83.07	81.47	79.23	77.73	75.38	74.01	72.39	71.42	70.28	69.61
2, Good	77.52	75.54	74.08	72.03	70.66	68.54	67.29	65.82	64.93	63.91	63.29
2 & 3	70.84	69.01	67.69	65.83	64.57	62.63	61.48	60.13	59.33	58.40	57.83
3, Average	63.92	62.28	61.08	59.41	58.27	56.50	55.47	54.27	53.55	52.69	52.20
3 & 4	57.06	55.58	54.52	53.01	52.00	50.44	49.52	48.44	47.80	47.04	46.58
4, Low	49.94	48.67	47.72	46.41	45.53	44.16	43.34	42.41	41.84	41.17	40.78

Second and Higher Stories
Square Foot Area

Quality Class	2,000	2,500	3,000	4,000	5,000	7,500	10,000	15,000	20,000	30,000	40,000
Exceptional	90.68	88.87	87.57	85.81	84.63	82.87	81.85	80.66	79.97	79.17	78.70
1, Best	82.88	81.20	80.02	78.41	77.34	75.72	74.78	73.71	73.06	72.33	71.90
1 & 2	75.80	74.29	73.22	71.74	70.75	69.27	68.42	67.43	66.85	66.18	65.79
2, Good	68.91	67.54	66.56	65.20	64.32	62.96	62.19	61.31	60.76	60.16	59.80
2 & 3	62.29	61.04	60.14	58.93	58.14	56.92	56.21	55.39	54.93	54.36	54.06
3, Average	55.47	54.37	53.58	52.50	51.79	50.71	50.08	49.36	48.93	48.43	48.15
3 & 4	49.19	48.22	47.51	46.54	45.91	44.95	44.40	43.75	43.38	42.93	42.70
4, Low	42.57	41.73	41.12	40.29	39.74	38.90	38.44	37.87	37.55	37.17	36.95

General Office Buildings - Wood Frame
Interior Suite Entrances, Length More Than 4 Times Width

Estimating Procedure

1. Use these figures to estimate general office buildings in which access to each suite is through an interior corridor. Medical and dental offices have smaller rooms and more plumbing fixtures than general offices and should be estimated with figures from the Medical and Dental Buildings Section. See page 118.
2. Establish the building quality class by applying the information on page 110.
3. Compute the first floor area. This should include everything within the exterior walls and all insets outside the main walls but under the main roof.
4. If the first floor wall height is more or less than 10 feet, add to or subtract from the first floor square foot cost below the appropriate amount from the Wall Height Adjustment Table on page 117.
5. Multiply the adjusted square foot cost by the first floor area.
6. Deduct, if appropriate, for common walls or no wall finish. Use the figures on page 117.
7. If there are second or higher floors, compute the square foot area on each floor. Locate the appropriate square foot cost from the table at the bottom of this page. Adjust this figure for a wall height more or less than 9 feet, using the figures on page 117. Multiply the adjusted cost by the square foot area on each floor. Use the figures on page 117 to deduct for common walls or no wall finish. Add the result to the cost from step 6 above.
8. Multiply the total cost by the location factor on page 7.
9. Add the cost of heating and air conditioning systems, elevators, fire sprinklers, exterior signs, paving and curbing, miscellaneous yard improvements, covered porches and garages. See page 201.

General Offices, Class 2

First Story
Square Foot Area

Quality Class	2,000	2,500	3,000	4,000	5,000	7,500	10,000	15,000	20,000	30,000	40,000
Exceptional	107.50	104.03	101.51	98.13	95.88	92.51	90.56	88.33	87.03	85.51	84.62
1, Best	98.20	95.01	92.73	89.63	87.58	84.44	82.71	80.68	79.49	78.10	77.31
1 & 2	89.35	86.46	84.38	81.55	79.68	76.88	75.26	67.81	72.32	71.07	70.34
2, Good	81.19	78.56	76.67	74.11	72.40	69.84	68.39	66.69	65.71	64.58	63.91
2 & 3	74.26	71.82	70.09	67.79	66.19	63.87	62.52	61.02	60.11	59.06	58.47
3, Average	67.10	64.93	63.36	61.25	59.84	57.73	56.51	55.12	54.32	53.36	52.82
3 & 4	59.85	57.91	56.50	54.62	53.37	51.51	50.41	49.17	48.45	47.59	47.12
4, Low	52.53	50.83	49.60	47.94	46.85	45.20	44.26	43.16	42.52	41.78	41.35

Second and Higher Stories
Square Foot Area

Quality Class	2,000	2,500	3,000	4,000	5,000	7,500	10,000	15,000	20,000	30,000	40,000
Exceptional	93.36	91.15	89.56	87.41	85.98	83.85	82.61	81.20	80.37	79.39	78.84
1, Best	85.48	83.46	82.01	80.04	78.74	76.79	75.65	74.35	73.59	72.71	72.20
1 & 2	78.36	76.51	75.18	73.37	72.17	70.38	69.34	68.14	67.45	66.65	66.18
2, Good	71.14	69.45	68.25	66.60	65.52	63.90	62.94	61.86	61.23	60.50	60.07
2 & 3	64.38	62.86	61.76	60.29	59.30	57.83	56.97	56.00	55.41	54.75	54.37
3, Average	57.42	56.07	55.09	53.78	52.89	51.57	50.81	49.93	49.44	48.84	48.49
3 & 4	50.97	49.75	48.90	47.73	46.94	45.78	45.10	44.33	43.88	43.35	43.05
4, Low	44.34	43.30	42.56	41.53	40.85	39.84	39.25	38.57	38.18	37.72	37.46

General Office Buildings - Wood Frame

Wall Height Adjustment

The square foot costs for general offices are based on the wall heights of 10 feet for first floors and 9 feet for higher floors. Add to or subtract from the amount listed in this table the square foot of floor cost for each foot of wall height more or less than 10 feet, if adjusting for a first floor, and 9 feet, if adjusting for upper floors.

Square Foot Area

Quality Class	1,000	1,500	2,000	3,000	4,000	5,000	7,500	10,000	15,000	20,000	40,000
1, Best	1.25	1.07	.93	.78	.67	.61	.47	.40	.32	.27	.13
2, Good	1.07	.91	.79	.67	.59	.49	.40	.35	.28	.23	.11
3, Average	.92	.78	.70	.57	.48	.44	.35	.30	.24	.21	.10
4, Low	.84	.70	.62	.51	.43	.39	.32	.27	.22	.18	.09

Perimeter Wall Adjustment

A common wall exists when two buildings share one wall. Adjust for common walls by deducting the linear foot cost below from the total structure cost. In some structures one or more walls are not owned at all. In this case, deduct the "No Ownership" cost per linear foot of wall not owned. Where a perimeter wall remains unfinished, deduct the "Lack of Exterior Finish" cost.

First Story

Class	For a Common Wall, Deduct Per L.F.	For No Wall Ownership, Deduct Per L.F.	For Lack of Exterior Finish, Deduct Per L.F.
1	$205.00	$410.00	$180.00
2	170.00	340.00	150.00
3	155.00	310.00	120.00
4	115.00	225.00	92.00

Second and Higher Stories

Class	For a Common Wall, Deduct Per L.F.	For No Wall Ownership, Deduct Per L.F.	For Lack of Exterior Finish, Deduct Per L.F.
1	$135.00	$265.00	$180.00
2	110.00	215.00	150.00
3	105.00	205.00	120.00
4	98.00	195.00	92.00

Medical-Dental Buildings - Masonry or Concrete

	Class 1 Best Quality	Class 2 Good Quality	Class 3 Average Quality	Class 4 Low Quality
First Floor Structure	Reinforced concrete slab on grade or standard wood frame.	Reinforced concrete slab on grade or standard wood frame.	Reinforced concrete slab on grade.	Reinforced concrete slab on grade.
Upper Floor Structure	Standard wood frame, plywood and 1-1/2" light-weight concrete subfloor.	Standard wood frame, plywood and 1-1/2" light-weight concrete subfloor.	Standard wood frame. 5/8" plywood subfloor.	Standard wood frame 5/8" plywood subfloor.
Walls	8" decorative concrete block or 6" concrete tilt-up.	8" decorative concrete block or 6" concrete tilt-up.	8" reinforced concrete block or 8" reinforced brick.	8" reinforced concrete block or clay tile.
Roof Structure	Standard wood frame, flat or low pitch.	Standard wood frame, flat or low pitch.	Standard wood frame, flat or low pitch.	Standard wood frame, flat or low pitch.
Exterior Finish Walls	Decorative block or large rock imbedded in tilt-up panels with 10 to 20% brick or stone veneer.	Decorative block or exposed aggregate and 10 to 20% brick or stone veneer.	Stucco or colored concrete block.	Painted.
Windows	Average number in good aluminum frame. Fixed plate glass in good frame on front.	Average number in good aluminum frame. Some fixed plate glass in front.	Average number of average aluminum sliding type.	Average number of low cost aluminum sliding type.
Roof Cover	5 ply built-up roofing on flat roofs. Heavy shake or tile on on sloping roofs.	5 ply built-up roofing on flat roofs. Average shake or composition, tar and large rock on sloping roofs.	4 ply built-up roofing on flat roofs. Wood shingle or composition, tar and pea gravel on sloping roofs.	3 ply built-up roofing on flat roofs. Composition shingle on sloping roofs.
Overhang	3' closed overhang, fully guttered.	2' closed overhang, fully guttered.	None on flat roofs. 18" open on sloping roofs, fully guttered.	None on flat roofs. 12" to 16" open on sloping roofs, gutters over entrances.
Business Offices	Good hardwood veneer paneling. Solid vinyl or carpet.	Hardwood paneling or vinyl wall cover. Resilient tile or carpet.	Gypsum wallboard, texture and paint. composition tile.	Gypsum wallboard, texture and paint. Minimum grade tile.
Corridors	Good hardwood veneer paneling. Solid vinyl or carpet.	Gypsum wallboard and vinyl wall cover. Resilient tile.	Gypsum wallboard, texture and paint. Composition tile.	Gypsum wallboard, texture and paint. Minimum grade tile.
Waiting Rooms	Good hardwood veneer paneling. Carpet.	Hardwood paneling. Carpet.	Gypsum wallboard, some paneling. Resilient tile or carpet.	Gypsum wallboard, texture and paint. Minimum grade tile.
Private Offices	Good hardwood veneer paneling. Carpet.	Textured wall cover and hardwood paneling. Carpet.	Gypsum wallboard and paper, wood paneling. Resilient tile or carpet.	Gypsum wallboard, texture and paint. Minimum grade tile.
Treatment Rooms	Gypsum wallboard and vinyl wall covering. Sheet vinyl or carpet.	Gypsum wallboard and enamel. Sheet vinyl.	Gypsum wallboard and enamel. Resilient tile.	Gypsum wallboard, texture and paint. Minimum grade tile.
Bathrooms	Gypsum wallboard and enamel with ceramic tile wainscot. Sheet vinyl or ceramic tile.	Gypsum wallboard and enamel or vinyl wall covering. Sheet vinyl or ceramic tile.	Gypsum wallboard and enamel. Resilient tile.	Gypsum wallboard, texture and paint. Minimum grade tile.
Ceiling Finish	Suspended "T" bar and acoustical tile.	Gypsum wallboard and acoustical tile.	Gypsum wallboard and acoustical tile.	Gypsum wallboard and paint.
Utilities Plumbing	Copper tubing, good fixtures.	Copper tubing, good fixtures.	Copper tubing, average fixtures.	Copper tubing, economy fixtures.
Lighting	Conduit wiring, good fixtures.	Conduit wiring, good fixtures.	Romex or conduit wiring, average fixtures.	Romex wiring, economy fixtures.
Cabinets	Formica faced with formica tops.	Good grade of hardwood with formica tops.	Average amount of painted wood or low grade hardwood with formica top.	Minimum amount of painted wood with formica top.

Square foot costs include the following components: Foundations as required for normal soil conditions. Floor, wall and roof structures. Interior floor, wall and ceiling finishes. Exterior wall finish and roof cover. Interior partitions. Cabinets, doors and windows. Basic electrical systems and lighting fixtures. Rough plumbing and fixtures. Permits and fees. Contractors' mark-up. In addition to the above components, costs for buildings with more than 10,000 feet include the cost of lead shielding for typical x-ray rooms.

Medical-Dental Buildings - Masonry or Concrete
Exterior Suite Entrances, Length Less Than Twice Width

Estimating Procedure

1. Use these figures to estimate medical, dental, psychiatric, optometry and similar professional buildings in which access to each office suite is through an exterior entrance. Buildings in this section have more plumbing fixtures per square foot of floor and smaller room sizes than general office buildings. Note also that buildings with more than 10,000 square feet are assumed to have lead shielded x-ray rooms.
2. Establish the building quality class by applying the information on page 118.
3. Compute the first floor area. This should include everything within the exterior walls and all insets outside the main walls but under the main roof.
4. If the first floor wall height is more or less than 10 feet, add to or subtract from the first floor square foot cost below the appropriate amount from the Wall Height Adjustment Table on page 125.
5. Multiply the adjusted square foot cost by the first floor area.
6. Deduct, if appropriate, for common walls or no wall finish. Use the figures on page 125.
7. If there are second or higher floors, compute the square foot area on each floor. Locate the appropriate square foot cost from the table at the bottom of this page. Adjust this figure for a wall height more or less than 9 feet, using the figures on page 125. Multiply the adjusted cost by the square foot area on each floor. Use the figures on page 125 to deduct for common walls or no wall finish. Add the result to the cost from step 6 above.
8. Multiply the total cost by the location factor listed on page 7.
9. Add the cost of heating and air conditioning systems, elevators, fire sprinklers, exterior signs, paving and curbing, miscellaneous yard improvements, covered porches and garages. See page 201.

Medical-Dental Building, Class 2

First Story
Square Foot Area

Quality Class	1,000	1,500	2,000	2,500	3,000	4,000	5,000	7,500	10,000	15,000	20,000
Exceptional	151.14	142.23	137.05	133.57	131.03	127.51	125.14	121.50	119.36	116.84	115.36
1, Best	137.86	129.71	124.99	121.81	119.50	116.31	114.13	110.82	108.86	106.57	105.22
1 & 2	130.07	122.39	117.95	114.95	112.76	109.74	107.69	104.57	102.72	100.57	99.27
2, Good	121.23	114.09	109.93	107.15	105.10	102.28	100.38	97.46	95.75	93.73	92.54
2 & 3	112.08	105.44	101.60	99.02	97.15	94.54	92.79	90.08	88.50	86.63	85.53
3, Average	102.58	96.61	93.08	90.73	89.00	85.74	85.00	82.53	81.07	79.36	78.36
3 & 4	96.24	90.56	87.27	85.05	83.45	81.20	79.68	77.37	76.01	74.41	73.45
4, Low	89.64	84.37	81.28	79.22	77.71	75.63	74.23	72.07	70.80	69.30	68.41

Second and Higher Stories
Square Foot Area

Quality Class	1,000	1,500	2,000	2,500	3,000	4,000	5,000	7,500	10,000	15,000	20,000
Exceptional	135.32	126.28	121.38	118.25	116.07	113.15	111.25	108.51	106.96	105.22	104.23
1, Best	130.21	121.50	116.81	113.79	111.68	108.88	107.06	104.40	102.91	101.25	100.29
1 & 2	123.11	114.88	110.44	107.59	105.60	102.94	101.22	98.71	97.30	95.73	94.82
2, Good	115.94	108.20	104.00	101.33	99.46	96.94	95.34	92.97	91.63	90.16	89.31
2 & 3	107.06	99.91	96.03	93.55	91.81	89.53	88.02	85.83	84.62	83.25	82.45
3, Average	98.11	91.55	88.00	85.73	84.16	82.03	80.67	78.66	77.54	76.28	75.57
3 & 4	91.92	85.79	82.46	80.33	78.85	76.87	75.58	73.70	76.15	71.48	70.80
4, Low	85.69	79.98	76.86	74.89	73.51	71.65	70.46	68.71	67.72	66.64	66.02

Medical-Dental Buildings - Masonry or Concrete
Exterior Suite Entrances, Length Between 2 and 4 Times Width

Estimating Procedure

1. Use these figures to estimate medical, dental, psychiatric, optometry and similar professional buildings in which access to each office suite is through an exterior entrance. Buildings in this section have more plumbing fixtures per square foot of floor and smaller room sizes than general office buildings. Note also that buildings with more than 10,000 square feet are assumed to have lead shielded x-ray rooms.
2. Establish the building quality class by applying the information on page 118.
3. Compute the first floor area. This should include everything within the exterior walls and all insets outside the main walls but under the main roof.
4. If the first floor wall height is more or less than 10 feet, add to or subtract from the first floor square foot cost below the appropriate amount from the Wall Height Adjustment Table on page 125.
5. Multiply the adjusted square foot cost by the first floor area.
6. Deduct, if appropriate, for common walls or no wall finish. Use the figures on page 125.
7. If there are second or higher floors, compute the square foot area on each floor. Locate the appropriate square foot cost from the table at the bottom of this page. Adjust this figure for a wall height more or less than 9 feet, using the figures on page 125. Multiply the adjusted cost by the square foot area on each floor. Use the figures on page 125 to deduct for common walls or no wall finish. Add the result to the cost from step 6 above.
8. Multiply the total cost by the location factor listed on page 7.
9. Add the cost of heating and air conditioning systems, elevators, fire sprinklers, exterior signs, paving and curbing, miscellaneous yard improvements, covered porches and garages. See page 201.

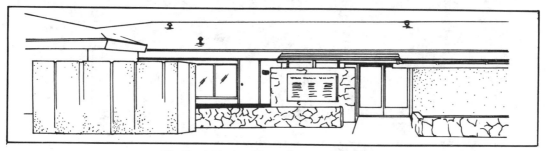

Medical Dental Building, Class 3 & 4

First Story
Square Foot Area

Quality Class	1,000	1,500	2,000	2,500	3,000	4,000	5,000	7,500	10,000	15,000	20,000
Exceptional	162.91	150.85	144.06	139.57	136.37	131.95	129.05	124.65	122.10	119.15	117.42
1, Best	148.85	137.84	131.61	127.51	124.58	120.56	117.90	113.87	111.55	108.86	107.29
1 & 2	139.32	129.01	123.21	119.38	116.62	112.86	110.37	106.61	104.42	101.89	100.44
2, Good	129.74	120.15	114.72	111.17	108.60	105.09	102.78	99.27	97.23	94.89	93.53
2 & 3	119.89	111.04	106.02	102.73	100.36	97.13	94.97	91.73	89.87	87.69	86.42
3, Average	109.88	101.76	97.16	94.15	91.97	89.01	87.05	84.07	82.35	80.36	79.21
3 & 4	102.96	95.34	91.04	88.20	86.17	83.41	81.55	78.78	77.15	75.30	74.21
4, Low	95.88	88.79	84.79	82.16	80.25	77.68	75.95	73.35	71.87	70.12	69.10

Second and Higher Stories
Square Foot Area

Quality Class	1,000	1,500	2,000	2,500	3,000	4,000	5,000	7,500	10,000	15,000	20,000
Exceptional	150.16	139.60	133.65	129.72	126.90	123.04	120.48	116.63	114.38	111.80	110.28
1, Best	137.18	127.65	122.10	118.49	115.92	112.39	110.07	106.54	104.48	102.11	100.74
1 & 2	129.67	120.54	115.39	112.01	109.57	106.23	104.03	100.69	98.76	96.52	95.21
2, Good	122.07	113.47	108.63	105.44	103.15	100.01	97.93	94.80	92.97	90.88	89.64
2 & 3	112.68	104.78	100.31	97.35	95.23	92.35	90.41	87.52	85.84	83.90	82.76
3, Average	103.25	95.99	91.89	89.19	87.25	84.60	82.85	80.18	78.65	76.88	75.84
3 & 4	96.80	90.01	86.16	83.63	81.82	79.32	77.68	75.18	73.75	72.06	71.09
4, Low	90.29	83.94	80.36	78.01	76.30	73.99	72.45	70.12	68.77	67.22	66.32

Medical-Dental Buildings - Masonry or Concrete

Exterior Suite Entrances, Length More Than 4 Times Width

Estimating Procedure

1. Use these figures to estimate medical, dental, psychiatric, optometry and similar professional buildings in which access to each office suite is through an exterior entrance. Buildings in this section have more plumbing fixtures per square foot of floor and smaller room sizes than general office buildings. Note also that buildings with more than 10,000 square feet are assumed to have lead shielded x-ray rooms.
2. Establish the building quality class by applying the information on page 118.
3. Compute the first floor area. This should include everything within the exterior walls and all insets outside the main walls but under the main roof.
4. If the first floor wall height is more or less than 10 feet, add to or subtract from the first floor square foot cost below the appropriate amount from the Wall Height Adjustment Table on page 125.
5. Multiply the adjusted square foot cost by the first floor area.
6. Deduct, if appropriate, for common walls or no wall finish. Use the figures on page 125.
7. If there are second or higher floors, compute the square foot area on each floor. Locate the appropriate square foot cost from the table at the bottom of this page. Adjust this figure for a wall height more or less than 9 feet, using the figures on page 125. Multiply the adjusted cost by the square foot area on each floor. Use the figures on page 125 to deduct for common walls or no wall finish. Add the result to the cost from step 6 above.
8. Multiply the total cost by the location factor listed on page 7.
9. Add the cost of heating and air conditioning systems, elevators, fire sprinklers, exterior signs, paving and curbing, miscellaneous yard improvements, covered porches and garages. See page 201.

Medical-Dental Building, Class 2 & 3

First Story
Square Foot Area

Quality Class	1,000	1,500	2,000	2,500	3,000	4,000	5,000	7,500	10,000	15,000	20,000
Exceptional	175.61	160.39	151.90	146.35	142.36	137.00	133.45	128.15	125.09	121.59	119.55
1, Best	160.43	146.53	138.76	133.69	130.07	125.15	121.91	117.06	114.27	111.08	109.21
1 & 2	149.90	136.89	129.63	124.90	121.53	116.91	113.89	109.36	106.76	103.77	102.03
2, Good	139.99	127.85	121.08	116.65	113.50	109.21	106.38	102.14	99.71	96.91	95.31
2 & 3	128.83	117.66	111.43	107.34	104.44	100.48	97.90	93.99	91.75	89.19	87.69
3, Average	117.92	107.69	101.99	98.26	95.59	91.98	89.61	86.04	83.99	81.63	80.27
3 & 4	110.47	100.87	95.53	92.05	89.55	86.17	83.94	73.63	78.66	76.48	75.19
4, Low	102.85	93.93	88.95	85.70	83.38	80.23	78.14	75.04	73.25	71.20	70.01

Second and Higher Stories
Square Foot Area

Quality Class	1,000	1,500	2,000	2,500	3,000	4,000	5,000	7,500	10,000	15,000	20,000
Exceptional	160.61	147.65	140.34	135.54	132.08	127.36	124.21	119.48	116.74	113.56	111.73
1, Best	146.12	134.33	127.69	123.30	120.16	115.85	113.01	108.69	106.21	103.33	101.66
1 & 2	138.59	127.43	121.12	116.98	113.97	109.90	107.19	103.11	100.75	98.02	96.42
2, Good	130.43	119.92	113.98	110.06	107.26	103.43	100.89	97.04	94.82	92.24	90.75
2 & 3	120.45	110.73	105.24	101.63	99.04	95.49	93.14	89.60	87.54	85.17	83.78
3, Average	110.32	101.44	96.41	93.11	90.74	87.49	85.33	82.08	80.19	78.03	76.76
3 & 4	103.58	95.21	90.51	87.41	85.17	82.13	80.12	77.06	75.29	73.24	72.05
4, Low	96.47	88.68	84.29	81.40	79.32	76.48	74.59	71.76	70.12	68.23	67.11

Interior Suite Entrances, Length Less Than Twice Width

Estimating Procedure

1. Use these figures to estimate medical, dental, psychiatric, optometry and similar professional buildings in which access to each office suite is through an interior corridor. Buildings in this section have more plumbing fixtures per square foot of floor and smaller room sizes than general office buildings. Note also that buildings with more than 10,000 square feet are assumed to have lead shielded x-ray rooms.
2. Establish the building quality class by applying the information on page 118.
3. Compute the first floor area. This should include everything within the exterior walls and all insets outside the main walls but under the main roof.
4. If the first floor wall height is more or less than 10 feet, add to or subtract from the first floor square foot cost below the appropriate amount from the Wall Height Adjustment Table on page 125.
5. Multiply the adjusted square foot cost by the first floor area.
6. Deduct, if appropriate, for common walls or no wall finish. Use the figures on page 125.
7. If there are second or higher floors, compute the square foot area on each floor. Locate the appropriate square foot cost from the table at the bottom of this page. Adjust this figure for a wall height more or less than 9 feet, using the figures on page 125. Multiply the adjusted cost by the square foot area on each floor. Use the figures on page 125 to deduct for common walls or no wall finish. Add the result to the cost from step 6 above.
8. Multiply the total cost by the location factor listed on page 7.
9. Add the cost of heating and air conditioning systems, elevators, fire sprinklers, exterior signs, paving and curbing, miscellaneous yard improvements, covered porches and garages. See page 201.

Medical-Dental Building, Class 3 & 4

First Story
Square Foot Area

Quality Class	2,000	2,500	3,000	4,000	5,000	7,500	10,000	15,000	20,000	30,000	40,000
Exceptional	127.76	124.30	121.78	118.33	116.02	112.48	110.41	107.99	106.57	104.92	103.94
1, Best	119.97	116.74	114.38	111.14	108.95	105.64	103.70	101.42	100.10	98.54	97.61
1 & 2	112.64	109.59	107.37	104.33	102.30	99.16	97.34	95.20	93.97	92.50	91.64
2, Good	105.20	102.35	100.29	97.45	95.52	92.61	90.92	88.93	87.76	86.39	85.58
2 & 3	97.51	94.87	92.96	90.32	88.55	85.84	84.28	82.42	81.34	80.06	79.32
3, Average	89.65	87.22	85.47	83.05	81.42	78.94	77.48	75.79	74.80	73.63	72.93
3 & 4	84.45	82.16	80.51	78.22	76.69	74.35	72.99	71.39	70.45	69.35	68.70
4, Low	79.16	77.02	75.47	73.31	71.89	69.70	68.41	66.93	66.04	65.02	64.41

Second and Higher Stories
Square Foot Area

Quality Class	2,000	2,500	3,000	4,000	5,000	7,500	10,000	15,000	20,000	30,000	40,000
Exceptional	120.78	117.69	115.46	112.38	110.31	107.16	105.33	103.19	101.93	100.45	99.57
1, Best	113.51	110.61	108.52	105.62	103.68	100.72	99.00	97.00	95.80	94.41	93.61
1 & 2	107.20	104.47	102.48	99.76	97.93	95.13	93.50	91.62	90.47	89.17	88.40
2, Good	100.80	98.23	96.37	93.79	92.06	89.45	87.91	86.12	85.07	83.85	83.11
2 & 3	93.19	90.81	89.09	86.73	85.13	82.70	81.29	79.63	78.65	77.52	76.84
3, Average	85.40	83.22	81.64	79.45	77.99	75.78	74.48	72.97	72.08	71.02	70.41
3 & 4	80.33	78.27	76.80	74.75	73.39	71.30	70.05	68.66	67.81	66.82	66.24
4, Low	75.28	73.35	71.96	70.03	68.75	66.80	65.66	64.33	63.54	62.61	62.07

Medical-Dental Buildings - Masonry or Concrete
Interior Suite Entrances, Length Between 2 and 4 Times Width

Estimating Procedure

1. Use these figures to estimate medical, dental, psychiatric, optometry and similar professional buildings in which access to each office suite is through an interior corridor. Buildings in this section have more plumbing fixtures per square foot of floor and smaller room sizes than general office buildings. Note also that buildings with more than 10,000 square feet are assumed to have lead shielded x-ray rooms.
2. Establish the building quality class by applying the information on page 118.
3. Compute the first floor area. This should include everything within the exterior walls and all insets outside the main walls but under the main roof.
4. If the first floor wall height is more or less than 10 feet, add to or subtract from the first floor square foot cost below the appropriate amount from the Wall Height Adjustment Table on page 125.
5. Multiply the adjusted square foot cost by the first floor area.
6. Deduct, if appropriate, for common walls or no wall finish. Use the figures on page 125.
7. If there are second or higher floors, compute the square foot area on each floor. Locate the appropriate square foot cost from the table at the bottom of this page. Adjust this figure for a wall height more or less than 9 feet, using the figures on page 125. Multiply the adjusted cost by the square foot area on each floor. Use the figures on page 125 to deduct for common walls or no wall finish. Add the result to the cost from step 6 above.
8. Multiply the total cost by the location factor listed on page 7.
9. Add the cost of heating and air conditioning systems, elevators, fire sprinklers, exterior signs, paving and curbing, miscellaneous yard improvements, covered porches and garages. See page 201.

Medical Dental Building, Class 3

First Story
Square Foot Area

Quality Class	2,000	2,500	3,000	4,000	5,000	7,500	10,000	15,000	20,000	30,000	40,000
Exceptional	133.74	129.48	126.40	122.18	119.38	115.13	112.65	109.78	108.09	106.12	104.96
1, Best	125.61	121.61	118.73	114.77	112.13	108.13	105.80	103.09	101.51	99.66	98.58
1 & 2	117.81	114.08	111.37	107.64	105.17	101.41	99.24	96.71	95.21	93.49	92.47
2, Good	109.93	106.45	103.93	100.45	98.16	94.65	92.61	90.24	88.86	87.24	86.30
2 & 3	101.83	98.59	96.23	93.05	90.91	87.64	85.76	83.58	82.30	80.80	79.92
3, Average	93.55	90.57	88.43	85.47	83.51	80.53	78.80	76.78	75.60	74.23	73.42
3 & 4	88.20	85.40	83.39	80.59	78.74	75.93	74.29	72.38	71.29	69.99	69.24
4, Low	82.55	79.94	78.03	75.42	73.69	71.07	69.55	67.76	66.73	65.51	64.79

Second and Higher Stories
Square Foot Area

Quality Class	2,000	2,500	3,000	4,000	5,000	7,500	10,000	15,000	20,000	30,000	40,000
Exceptional	126.07	122.26	119.53	115.77	113.26	109.47	107.28	104.71	103.21	101.48	100.44
1, Best	118.40	114.82	112.25	108.72	106.36	102.82	100.73	98.33	96.92	95.30	94.32
1 & 2	111.78	108.42	105.98	102.65	100.44	97.05	95.11	92.84	91.50	89.97	89.06
2, Good	105.12	101.96	99.66	96.53	94.45	91.29	89.46	87.33	86.07	84.61	83.75
2 & 3	97.05	94.13	92.02	89.14	87.20	84.29	82.59	80.61	79.47	78.11	77.34
3, Average	88.86	86.20	84.24	81.61	79.84	77.17	75.60	73.81	72.76	71.52	70.79
3 & 4	83.76	81.21	79.39	76.90	75.24	72.71	71.26	69.56	68.56	67.39	66.71
4, Low	78.49	76.12	74.42	72.08	70.52	68.15	66.78	65.19	64.26	63.18	62.53

Interior Suite Entrances, Length More Than 4 Times Width

Estimating Procedure

1. Establish the building quality class by applying the information on page 118.
3. Compute the first floor area. This should include everything within the exterior walls and all insets outside the main walls but under the main roof.
4. If the first floor wall height is more or less than 10 feet, add to or subtract from the first floor square foot cost below the appropriate amount from the Wall Height Adjustment Table on page 125.
5. Multiply the adjusted square foot cost by the first floor area.
6. Deduct, if appropriate, for common walls or no wall finish. Use the figures on page 125.
7. If there are second or higher floors, compute the square foot area on each floor. Locate the appropriate square foot cost from the table at the bottom of this page. Adjust this figure for a wall height more or less than 9 feet, using the figures on page 125. Multiply the adjusted cost by the square foot area on each floor. Use the figures on page 125 to deduct for common walls or no wall finish. Add the result to the cost from step 6 above.
8. Multiply the total cost by the location factor listed on page 7.
9. Add the cost of heating and air conditioning systems, elevators, fire sprinklers, exterior signs, paving and curbing, miscellaneous yard improvements, covered porches and garages. See page 201.

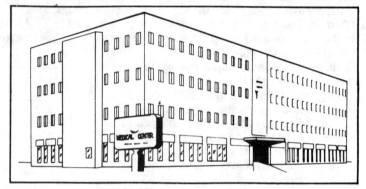

Medical-Dental Building, Class 2 & 3

First Story
Square Foot Area

Quality Class	2,000	2,500	3,000	4,000	5,000	7,500	10,000	15,000	20,000	30,000	40,000
Exceptional	141.41	136.07	132.24	127.05	123.62	118.47	115.50	112.08	110.10	107.80	106.45
1, Best	132.81	127.79	124.22	119.33	116.09	111.25	108.48	105.26	103.39	101.25	99.97
1 & 2	124.47	119.77	116.41	111.83	108.81	104.28	101.67	98.67	96.92	94.88	93.70
2, Good	116.06	111.69	108.55	104.28	101.46	97.22	94.80	91.99	90.36	88.47	87.37
2 & 3	107.36	103.31	100.41	96.47	93.86	89.95	87.70	85.11	83.60	81.86	80.83
3, Average	98.53	94.82	92.15	88.52	86.12	82.54	80.47	78.10	76.71	75.11	74.17
3 & 4	92.78	89.27	86.76	83.38	81.12	77.72	75.79	73.54	72.24	70.72	69.85
4, Low	86.89	83.60	81.26	78.07	75.96	72.80	70.96	68.88	67.65	66.24	65.41

Second and Higher Stories
Square Foot Area

Quality Class	2,000	2,500	3,000	4,000	5,000	7,500	10,000	15,000	20,000	30,000	40,000
Exceptional	132.59	127.85	124.44	119.81	116.77	112.20	109.56	106.54	104.79	102.76	101.57
1, Best	124.60	120.13	116.92	112.58	109.72	105.42	102.95	100.12	98.45	96.54	95.42
1 & 2	117.58	113.37	110.36	106.25	103.54	99.48	97.17	94.46	92.92	91.11	90.06
2, Good	110.54	106.58	103.74	99.89	97.35	93.54	91.33	88.82	87.36	85.65	84.67
2 & 3	101.95	98.30	95.68	92.14	89.78	86.28	84.25	81.94	80.58	79.00	78.10
3, Average	93.48	90.13	87.74	84.48	82.32	79.11	77.25	75.11	73.87	72.45	71.59
3 & 4	88.07	84.91	82.64	79.58	77.54	74.53	72.76	70.76	69.60	68.24	67.45
4, Low	82.53	79.59	77.44	74.57	72.68	69.84	68.19	66.32	65.21	63.95	63.21

Medical-Dental Buildings - Masonry or Concrete

Wall Height Adjustment

The square foot costs for medical-dental buildings are based on the wall heights of 10 feet for first floors and 9 feet for higher floors. The main or first floor height is the distance from the bottom of the floor slab or joists to the top of the roof slab or ceiling joists. Second and higher floors are measured from the top of the floor slab or floor joists to the top of the roof slab or ceiling joists. Add or subtract the amount listed in this table to the square foot of floor cost for each foot of wall height more or less than 10 feet, if adjusting for a first floor, and 9 feet, if adjusting for upper floors.

Square Foot Area

Quality Class	1,000	1,500	2,000	3,000	4,000	5,000	7,500	10,000	15,000	20,000	40,000
1 Best	3.76	2.93	2.48	1.94	1.64	1.46	1.16	1.00	.80	.68	.47
2, Good	3.16	2.48	2.08	1.65	1.39	1.25	.99	.82	.68	.58	.40
3, Average	2.70	2.11	1.77	1.41	1.20	1.05	.85	.71	.59	.50	.35
4, Low	2.36	1.86	1.56	1.23	1.04	.92	.73	.64	.51	.42	.30

Perimeter Wall Adjustment

A common wall exists when two buildings share one wall. Adjust for common walls by deducting the linear foot costs below from the total structure cost. In some structures, one or more walls are not owned at all. In this case, deduct the "No Ownership" cost per linear foot of wall not owned. If a wall has no exterior finish, deduct the "Lack of Exterior Finish" cost.

First Story

Class	For a Common Wall, Deduct Per L.F.	For No Wall Ownership, Deduct Per L.F.	For Lack of Exterior Finish, Deduct Per L.F.
1	$213.00	$428.00	$162.00
2	162.00	284.00	103.00
3	116.00	232.00	54.00
4	89.00	177.00	33.00

Second and Higher Stories

Class	For a Common Wall, Deduct Per L.F.	For No Wall Ownership, Deduct Per L.F.	For Lack of Exterior Finish, Deduct Per L.F.
1	$191.00	$382.00	$162.00
2	155.00	310.00	103.00
3	118.00	236.00	54.00
4	103.00	207.00	33.00

Medical-Dental Buildings - Wood Frame

	Class 1 Best Quality	Class 2 Good Quality	Class 3 Average Quality	Class 4 Low Quality
Foundation	Reinforced concrete.	Reinforced concrete.	Reinforced concrete.	Reinforced concrete.
First Floor Structure	Reinforced concrete slab on grade or standard wood frame.	Reinforced concrete slab on grade or standard wood frame.	Reinforced concrete slab on grade or 4" x 6" girders with 2" T&G subfloor.	Reinforced concrete slab on grade.
Upper Floor Structure	Standard wood frame, plywood and 1-1/2" light-weight concrete subfloor.	Standard wood frame, plywood and 1-1/2" light-weight concrete subfloor.	Standard wood frame, 5/8" plywood subfloor.	Standard wood frame, 5/8" plywood subfloor.
Walls	Standard wood frame.	Standard wood frame.	Standard wood frame.	Standard wood frame.
Roof	Standard wood frame, flat or low pitch.	Standard wood frame, flat or low pitch.	Standard wood frame, flat or low pitch.	Standard wood frame, flat or low pitch.
Exterior Finish Walls	Good wood siding with 10 to 20% brick or stone veneer.	Average wood siding or stucco and 10 to 20% brick or stone veneer.	Stucco with some wood trim or cheap wood siding.	Stucco.
Windows	Average number in good aluminum frame. Fixed plate glass in good frame on front.	Average number in good aluminum frame. Some fixed plate glass in front.	Average amount of average aluminum sliding type.	Average number of low cost aluminum sliding type.
Roof Cover	5 ply built-up roofing on flat roofs. Heavy shake or tile on sloping roofs.	5 ply built-up roofing on flat roofs. Average shake or composition, tar and large rock on	4 ply built-up roofing on flat roofs. Wood shingle or composition, tar and pea gravel on sloping roofs.	3 ply built-up roofing on flat roofs. Composition shingle on sloping roofs.
Overhang	3' sealed overhang, fully guttered.	2' sealed overhang, fully guttered.	None on flat roofs. 18" unsealed on sloping roofs, fully guttered.	None on flat roofs. 12" to 16" unsealed on sloping roofs, gutters over entrances.
Floor Finish Business Offices	Solid vinyl tile or carpet.	Resilient tile or carpet.	Composition tile.	Minimum tile.
Corridors	Solid vinyl tile or carpet.	Resilient tile or carpet.	Composition tile.	Minimum tile.
Waiting Rooms	Carpet.	Carpet.	Composition tile or carpet.	Minimum tile.
Private Offices	Carpet.	Carpet.	Composition tile or carpet.	Minimum tile.
Treatment Room	Sheet vinyl or carpet.	Sheet vinyl.	Composition tile.	Minimum tile.
Bathrooms	Sheet vinyl or ceramic tile.	Sheet vinyl or ceramic tile.	Composition tile.	Minimum tile.
Interior Wall Finish Business Offices	Good hardwood veneer paneling.	Hardwood paneling or vinyl wall cover.	Gypsum wallboard, texture and paint.	Gypsum wallboard, texture and paint.
Corridors	Good hardwood veneer paneling.	Gypsum wallboard and vinyl wall cover.	Gypsum wallboard, texture and paint.	Gypsum wallboard, texture and paint.
Waiting Room	Good hardwood veneer paneling.	Hardwood paneling.	Gypsum wallboard and paper, some wood paneling.	Gypsum wallboard, texture and paint.
Treatment Room	Gypsum wallboard and vinyl wall covering.	Gypsum wallboard and enamel.	Gypsum wallboard and enamel.	Gypsum wallboard, texture and paint.
Bathrooms	Gypsum wallboard and enamel with ceramic tile wainscot.	Gypsum wallboard and enamel or vinyl wall covering.	Gypsum wallboard and enamel.	Gypsum wallboard, texture and paint.
Ceiling Finish	Suspended "T" bar and acoustical tile.	Gypsum wallboard and acoustical tile.	Gypsum wallboard and acoustical tile.	Gypsum wallboard and paint.
Utilities Plumbing	Copper tubing, good fixtures.	Copper tubing, good fixtures.	Copper tubing, average fixtures.	Copper tubing, economy fixtures.
Lighting	Conduit wiring, good fixtures.	Conduit wiring, good fixtures.	Romex or conduit wiring, average fixtures.	Romex wiring, economy fixtures.
Cabinets	Formica faced with formica tops.	Good grade of hardwood with formica tops.	Average amount of painted wood or low grade hardwood with formica top.	Minimum amount of painted wood with formica top.

Square foot costs include the following components: Foundations as required for normal soil conditions. Floor, wall and roof structures. Interior floor, wall and ceiling finishes. Exterior wall finish and roof cover. Interior partitions. Cabinets, doors and windows. Basic electrical systems and lighting fixtures. Rough plumbing and fixtures. Permits and fees. Contractors' mark-up. In addition to the above components, costs for buildings with more than 10,000 feet include the cost of lead shielding for typical x-ray rooms.

Medical-Dental Buildings - Wood Frame
Exterior Suite Entrances, Length Less Than Twice Width

Estimating Procedure

1. Use these figures to estimate medical, dental, psychiatric, optometry and similar professional buildings in which access to each office suite is through an exterior entrance. Buildings in this section have more plumbing fixtures per square foot of floor and smaller room sizes than general office buildings. Note also that buildings with more than 10,000 square feet are assumed to have lead shielded x-ray rooms.
2. Establish the building quality class by applying the information on page 126.
3. Compute the first floor area. This should include everything within the exterior walls and all insets outside the main walls but under the main roof.
4. If the first floor wall height is more or less than 10 feet, add to or subtract from the first floor square foot cost below the appropriate amount from the Wall Height Adjustment Table on page 133.
5. Multiply the adjusted square foot cost by the first floor area.
6. Deduct, if appropriate, for common walls or no wall finish. Use the figures on page 133.
7. If there are second or higher floors, compute the square foot area on each floor. Locate the appropriate square foot cost from the table at the bottom of this page. Adjust this figure for a wall height more or less than 9 feet, using the figures on page 133. Multiply the adjusted cost by the square foot area on each floor. Use the figures on page 133 to deduct for common walls or no wall finish. Add the result to the cost from step 6 above.
8. Multiply the total cost by the location factor listed on page 7.
9. Add the cost of heating and air conditioning systems, elevators, fire sprinklers, exterior signs, paving and curbing, miscellaneous yard improvements, covered porches and garages. See page 201.

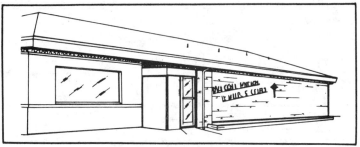

Medical-Dental Building, Class 2 & 3

First Story
Square Foot Area

Quality Class	1,000	1,500	2,000	2,500	3,000	4,000	5,000	7,500	10,000	15,000	20,000
Exceptional	134.92	129.68	126.67	124.68	123.24	121.27	119.95	117.95	116.79	115.42	114.62
1, Best	123.25	118.45	115.73	113.90	112.61	110.79	109.60	107.76	106.68	105.45	104.71
1 & 2	115.63	111.14	108.59	106.88	105.66	103.96	102.82	101.10	100.09	98.92	98.24
2, Good	107.80	103.62	101.22	99.62	98.49	96.90	95.85	94.25	93.32	92.22	91.59
2 & 3	100.10	96.21	93.98	92.52	91.45	89.96	89.01	87.50	86.64	85.63	85.03
3, Average	92.33	88.75	86.71	85.33	84.37	83.01	82.11	80.73	79.92	79.00	78.46
3 & 4	86.31	82.96	81.03	79.77	78.85	77.57	76.75	75.46	74.71	73.84	73.32
4, Low	80.22	77.09	75.33	74.14	73.29	72.13	71.33	70.15	69.45	68.63	68.15

Second and Higher Stories
Square Foot Area

Quality Class	1,000	1,500	2,000	2,500	3,000	4,000	5,000	7,500	10,000	15,000	20,000
Exceptional	120.28	117.23	115.46	114.26	113.38	112.13	111.30	110.00	109.23	108.34	107.80
1 Best	110.03	107.24	105.62	104.52	103.71	102.59	101.82	100.63	99.94	99.11	98.61
1 & 2	102.42	99.84	98.32	97.30	96.54	95.49	94.78	93.69	93.03	92.25	91.80
2, Good	98.31	95.84	94.40	93.41	92.68	91.67	91.00	89.94	85.92	88.56	88.12
2 & 3	90.84	88.54	87.21	86.30	85.63	84.69	84.06	83.08	82.50	81.83	81.42
3, Average	83.31	81.52	79.97	79.13	78.53	77.68	77.10	76.20	75.66	75.06	74.68
3 & 4	77.86	75.89	74.75	73.97	73.39	72.61	72.05	71.23	70.73	70.13	69.80
4, Low	72.48	70.66	69.59	68.86	68.32	67.60	67.07	66.31	65.85	65.30	64.97

Exterior Suite Entrances, Length Between 2 and 4 Times Width

Estimating Procedure

1. Use these figures to estimate medical, dental, psychiatric, optometry and similar professional buildings in which access to each office suite is through an exterior entrance. Buildings in this section have more plumbing fixtures per square foot of floor and smaller room sizes than general office buildings. Note also that buildings with more than 10,000 square feet are assumed to have lead shielded x-ray rooms.
2. Establish the building quality class by applying the information on page 126.
3. Compute the first floor area. This should include everything within the exterior walls and all insets outside the main walls but under the main roof.
4. If the first floor wall height is more or less than 10 feet, add to or subtract from the first floor square foot cost below the appropriate amount from the Wall Height Adjustment Table on page 133.
5. Multiply the adjusted square foot cost by the first floor area.
6. Deduct, if appropriate, for common walls or no wall finish. Use the figures on page 133.
7. If there are second or higher floors, compute the square foot area on each floor. Locate the appropriate square foot cost from the table at the bottom of this page. Adjust this figure for a wall height more or less than 9 feet, using the figures on page 133. Multiply the adjusted cost by the square foot area on each floor. Use the figures on page 133 to deduct for common walls or no wall finish. Add the result to the cost from step 6 above.
8. Multiply the total cost by the location factor listed on page 7.
9. Add the cost of heating and air conditioning systems, elevators, fire sprinklers, exterior signs, paving and curbing, miscellaneous yard improvements, covered porches and garages. See page 201.

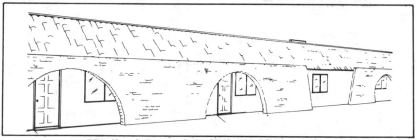

Medical-Dental Building, Class 1 & 2

First Story
Square Foot Area

Quality Class	1,000	1,500	2,000	2,500	3,000	4,000	5,000	7,500	10,000	15,000	20,000
Exceptional	137.76	131.26	127.61	125.20	123.48	121.10	119.54	117.17	115.80	114.21	113.28
1, Best	125.84	119.91	116.57	114.37	112.79	110.63	109.21	107.05	105.79	104.34	103.49
1 & 2	117.91	112.35	109.22	107.16	105.69	103.66	102.32	100.30	99.12	97.77	104.10
2, Good	109.74	104.69	101.75	99.84	98.46	96.57	95.33	93.44	92.35	91.08	90.35
2 & 3	101.95	97.17	94.43	92.66	91.38	89.62	88.47	86.73	85.70	84.54	83.86
3, Average	94.01	89.60	87.10	85.46	84.27	82.65	81.58	79.97	79.03	77.93	77.32
3 & 4	87.80	83.68	81.35	79.83	78.73	77.22	76.22	74.70	73.84	72.83	72.23
4, Low	81.62	77.76	75.61	74.17	73.15	71.76	70.82	69.41	68.61	67.67	67.11

Second and Higher Stories
Square Foot Area

Quality Class	1,000	1,500	2,000	2,500	3,000	4,000	5,000	7,500	10,000	15,000	20,000
Exceptional	121.69	117.81	115.59	114.11	113.03	111.56	110.56	109.06	108.18	107.15	106.55
1, Best	111.17	107.62	105.60	104.24	103.28	101.91	101.02	99.63	98.83	97.88	97.34
1 & 2	105.13	101.78	99.87	98.57	97.67	96.38	95.53	94.22	93.46	92.58	92.06
2, Good	99.44	96.27	94.46	93.25	92.37	91.16	90.36	89.13	88.41	87.56	87.07
2 & 3	91.53	88.60	86.95	85.81	84.99	83.89	83.16	82.01	81.35	80.58	80.13
3, Average	83.97	81.30	79.77	78.75	78.02	76.98	76.31	75.25	74.65	73.95	73.52
3 & 4	78.55	76.07	74.61	73.65	72.96	72.01	71.36	70.39	69.84	69.18	68.78
4, Low	72.90	70.58	69.26	68.36	67.73	66.83	66.23	65.35	64.82	64.20	63.83

Medical-Dental Buildings - Wood Frame
Exterior Suite Entrances, Length More Than 4 Times Width

Estimating Procedure

1. Use these figures to estimate medical, dental, psychiatric, optometry and similar professional buildings in which access to each office suite is through an exterior entrance. Buildings in this section have more plumbing fixtures per square foot of floor and smaller room sizes than general office buildings. Note also that buildings with more than 10,000 square feet are assumed to have lead shielded x-ray rooms.
2. Establish the building quality class by applying the information on page 126.
3. Compute the first floor area. This should include everything within the exterior walls and all insets outside the main walls but under the main roof.
4. If the first floor wall height is more or less than 10 feet, add to or subtract from the first floor square foot cost below the appropriate amount from the Wall Height Adjustment Table on page 133.
5. Multiply the adjusted square foot cost by the first floor area.
6. Deduct, if appropriate, for common walls or no wall finish. Use the figures on page 133.
7. If there are second or higher floors, compute the square foot area on each floor. Locate the appropriate square foot cost from the table at the bottom of this page. Adjust this figure for a wall height more or less than 9 feet, using the figures on page 133. Multiply the adjusted cost by the square foot area on each floor. Use the figures on page 133 to deduct for common walls or no wall finish. Add the result to the cost from step 6 above.
8. Multiply the total cost by the location factor listed on page 7.
9. Add the cost of heating and air conditioning systems, elevators, fire sprinklers, exterior signs, paving and curbing, miscellaneous yard improvements, covered porches and garages. See page 201.

Medical-Dental Building, Class 1

First Story
Square Foot Area

Quality Class	1,000	1,500	2,000	2,500	3,000	4,000	5,000	7,500	10,000	15,000	20,000
Exceptional	144.03	136.05	131.57	128.63	126.54	123.68	121.80	118.95	117.31	115.42	114.33
1, Best	131.86	124.55	120.47	117.79	115.86	113.24	111.52	108.92	107.43	105.67	104.67
1 & 2	123.41	116.57	112.73	110.22	108.41	105.98	104.36	101.93	100.51	98.89	97.96
2, Good	114.82	108.47	104.90	102.57	100.89	98.61	97.11	94.83	93.52	92.02	91.15
2 & 3	106.54	100.61	97.32	95.15	93.61	91.48	90.09	87.99	86.76	85.37	84.56
3, Average	98.18	92.76	89.69	87.70	86.27	84.31	83.02	81.09	79.97	78.68	77.93
3 & 4	91.66	86.60	83.75	81.88	80.53	78.72	77.51	75.72	74.66	73.47	72.77
4, Low	85.15	80.43	77.78	76.05	74.80	73.11	72.00	70.31	69.34	68.24	67.60

Second and Higher Stories
Square Foot Area

Quality Class	1,000	1,500	2,000	2,500	3,000	4,000	5,000	7,500	10,000	15,000	20,000
Exceptional	125.79	120.82	118.02	116.20	114.89	113.11	111.91	110.14	109.11	107.93	107.24
1, Best	114.97	110.44	107.89	106.22	105.01	103.38	102.31	100.68	99.74	98.64	98.03
1 & 2	108.59	104.29	101.90	100.32	99.19	97.63	96.62	95.08	94.19	93.18	92.59
2, Good	102.35	98.31	96.05	94.55	93.48	92.03	91.07	89.62	88.79	87.83	87.26
2 & 3	94.60	90.86	88.76	87.39	86.39	85.06	84.18	82.83	82.04	81.18	80.64
3, Average	86.68	83.26	81.34	80.08	79.18	77.94	77.13	75.89	75.19	74.38	73.91
3 & 4	81.10	77.88	76.11	74.92	74.07	72.92	72.16	71.01	70.35	69.60	69.14
4, Low	75.36	72.40	70.70	69.63	68.84	67.76	67.05	66.00	65.37	64.66	64.24

Medical-Dental Buildings - Wood Frame
Interior Suite Entrances, Length Less Than Twice Width

Estimating Procedure

1. Use these figures to estimate medical, dental, psychiatric, optometry and similar professional buildings in which access to each office suite is through an interior corridor. Buildings in this section have more plumbing fixures per square foot or floor and smaller room sizes than general office buildings. Note also that buildings with more than 10,000 square feet are assumed to have lead shielded x-ray rooms.
2. Establish the building quality class by applying the information on page 126.
3. Compute the first floor area. This should include everything within the exterior walls and all insets outside the main walls but under the main roof.
4. If the first floor wall height is more or less than 10 feet, add or subtract from the first floor square foot cost below the appropriate amount from the Wall Height Adjustment Table on page 133.
5. Multiply the adjusted square foot cost by the first floor area.
6. Deduct, if appropriate, for common walls or no wall finish. Use the figures on page 133.
7. If there are second or higher floors, compute the square foot area on each floor. Locate the appropriate square foot cost from the table at the bottom of this page. Adjust this figure for a wall height more or less than 9 feet, using the figures on page 133. Multiply the adjusted cost by the square foot area on each floor. Use the figures on page 133 to deduct for common walls or no wall finish. Add the result to the cost from step 6 above.
8. Multiply the total cost by the location factor listed on page 7.
9. Add the cost of heating and air conditioning systems, elevators, fire sprinklers, exterior signs, paving and curbing, miscellaneous yard improvements, covered porches and garages. See page 201.

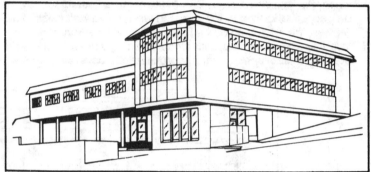

Medical-Dental Building, Class 3

First Story
Square Foot Area

Quality Class	2,000	2,500	3,000	4,000	5,000	7,500	10,000	15,000	20,000	30,000	40,000
Exceptional	114.23	112.36	111.01	109.13	107.90	106.00	104.88	103.58	102.82	101.92	101.41
1, Best	107.62	105.86	104.57	102.83	101.63	99.85	98.81	97.59	96.87	96.02	95.53
1 & 2	101.15	99.51	98.30	96.66	95.54	93.86	92.90	91.73	91.05	90.25	89.80
2, Good	94.56	93.03	91.91	90.36	89.32	87.75	86.84	85.77	85.13	84.40	83.95
2 & 3	87.83	86.39	85.36	83.93	82.96	81.50	80.63	79.62	79.06	78.37	77.98
3, Average	81.19	79.86	78.89	77.56	76.68	75.33	74.54	73.62	73.08	72.46	72.08
3 & 4	76.35	75.10	74.21	72.94	72.13	70.85	70.10	69.24	68.72	68.13	67.77
4, Low	71.47	70.30	69.45	68.28	67.49	66.32	65.63	64.82	64.31	63.78	63.45

Second and Higher Stories
Square Foot Area

Quality Class	2,000	2,500	3,000	4,000	5,000	7,500	10,000	15,000	20,000	30,000	40,000
Exceptional	104.73	103.58	102.76	101.60	100.84	99.64	98.94	98.13	97.65	97.10	96.76
1, Best	98.82	97.74	96.96	95.86	95.14	94.02	93.36	92.59	92.15	91.61	91.29
1 & 2	93.37	92.35	91.60	90.57	89.88	88.83	88.22	87.49	87.05	86.55	80.65
2, Good	87.87	86.90	86.20	85.24	84.58	83.59	83.02	82.32	81.93	81.45	81.17
2 & 3	81.19	80.29	79.65	78.76	78.15	77.25	76.70	76.07	75.70	75.27	75.01
3, Average	74.43	73.62	73.02	72.20	71.65	70.81	70.31	69.74	69.40	68.99	68.77
3 & 4	69.96	69.21	68.65	57.53	67.36	66.58	66.11	65.55	65.23	64.86	64.64
4, Low	65.41	64.69	64.17	63.45	62.97	62.23	61.80	61.29	60.98	60.64	60.43

Medical-Dental Buildings - Wood Frame
Interior Suite Entrances, Length Between 2 and 4 Times Width

Estimating Procedure

1. Use these figures to estimate medical, dental, psychiatric, optometry and similar professional buildings in which access to each office suite is through an interior corridor. Buildings in this section have more plumbing fixtures per square foot of floor and smaller room sizes than general office buildings. Note also that buildings with more than 10,000 square feet are assumed to have lead shielded x-ray rooms.
2. Establish the building quality class by applying the information on page 126.
3. Compute the first floor area. This should include everything within the exterior walls and all insets outside the main walls but under the main roof.
4. If the first floor wall height is more or less than 10 feet, add to or subtract from the first floor square foot cost below the appropriate amount from the Wall Height Adjustment Table on page 133.
5. Multiply the adjusted square foot cost by the first floor area.
6. Deduct, if appropriate, for common walls or no wall finish. Use the figures on page 133.
7. If there are second or higher floors, compute the square foot area on each floor. Locate the appropriate square foot cost from the table at the bottom of this page. Adjust this figure for a wall height more or less than 9 feet, using the figures on page 133. Multiply the adjusted cost by the square foot area on each floor. Use the figures on page 133 to deduct for common walls or no wall finish. Add the result to the cost from step 6 above.
8. Multiply the total cost by the location factor listed on page 7.
9. Add the cost of heating and air conditioning systems, elevators, fire sprinklers, exterior signs, paving and curbing, miscellaneous yard improvements, covered porches and garages. See page 201.

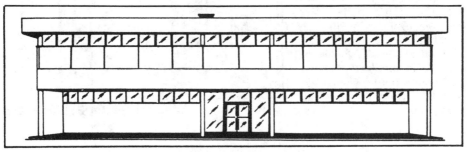

Medical-Dental Building, Class 1 & 2

First Story
Square Foot Area

Quality Class	2,000	2,500	3,000	4,000	5,000	7,500	10,000	15,000	20,000	30,000	40,000
Exceptional	117.09	114.81	113.18	110.94	109.45	107.19	105.88	104.34	103.46	102.41	101.81
1, Best	110.70	108.55	107.00	104.89	103.46	101.34	100.08	98.63	97.81	96.81	96.24
1 & 2	103.88	101.87	100.42	98.44	97.11	95.11	93.92	92.58	91.79	90.87	90.30
2, Good	97.03	95.16	93.80	91.95	90.71	88.83	87.74	86.48	85.73	84.87	84.37
2 & 3	90.16	88.41	87.14	85.42	84.27	82.53	81.53	80.33	79.65	78.85	78.39
3, Average	83.24	81.62	80.47	78.86	77.81	76.20	75.25	74.17	73.55	72.80	72.36
3 & 4	78.18	76.67	75.57	74.07	73.09	71.57	70.69	69.67	69.06	68.37	67.97
4, Low	73.24	71.82	70.80	69.38	68.45	67.04	66.21	65.26	64.71	64.07	63.67

Second and Higher Stories
Square Foot Area

Quality Class	2,000	2,500	3,000	4,000	5,000	7,500	10,000	15,000	20,000	30,000	40,000
Exceptional	107.00	105.44	104.31	102.81	101.83	100.36	99.53	98.59	98.05	97.41	97.04
1, Best	100.98	99.49	98.45	97.02	96.09	94.72	93.92	93.03	92.51	91.91	91.59
1 & 2	95.38	93.97	92.98	91.64	90.77	89.46	88.72	87.88	87.39	86.84	86.51
2, Good	89.72	88.41	87.47	86.21	85.38	84.16	83.46	82.66	82.22	81.69	81.38
2 & 3	83.21	81.99	81.12	79.95	79.18	78.05	77.40	76.66	76.25	75.75	75.47
3, Average	76.02	74.90	74.12	73.04	72.34	71.30	70.72	70.05	69.66	69.21	68.95
3 & 4	71.47	70.41	69.66	68.66	68.01	67.03	66.48	65.85	65.47	65.05	64.80
4, Low	66.82	65.84	65.14	64.20	63.59	62.66	62.16	61.56	61.24	60.85	60.61

Medical-Dental Buildings - Wood Frame
Interior Suite Entrances, Length More Than 4 Times Width

Estimating Procedure

1. Use these figures to estimate medical, dental, psychiatric, optometry and similar professional buildings in which access to each office suite is through an interior corridor. Buildings in this section have more plumbing fixtures per square foot of floor and smaller room sizes than general office buildings. Note also that buildings with more than 10,000 square feet are assumed to have lead shielded x-ray rooms.
2. Establish the building quality class by applying the information on page 126.
3. Compute the first floor area. This should include everything within the exterior walls and all insets outside the main walls but under the main roof.
4. If the first floor wall height is more or less than 10 feet, add to or subtract from the first floor square foot cost below the appropriate amount from the Wall Height Adjustment Table on page 133.
5. Multiply the adjusted square foot cost by the first floor area.
6. Deduct, if appropriate, for common walls or no wall finish. Use the figures on page 133.
7. If there are second or higher floors, compute the square foot area on each floor. Locate the appropriate square foot cost from the table at the bottom of this page. Adjust this figure for a wall height more or less than 9 feet, using the figures on page 133. Multiply the adjusted cost by the square foot area on each floor. Use the figures on page 133 to deduct for common walls or no wall finish. Add the result to the cost from step 6 above.
8. Multiply the total cost by the location factor listed on page 7.
9. Add the cost of heating and air conditioning systems, elevators, fire sprinklers, exterior signs, paving and curbing, miscellaneous yard improvements, covered porches and garages. See page 201.

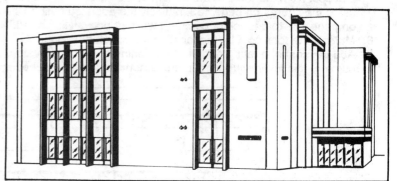

Medical-Dental Building, Class 2

First Story
Square Foot Area

Quality Class	2,000	2,500	3,000	4,000	5,000	7,500	10,000	15,000	20,000	30,000	40,000
Exceptional	121.63	118.93	116.95	114.18	112.30	109.39	107.66	105.63	104.40	102.97	102.12
1, Best	114.75	112.22	110.35	107.74	105.96	103.21	101.57	99.65	98.51	97.16	96.35
1 & 2	107.63	105.25	103.49	101.05	99.37	96.80	95.26	93.46	92.39	91.13	90.37
2, Good	100.42	98.20	96.56	94.28	92.72	90.32	88.89	87.21	86.20	85.01	84.31
2 & 3	93.26	91.20	89.69	87.55	86.12	83.89	82.55	80.99	80.07	78.95	78.32
3, Average	86.09	84.19	82.78	80.83	79.50	77.44	76.21	74.77	73.89	72.88	72.28
3 & 4	83.08	81.24	79.87	78.01	76.71	74.72	73.55	72.15	71.31	70.34	69.75
4, Low	75.64	73.96	72.73	71.00	69.85	68.03	66.96	65.70	64.92	64.05	63.50

Second and Higher Stories
Square Foot Area

Quality Class	2,000	2,500	3,000	4,000	5,000	7,500	10,000	15,000	20,000	30,000	40,000
Exceptional	109.21	107.44	106.17	104.44	103.30	101.55	100.58	99.43	98.76	97.98	97.54
1, Best	103.05	101.39	100.18	98.54	97.45	95.84	94.89	93.82	93.18	92.46	92.02
1 & 2	97.32	95.73	94.59	93.05	92.03	90.49	89.60	88.59	87.99	87.29	86.89
2, Good	91.53	90.05	88.98	87.54	86.56	85.13	84.29	83.33	82.77	82.12	81.74
2 & 3	84.60	83.22	82.24	80.90	80.01	78.68	77.88	77.01	76.49	75.91	75.56
3, Average	77.56	76.31	75.40	74.16	73.35	72.13	71.43	70.62	70.14	69.59	69.26
3 & 4	72.89	71.70	70.86	69.70	68.94	67.79	67.11	66.36	57.33	65.40	65.09
4, Low	68.11	67.01	66.23	65.14	64.43	63.36	62.74	62.01	61.60	61.12	60.84

Medical-Dental Buildings - Wood Frame

Wall Height Adjustment

The square foot costs for medical and dental buildings are based on the wall heights of 10 feet for first floors and 9 feet for higher floors. The first floor height is the distance from the bottom of the floor slab or joists to the top of the roof slab or ceiling joists. Second and higher floors are measured from the top of the floor slab or floor joists to the top of the roof slab or ceiling joists. Add or subtract the amount listed in this table to the square foot of floor cost for each foot of wall height more or less than 10 feet, if adjusting for first floor, and 9 feet, if adjusting for upper floors.

Square Foot Area

Quality Class	1,000	1,500	2,000	3,000	4,000	5,000	7,500	10,000	15,000	20,000	40,000
1, Best	1.26	1.01	.85	.68	.59	.51	.41	.36	.30	.27	.19
2, Good	1.08	.84	.72	.59	.49	.43	.35	.31	.26	.23	.18
3, Average	.94	.72	.62	.50	.42	.37	.31	.27	.23	.20	.11
4, Low	.83	.65	.55	.43	.37	.34	.27	.24	.20	.17	.11

Perimeter Wall Adjustment

A common wall exists when two buildings share one wall. Adjust for common walls by deducting the linear foot costs below from the total structure cost. In some structures, one or more walls are not owned at all. In this case, deduct the "No Ownership" cost per linear foot of wall not owned. If a wall has no exterior finish, deduct the "Lack of Exterior Finish" cost.

First Story

Class	For a Common Wall, Deduct Per L.F.	For No Wall Ownership, Deduct Per L.F.	For Lack of Exterior Finish, Deduct Per L.F.
1	$154.00	$308.00	$134.00
2	124.00	247.00	103.00
3	98.00	196.00	72.00
4	78.00	155.00	52.00

Second and Higher Stories

Class	For a Common Wall, Deduct Per L.F.	For No Wall Ownership, Deduct Per L.F.	For Lack of Exterior Finish, Deduct Per L.F.
1	$103.00	$206.00	$134.00
2	77.00	154.00	103.00
3	72.00	144.00	72.00
4	67.00	135.00	52.00

Quality Classification

	Class 1 Best Quality	Class 2 Good Quality	Class 3 Average Quality	Class 4 Low Quality
Foundation	Reinforced concrete.	Reinforced concrete.	Reinforced concrete.	Reinforced concrete.
Floor Structure	Reinforced concrete.	Reinforced concrete.	Reinforced concrete.	Reinforced concrete.
Roof	Standard wood frame.	Standard wood frame.	Standard wood frame.	Standard wood frame.
Exterior Finish **Walls**	Furred out gypsum wall-board interior, colored block exterior.	Furred out gypsum wall-board interior, colored block exterior.	Colored block with stucco interior and painted exterior	Plain 8" concrete block painted on the inside.
Front	Brick or stone veneer and plate glass in aluminum frame.	Stone or brick veneer. Small amount of plate glass in aluminum frame.	Rustic siding or some brick or stone veneer.	Stucco and small amount siding.
Windows	Good aluminum sliding glass doors or anodized aluminum casement windows.	Good aluminum sliding glass doors with aluminum jalousie window for vent.	Average aluminum sliding type. 6' sliding glass door in bedrooms.	Low cost aluminum sliding type.
Doors	Anodized aluminum entry doors.	Aluminum and glass entry doors.	Good hardwood slab door with side lights or aluminum and glass entry door.	Wood slab door.
Roof	Composition with large colored rock or tile. 4' to 6' sealed overhang.	Composition with large colored rock or asbestos shingles. 3' to 4' sealed overhang.	Composition and small colored rock. 4' exposed or 3' sealed overhang.	Composition and pea gravel or composition shingles. 2' exposed overhang.
Interior Finish **Floors**	Resilient tile. Good carpet in entry and day room. Quarry tile or epoxy type floor in kitchen.	Resilient tile. Good carpet in day room. May have coved sheet vinyl in kitchen.	Composition tile. Day room may have average grade carpet.	Low cost tile. Composition tile in kitchen. Concrete in storage areas.
Walls	Gypsum wallboard and enamel. Large amount of plastic wall cover. Some wood paneling or glass walls in entry and day rooms. 4' ceramic tile wainscot in kitchen. Gypsum wallboard and paint in storage areas.	Gypsum wallboard and enamel. Gypsum wallboard and some plastic wall cover or wood paneling in entry, bedrooms, dining room, day room, and kitchen. Gypsum wallboard and paint in storage areas.	Gypsum wallboard and enamel. Some plastic wall cover or wood paneling in entry and day room. Gypsum wallboard and paint in storage areas.	Gypsum wallboard and texture. Gypsum wallboard and enamel in kitchen. Unfinished gypsum wallboard in storage areas.
Ceilings	Acoustical tile. Washable acoustical tile in kitchen. Gypsum wallboard and paint in storage areas.	Gypsum wallboard and enamel. Acoustical tile in corridor. Open beam ceilings with wood decking in entry and day room. Gypsum wallboard and paint in storage areas.	Gypsum wallboard & enamel or acoustical texture. Open beams with wood decking. Gypsum wallboard and paint in storage areas.	Gypsum wallboard and texture. Gypsum wallboard and acoustical texture in entry. Gypsum wallboard and enamel in kitchen. Unfinished gypsum in storage areas.
Bath Finish Detail **Main Bath** **& Showers**	Ceramic tile floors. Ceramic tile walls. Gypsum wallboard and enamel ceilings.	Ceramic tile floors. 6' ceramic tile wainscot. Gypsum wallboard and enamel walls and ceilings.	Ceramic tile floors. 4' ceramic tile wainscot. Gypsum wallboard and enamel walls and ceilings.	Resilient tile floors. Gypsum wallboard and enamel walls and ceilings.
Toilet Rooms	Sheet vinyl or ceramic tile floors. 4' ceramic tile wainscot. Gypsum wallboard and enamel walls and ceilings.	Sheet vinyl floors. 4' plastic or ceramic tile wainscot.	Resilient tile or linoleum floors. 4' plastic or marlite wainscot.	Composition tile floors. Gypsum wallboard and enamel walls and ceilings.

Square foot costs include the following components: Foundations as required for normal soil conditions. Floor, wall and roof structures. Interior floor, wall and ceiling finishes. Exterior wall finish and roofing. Interior partitions. Cabinets, doors and windows. Basic electrical systems and all plumbing. Permits and fees. Contractors' mark-up. Cost of nurses' built-in station desks and cabinets. Nurses' call system. Emergency lighting system. Cubicle curtain tracks. Kitchen range hood.

The in-place cost of these extra components should be added to the basic building cost to arrive at the total structure cost. See the section "Additional Costs for Commercial Buildings" on page 201. Heating and air conditioning systems. Elevators. Fire sprinklers. Dumbwaiters. Fireplaces. *Square foot costs do not include the following items:* Drapes. Cubicle curtains. Incinerators. Laundry or kitchen equipment, other than the range hood.

Convalescent Hospitals - Masonry or Concrete

Estimating Procedure

1. Use these figures to estimate institutions such as nursing or rest homes which have facilities limited to bed care for people unable to care for themselves.
2. Establish the building quality class by applying the information on page 134.
3. Compute the floor area. This should include everything within the exterior walls and all insets outside the main walls but under the main roof. Areas such as storage rooms which have a substantially inferior finish or any area with a slightly interior finish and which does not perform a function related to the hospital operation should be estimated at 1/2 to 3/4 of the square foot costs listed.
4. Estimate second stories and basements at 100% of the first floor cost if they are of the same general quality. A storage area of the same or slightly inferior quality located in the basement, second floor or separate building should be estimated at 100% of the building square foot cost.
5. Multiply the appropriate square foot cost below by the floor area.
6. Multiply the total cost by the location factor on page 7.
7. Add the cost of heating and air conditioning systems, fire sprinklers, drapes, therapy facilities, cubicle curtains, incinerators, elevators, dumbwaiters, fireplaces, laundry equipment and kitchen equipment other than a range hood. See the section beginning on page 201.

Convalescent Hospital, Class 2

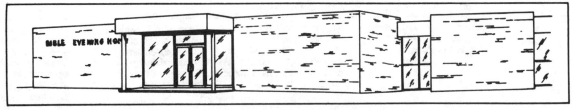

Convalescent Hospital, Class 2

Square Foot Area

Quality Class	5,000	6,000	7,000	8,000	9,000	10,000	12,500	15,000	17,500	20,000	25,000	30,000
Exceptional	115.33	111.71	109.10	107.16	105.65	104.45	102.30	100.87	99.86	99.10	98.06	97.37
1, Best	108.35	104.91	102.48	100.65	99.24	98.11	96.10	94.75	93.80	93.09	92.10	91.45
1 & 2	101.64	98.42	96.13	94.42	93.11	92.05	90.16	88.89	88.01	87.33	86.39	85.80
2, Good	94.48	91.50	89.37	87.78	86.55	85.57	83.81	82.65	81.82	81.19	80.33	79.76
2 & 3	87.25	84.49	82.53	81.06	79.93	79.00	77.39	76.32	75.53	74.97	74.17	73.65
3, Average	81.12	78.56	76.73	75.37	74.32	73.47	71.94	70.96	70.24	69.71	68.96	68.48
3 & 4	74.31	71.97	70.31	69.04	68.08	67.31	65.93	65.00	64.34	63.85	63.18	62.73
4, Low	67.69	65.54	64.02	62.89	61.99	61.29	60.03	59.20	58.61	58.14	57.56	57.13

Convalescent Hospitals - Wood Frame

Quality Classification

	Class 1 Best Quality	Class 2 Good Quality	Class 3 Average Quality	Class 4 Low Quality
Foundation	Reinforced concrete.	Reinforced concrete.	Reinforced concrete.	Reinforced concrete.
Floor Structure	Reinforced concrete.	Reinforced concrete.	Reinforced concrete.	Reinforced concrete.
Wall Structure	Standard wood frame.	Standard wood frame.	Standard wood frame.	Standard wood frame.
Roof	Standard wood frame.	Standard wood frame.	Standard wood frame.	Standard wood frame.
Exterior Finish **Walls**	Wood siding or stucco with extensive trim.	Stucco or wood siding.	Stucco or good plywood siding.	Stucco.
Front	Brick or stone veneer and plate glass in aluminum frame.	Stone or brick veneer. Small amount of plate glass in aluminum frame.	Rustic siding or some brick or stone veneer.	Stucco and small amount siding.
Windows	Good aluminum sliding glass doors or anodized aluminum casement windows.	Good aluminum sliding glass doors with aluminum jalousie window for vent.	Average aluminum sliding type. 6' sliding glass door in bedrooms.	Low cost aluminum sliding type.
Doors	Anodized aluminum entry doors.	Aluminum and glass entry doors.	Good hardwood slab door with side lights or aluminum and glass entry door.	Wood slab door.
Roof	Composition with large colored rock or tile. 4' to 6' sealed overhang.	Composition with large colored rock or asbestos shingles. 3' to 4' sealed overhang.	Composition and small colored rock. 4' exposed or 3' sealed overhang.	Composition and pea gravel or composition shingles. 2' exposed overhang.
Interior Finish **Floors**	Resilient tile. Good carpet in entry and day room. Quarry tile or epoxy type floor in kitchen.	Resilient tile. Good carpet in day room. May have coved sheet vinyl in kitchen.	Composition tile. Day room may have average grade carpet.	Low cost tile. Better resilient tile in kitchen. Concrete in storage areas.
Walls	Gypsum wallboard and enamel. Large amount of plastic wall cover. Some wood paneling or glass walls in entry and day rooms. 4' ceramic tile wainscot in kitchen. Gypsum wallboard and paint in storage areas.	Gypsum wallboard and enamel. Gypsum wallboard and some plastic wall cover or wood paneling in entry, bedrooms, dining room, day room, and kitchen. Gypsum wallboard and paint in storage areas.	Gypsum wallboard and enamel. Some plastic wall cover or wood paneling in entry and day room. Gypsum wallboard and paint in storage areas.	Gypsum wallboard and texture. Gypsum wallboard and enamel in kitchen. Unfinished gypsum wallboard in storage areas.
Ceilings	Acoustical tile. Washable acoustical tile in kitchen. Gypsum wallboard and paint in storage areas.	Gypsum wallboard and enamel. Acoustical tile in corridor. Open beam ceilings with wood decking in entry and day room. Gypsum wallboard and paint in storage areas.	Gypsum wallboard and enamel or acoustical texture. Open beams with wood decking. Gypsum wallboard and paint in storage areas.	Gypsum wallboard and texture. Gypsum wallboard and acoustical texture in entry. Gypsum wallboard and enamel in kitchen. Unfinished gypsum wallboard in storage areas.
Bath Finish Detail **Main Bath** **& Showers**	Ceramic tile floors. Ceramic tile walls. Gypsum wallboard and enamel ceilings.	Ceramic tile floors. 6' ceramic tile wainscot. Gypsum wallboard and enamel walls and ceilings.	Ceramic tile floors. 4' ceramic tile wainscot. Gypsum wallboard and enamel walls and ceilings.	Composition tile floors. Gypsum wallboard and enamel walls and ceilings.
Toilet Rooms	Vinyl or ceramic tile floors. 4' ceramic tile wainscot. Gypsum wallboard and enamel walls and ceilings.	Vinyl floors. 4' plastic or ceramic tile wainscot.	Resilient tile floors. 4' plastic wainscot.	Composition tile floors. Gypsum wallboard and enamel walls and ceilings

Square foot costs include the following components: Foundations as required for normal soil conditions. Floor, wall and roof structures. Interior floor, wall and ceiling finishes. Exterior wall finish and roof cover. Interior partitions. Cabinets, doors and windows. Basic electrical systems and plumbing systems. Permits and fees. Contractors' mark-up. Cost of nurses' built-in station desks and cabinets. Nurses' call system. Emergency lighting system. Cubicle curtain tracks. Kitchen range hood.

The in-place cost of these extra components should be added to the basic building cost to arrive at the total structure cost. See the section "Additional Costs for Commercial Buildings" on page 201. Heating and air conditioning systems. Elevators. Fire sprinklers. Dumbwaiters. Fireplaces. *Square foot costs do not include the following items:* Drapes. Cubicle curtains. Incinerators. Laundry or kitchen equipment, other than the range hood.

Convalescent Hospitals - Wood Frame

Estimating Procedure

1. Use these figures to estimate institutions such as nursing or rest homes which have facilities limited to bed care for people unable to care for themselves.
2. Establish the building quality class by applying the information on page 136.
3. Compute the floor area. This should include everything within the exterior walls and all insets outside the main walls but under the main roof. Areas such as storage rooms which have a substantially inferior finish or any area with a slightly interior finish and which does not perform a function related to the hospital operation should be estimated at 1/2 to 3/4 of the square foot costs listed.
4. Estimate second stories and basements at 100% of the first floor cost if they are of the same general quality. A storage area of the same or slightly inferior quality located in a basement, second floor or separate building should be estimated at 100% of the building square foot cost.
5. Multiply the appropriate square foot cost below by the floor area.
6. Multiply the total cost by the location factor on page 7.
7. Add the cost of heating and air conditioning systems, fire sprinklers, drapes, therapy facilities, cubicle curtains, incinerators, elevators, dumbwaiters, fireplaces, laundry equipment and kitchen equipment other than a range hood. See the section beginning on page 201.

Convalescent Hospital, Class 4

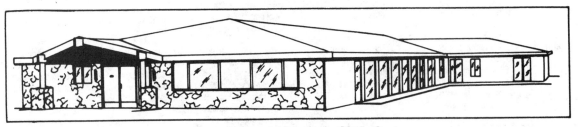

Convalescent Hospital, Class 3

Square Foot Area

Quality Class	5,000	6,000	7,000	8,000	9,000	10,000	12,500	15,000	17,500	20,000	25,000	30,000
Exceptional	103.07	99.84	97.49	95.70	94.30	93.19	91.14	89.76	88.74	87.97	86.88	86.13
1, Best	97.02	93.96	91.76	90.09	88.78	87.71	85.78	84.48	83.54	82.80	81.77	81.08
1 & 2	90.75	87.88	85.81	84.27	83.02	82.04	80.24	79.00	78.13	77.45	76.48	70.22
2, Good	84.58	81.92	79.99	78.54	77.38	76.47	74.79	73.65	72.81	72.19	71.29	70.67
2 & 3	79.00	76.50	74.71	73.34	72.27	71.43	69.84	68.79	68.02	67.42	66.58	66.01
3, Average	72.64	70.36	68.71	67.44	66.47	65.69	64.23	63.25	62.53	62.01	61.24	60.72
3 & 4	65.35	63.31	61.80	60.68	59.80	59.08	57.79	56.90	56.26	55.78	55.08	54.61
4, Low	60.92	58.99	57.59	56.54	55.73	55.08	53.87	53.04	52.43	51.99	51.34	50.89

Funeral Homes

Quality Classification

	Class 1 Best Quality	Class 2 Good Quality	Class 3 High Average Quality	Class 4 Low Average Quality	Class 5 Low Quality
Foundation	Reinforced concrete.	Reinforced concrete.	Reinforced concrete.	Reinforced concrete.	Reinforced concrete.
Floor Structure	Standard wood frame or reinforced concrete.	Standard wood frame or reinforced concrete.	Standard wood frame or reinforced concrete.	Standard wood frame or reinforced concrete.	Standard wood frame or reinforced concrete.
Wall Structure*	Standard wood frame.	Standard wood frame.	Standard wood frame.	Standard wood frame.	Standard wood frame.
Roof	Standard wood frame.	Standard wood frame.	Standard wood frame.	Standard wood frame.	Standard wood frame.
Exterior Finish Walls	Brick or stone veneer.	Good wood siding, brick or stone veneer.	Wood siding or stucco with extensive wood trim.	Stucco or wood siding.	Stucco.
Front	Brick or stone veneer. Some tinted plate glass in bronze frames.	Brick or stone veneer. Some tinted plate glass in anodized aluminum frames.	Brick or stone veneer. Some tinted plate glass in anodized aluminum frames.	Stone or brick veneer. Some tinted crystal or plate glass in wood frame.	Wood siding or stucco with stone brick veneer. Some colored veneer windows with wood frame (bottle glass type).
Windows	Good anodized aluminum windows. Extensive stained glass in chapel.	Good aluminum sliding type. Stained glass in chapel.	Good aluminum sliding type. Leaded glass in chapel.	Good aluminum sliding type. Mock leaded glass in chapel.	Average aluminum sliding type. Colored plastic in chapel.
Doors	Elaborate hardwood entry doors.	Custom-built hardwood entry doors.	Custom hardwood or anodized aluminum and glass.	Good hardwood or anodized aluminum and glass doors.	Good hardwood slab or aluminum and glass entry doors.
Roof Cover	Slate, 3' sealed overhang.	Bar tile, 3' sealed overhang.	Heavy shake, 3' sealed overhang.	Composition and large colored rock, asbestos shingles or medium shake, 3' exposed overhang.	Compositon and small colored rock or good composition shingles, 2' exposed overhang.
Interior Finish Floors	Excellent carpet. Good tile in kitchen.	Excellent carpet. Good carpet in living quarters. Good tile in kitchen.	Good carpet. Good tile in kitchen.	Good carpet. Average carpet in living quarters. Tile in kitchen.	Average carpet. Composition tile in hallways and living quarters.
Walls	Hardwood paneling, extensive brick or stone work in lobby and chapel. Gypsum wallboard and texture in living quarters.	Good hardwood veneer with brick or stone work in lobby and chapel. Gypsum wallboard and texture in living quarters.	Gypsum wallboard and embossed wallpaper or fabric textured wallcover. Hardwood in chapel. Gypsum wallboard and texture in living quarters.	Gypsum wallboard and wallpaper. Hardwood veneer in chapel.	Gypsum wallboard and texture. Some hardwood veneer in chapel.
Ceilings	Acoustical plaster. Exposed ornamental beams with ornamental hardwood decking or ornamental acoustical plaster in chapel.	Acoustical plaster. Exposed ornamental beams with ornamental wood decking or acoustical plaster in chapel.	Gypsum wallboard and heavy acoustical texture. Exposed architectural glu-lams w/ deck in chapel. Gypsum wallboard and texture in living quarters.	Gypsum wallboard and acoustical texture. Exposed wood beam with wood deck in chapel.	Gypsum wallboard and acoustical texture. Exposed beam with wood deck in chapel.
Special Lighting Fixtures	Two expensive chandeliers in lobby (4' to 6' diameter). Recessed and spotlighting in chapel.	Two expensive chandeliers in lobby (4' to 6' diameter). Recessed and spotlighting in chapel.	Two good chandeliers in lobby (4' diameter). Recessed and spotlighting in chapel.	Two average chandeliers in lobby (3' diameter).	One average chandelier in lobby.
Bath Detail	Ceramic tile floors and walls. Gypsum wallboard and enamel ceilings. Excellent fixtures.	Ceramic tile floors. 4' ceramic tile wainscot. Gypsum wallboard and enamel ceilings. Excellent fixtures.	Ceramic tile floors. Gypsum wallboard and enamel walls and ceilings. Good fixtures.	Good tile floors. Gypsum wallboard and enamel walls and ceilings. Average fixtures.	Tile floors. Gypsum wallboard and enamel walls and ceilings. Average fixtures.

*Add 9% to the square foot cost if the exterior walls are concrete block or brick. Add 13% if the exterior walls are concrete.

Square foot costs include the following components: Foundations as required for normal soil conditions. Floor, wall and roof structures. Interior floor, wall and ceiling finishes. Exterior wall finish and roof cover. Interior partitions. Cabinets, doors and windows. Basic electrical system and plumbing system. Special lighting fixtures. Bath detail. Permits and fees. Contractors' mark-up.

Funeral Homes - Wood Frame

Estimating Procedure

1. Use these figures to estimate mortuaries and buildings designed for care of the dead and funeral services.
2. Establish the building quality class by applying the information on page 138.
3. Compute the first floor area under the main roof and inside the exterior walls. Include lobby, chapel and living sections, but exclude garages, whether internal, external or attached.
4. Add to the first floor area between 1/3 and 1/4 of the actual garage area. Use the higher value for better quality garages.
5. Add to the first floor area all finished basement and second floor area if it is similar in quality to the first floor area. Inferior quality second floor area should be reclassified as to quality and calculated separately. Unfinished funeral home basements should be estimated from the figures on page 202.
6. Finished space within the area formed by the roof but with an exterior wall height less than 7 feet (half story area) should be included at 1/3 to 1/2 of the actual floor area. Exclude half story area when selecting a square foot cost from the table.
7. Multiply the sum from steps 3 to 6 above by the appropriate square foot cost below.
8. Add 8% if the exterior walls are brick or block and 12% if the exterior walls are concrete. Add 2% if the building has more than 16 corners or if the length is more than twice with width.
9. Multiply the building cost by the location factor listed on page 7.
10. Add the cost of heating and cooling systems, elevators, fireplaces, porches, pews, drapes, signs and yard improvement. See the section beginning on page 201.

Funeral Home, Class 5

Square Foot Area

Quality	2,000	2,500	3,000	4,000	5,000	7,500	10,000	12,500	15,000	20,000	25,000	30,000
1, Best	--	--	--	--	90.55	87.25	85.51	84.40	83.62	82.59	81.93	81.47
1 & 2	--	--	--	--	84.31	81.26	79.62	78.58	77.86	76.90	76.29	75.86
2, Good	--	--	--	--	78.75	75.90	74.37	73.40	72.74	71.84	71.27	70.86
2 & 3	--	--	81.91	76.73	73.93	69.85	69.53	68.97	68.75	68.60	68.69	68.80
2, High	--	--	76.66	71.81	69.20	65.36	65.08	64.56	64.33	64.21	64.27	64.40
3 & 4	78.57	74.37	71.54	68.00	65.85	62.95	61.47	60.57	59.96	59.20	58.73	58.41
4, Low	73.93	69.97	67.32	63.98	61.96	59.23	57.85	56.99	56.43	55.70	55.27	54.96
4 & 5	69.38	65.67	63.17	60.05	58.14	55.58	54.28	53.48	52.94	52.27	51.86	51.58
5, Low	64.37	60.93	58.61	55.70	53.95	51.57	50.37	49.62	49.13	48.50	48.11	47.84

Self Service Restaurants

Quality Classification

	Class 1 Best Quality	Class 2 Good Quality	Class 3 Average Quality	Class 4 Low Quality
Foundation	Reinforced concrete.	Reinforced concrete.	Reinforced concrete.	Reinforced concrete.
Floor Structure	Reinforced concrete or standard wood frame.	Reinforced concrete or standard wood frame.	Reinforced concrete or standard wood frame.	Reinforced concrete or standard wood frame.
Walls, *Wood Frame Construction*	Standard wood frame.	Standard wood frame.	Standard wood frame.	Standard wood frame.
Walls, *Masonry or Concrete Const.*	8" reinforced decorative concrete block or brick.	8" reinforced decorative concrete block or brick.	8" reinforced concrete block.	8" reinforced concrete block.
Roof Structure	Complex angular structure.	Standard wood frame structure, low or medium pitch.	Standard wood frame structure, low pitch shed or gable type.	Standard wood frame structure, low pitch shed or gable type.
Roof Cover	Concrete tile, mission tile, or heavy shake.	Composition tar and gravel or medium shake.	Composition tar and gravel.	Composition or composition shingle.
Floor Finish	Terrazzo in dining areas. Quarry tile in kitchen.	Vinyl tile in dining area. Terrazzo or quarry tile in kitchen.	Resilient tile in dining area. Grease proof tile in kitchen.	Composition tile in dining area. Grease proof tile in kitchen.
Walls, *Exterior Finish*	Very good wood siding; ornamental brick or natural stone veneer. Minimum 60% of front and side walls have plate glass in heavy metal frames or rabbeted timbers.	Good wood siding or plywood; ornamental brick or concrete block; good stucco with color finish and with 2' to 4' brick, or natural stone veneer bulkheads. 50% to 75% of front and side walls have plate glass set in good wood or metal frames.	Stucco or average wood siding with 2' to 4' brick or stone veneer bulkheads. Plate glass front from waist high to roof line running along service front. Glassed area from 40% to 70% of front and side walls.	Stucco, low cost wood siding or plywood. 4' to 6' fixed glass waist high with sliding aluminum window opening running along service front. Glassed area from 30% to 40% of front and side walls.
Interior Finish	Plaster with putty coat finish; plastic coated paneling.	Plaster with putty coat finish; plastic coated paneling.	Plaster putty coat finish; textured gypsum wallboard with portions of wall finished with natural wood or plastic coated veneers.	Painted plaster or gypsum wallboard.
Ceilings	Complex angular structure with acoustical plaster.	Multi-level structure with good plaster or acoustical tile.	Flat or low slope with textured gypsum wallboard or acoustical tile.	Flat or low slope with textured gypsum wallboard or acoustical tile.
Plumbing	Good fixtures with metal or marble toilet partitions. Full wall height ceramic tile wainscot. Numerous floor drains.	Commercial fixtures with metal toilet partitions. Ceramic tile wainscot. Numerous floor drains in kitchen and dishwasher area.	4 to 6 standard commercial fixtures with metal toilet partitions. Floor drains in kitchen and dishwashing area.	4 to 6 standard commercial fixtures with wood toilet partitions. Minimum floor drainage in kitchen area.
Lighting	Triple encased fluorescent fixtures or decorative and ornate incandescent fixtures.	Triple encased fluorescent fixtures or decorative incandescent fixtures.	Triple open or double encased louvered fluorescent fixtures with some decorative incandescent fixtures.	Double open strip fluorescent fixtures or low cost incandescent fixtures.

Square foot costs include the following components: Foundations as required for normal soil conditions. Floor, wall and roof structures. Interior floor, wall and ceiling finishes. Exterior wall finish and roof cover. All glass and glazing. Interior partitions. Basic electrical systems and lighting fixtures. Rough and finish plumbing. Permits and fees. Contractors' mark-up.

The in-place cost of these extra components should be added to the basic building cost to arrive at the total structure cost. See the section beginning on page 201. Heating and air conditioning systems. Booths and counters. Kitchen equipment. Fire sprinklers. Exterior signs. Paving and curbing. Yard improvements. Canopies and overhang.

Self Service Restaurants - Masonry

Estimating Procedure

1. Use these figures to estimate the cost of self-service or drive-in restaurants. These restaurants specialize in rapid service, may or may not have an interior eating area, have a large amount of glass on the front and side walls and usually are well lighted.
2. Establish the structure quality class by applying the information on page 140.
3. Compute the building floor area. This should include everything within the exterior walls and all insets outside the main walls but under the main roof.
4. Multiply the floor area by the appropriate cost below.
5. Multiply the total cost by the factor listed on page 7.
6. Add the cost of heating and air conditioning systems, booths and counters, kitchen equipment, fire sprinklers, exterior signs, paving and curbing, yard improvements and canopies. See the section beginning on page 201.

Self Service Restaurant, Class 2

Self Service Restaurant, Class 3

Square Foot Area

Quality Class	400	500	600	700	800	900	1,000	1,100	1,200	1,300	1,400
1, Best	249.59	230.40	216.11	204.90	195.82	188.24	181.81	176.25	171.40	167.12	163.26
1 & 2	215.93	199.35	186.96	177.26	169.40	162.86	157.30	152.47	148.29	144.56	141.25
2, Good	186.79	172.46	161.74	153.36	146.53	140.88	136.07	131.93	128.27	125.06	122.19
2 & 3	161.57	149.18	139.91	132.66	126.76	121.87	117.70	114.11	110.97	108.17	105.70
3, Average	139.78	129.05	121.04	114.76	109.67	105.42	101.84	98.72	96.00	93.59	91.43
3 & 4	120.91	111.63	104.69	99.26	94.86	91.20	88.08	85.39	83.03	80.95	79.08
4, Low	104.59	96.55	90.55	85.87	82.05	78.87	76.19	73.86	71.82	70.01	68.41

Square Foot Area

Quality Class	1,500	1,600	1,700	1,800	2,000	2,200	2,400	2,600	2,800	3,000	3,400
1, Best	160.21	158.21	156.31	154.52	150.63	148.24	145.54	143.08	140.82	138.75	135.05
1 & 2	138.66	136.91	135.28	133.73	130.38	128.30	125.96	123.83	121.87	120.09	116.88
2, Good	119.88	118.38	116.97	115.63	112.72	110.95	108.92	107.07	105.39	103.84	101.06
2 & 3	103.71	102.41	101.19	100.03	97.52	95.97	94.21	92.63	91.16	89.83	87.43
3, Average	89.71	88.59	87.54	86.53	84.36	83.02	81.49	80.12	78.86	77.70	75.63
3 & 4	77.62	76.64	75.72	74.85	72.97	71.81	70.52	69.31	68.23	67.22	65.42
4, Low	67.13	66.30	65.49	64.75	63.13	62.11	61.00	59.97	59.01	58.14	56.59

Self Service Restaurants - Wood Frame

Estimating Procedure

1. Use these figures to estimate the cost of self-service or drive-in restaurants. These restaurants specialize in rapid service, may or may not have an interior eating area, have a large amount of glass on the front and side walls and usually are well lighted.
2. Establish the structure quality class by applying the information on page 140.
3. Compute the building floor area. This should include everything within the exterior walls and all insets outside the main walls but under the main roof.
4. Multiply the floor area by the appropriate cost below.
5. Multiply the total cost by the factor listed on page 7.
6. Add the cost of heating and air conditioning systems, booths and counters, kitchen equipment, fire sprinklers, exterior signs, paving and curbing, yard improvements and canopies. See the section beginning on page 201.

Self Service Restaurant, Class 1 & 2

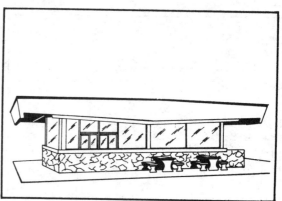

Self Service Restaurant, Class 1 & 2

Square Foot Area

Quality Class	400	500	600	700	800	900	1,000	1,100	1,200	1,300	1,400
1, Best	254.08	234.56	219.99	208.58	199.33	191.62	185.09	179.43	174.49	170.13	166.21
1 & 2	219.80	202.91	190.31	180.43	172.43	165.78	160.10	155.24	150.94	147.17	143.79
2, Good	190.07	175.46	164.57	156.01	149.11	143.36	138.46	134.24	130.53	127.27	124.35
2 & 3	164.32	151.68	142.28	134.89	128.91	123.92	119.69	116.04	112.85	110.02	107.51
3, Average	141.98	131.06	122.93	116.56	111.38	107.08	103.42	100.28	97.51	95.05	92.88
3 & 4	122.96	113.50	106.47	100.95	96.44	92.73	89.56	86.84	84.46	82.32	80.43
4, Low	106.30	98.13	92.03	87.27	83.39	80.18	77.45	75.07	73.01	71.18	69.54

Square Foot Area

Quality Class	1,500	1,600	1,700	1,800	2,000	2,200	2,400	2,600	2,800	3,000	3,400
1, Best	163.78	161.29	159.04	156.99	153.37	150.29	147.65	145.32	143.26	141.42	138.26
1 & 2	141.54	139.38	137.43	135.66	132.53	129.89	127.60	125.58	123.80	122.22	119.48
2, Good	122.52	120.64	118.97	117.42	114.73	112.43	110.44	108.71	107.16	105.78	103.43
2 & 3	105.95	104.33	102.88	101.54	99.22	97.23	95.51	93.99	92.66	91.47	89.43
3, Average	91.61	90.23	88.96	87.82	85.80	84.08	82.60	81.30	80.13	79.10	77.35
3 & 4	79.26	78.05	76.96	75.97	74.21	72.73	71.45	70.33	69.30	68.45	66.91
4, Low	68.53	67.48	66.55	65.68	64.17	62.89	61.77	60.81	59.95	59.18	57.86

Coffee Shop Restaurants

Quality Classification

	Class 1 Best Quality	Class 2 Good Quality	Class 3 Average Quality	Class 4 Low Quality
Foundation	Reinforced concrete.	Reinforced concrete.	Reinforced concrete.	Reinforced concrete.
Floor Structure	Reinforced concrete or standard wood frame.	Reinforced concrete or standard wood frame.	Reinforced concrete or standard wood frame.	Reinforced concrete or standard wood frame.
Walls, *Wood Frame Construction*	Standard wood frame.	Standard wood frame.	Standard wood frame.	Standard wood frame.
Walls, *Masonry or Concrete Construction*	8" reinforced decorative concrete block or brick.	8" reinforced decorative concrete block or brick.	8" reinforced concrete block.	8" reinforced concrete block.
Roof Structure	Complex angular structure.	Standard wood frame structure, low or medium pitch.	Standard wood frame, low pitch shed or gable. type.	Standard wood frame, low pitch shed or gable type.
Roof Cover	Concrete tile, mission tile, or heavy shake.	Composition tar and gravel or medium shake.	Composition tar and gravel.	Composition or composition shingle.
Floor Finish	Terrazzo or good carpeting in dining area. Quarry tile in kitchen. Terrazzo or natural flagstone in entry vestibule.	Good tile or average carpeting in dining area. Terrazzo or quarry tile in kitchen. Terrazzo entry vestibule.	Good tile in dining area. Grease proof tile in kitchen area. Terrazzo in entry vestibule.	Composition tile in dining area. Grease proof tile in kitchen area.
Exterior Wall Finish	Very good wood siding; ornamental brick or natural stone veneer. Minimum 60% of front and side walls in plate glass in heavy metal frames or rabbeted timbers. Entry vestibule with decorative screens of ornamental concrete block or natural stone panels.	Good wood siding; ornamental brick or concrete blocks. Good stucco with color finish and with 2' to 4' brick, or natural stone veneer bulkheads. 50% to 75% of front and side walls have plate glass set in good wood or metal frames. Double entry doors, recessed entries with decorative screen of concrete block.	Stucco or average wood siding with 2' to 4' brick or stone veneer bulkheads. 50% to 60% of front and side walls in plate glass set in average quality wood or aluminum frames. Double entry doors of metal or wood.	Stucco, low cost wood siding or plywood. 40% to 60% of front and side walls in crystal or plate glass set in low-cost wood or metal frames. Double entry doors.
Interior Finish	Select and matched wood paneling, decorative wallpaper or synthetic fabrics.	Plaster with putty coat finish. Wood or plastic coated veneer paneling. Decorative wallpapers or fabrics.	Plaster putty coat finish textured gypsum wallboard with portions of wall finished with natural wood or plastic coated veneers.	Painted plaster or gypsum wallboard.
Ceilings	Complex angular structure with acoustical plaster.	Multi-level structure with good plaster or acoustical tile.	Flat or low slope with textured gypsum wallboard or acoustical tile.	Flat or low slope with textured gypsum wallboard or acoustical tile.
Plumbing	Good fixtures with metal or marble toilet partitions. Full wall height ceramic tile wainscot. Numerous floor drains.	Commercial fixtures with metal toilet partitions; ceramic tile wainscot. Numerous floor drains in kitchen and dishwasher area.	4 to 6 standard commercial fixtures with metal toilet partitions. Floor drains in kitchen and dishwashing area.	4 to 6 standard commercial fixtures with wood toilet partitions. Minimum floor drainage in kitchen.
Lighting	Triple encased fluorescent fixtures or decorative and ornate incandescent fixtures.	Triple encased fluorescent fixtures or decorative incandescent fixtures.	Triple open or double encased louvered fluorescent fixtures with some decorative incandescent fixtures.	Double open strip fluorescent fixtures or low cost incandescent fixtures.

Square foot costs include the following components: Foundations as required for normal soil conditions. Floor, wall and roof structures. Interior floor, wall and ceiling finishes. Exterior wall finish and roof cover. All glass and glazing. Interior partitions. Basic electrical systems and lighting fixtures. Rough and finish plumbing. Permits and fees. Contractors' mark-up.

Coffee Shop Restaurants - Masonry

Estimating Procedure

1. Use these figures to estimate the cost of restaurants with counters, or with counter, booth and table seating. A lounge or bar may be part of the building. Higher quality coffee shops usually have cut-up or complex sloping ceiling and roof structures. Coffee shops have extensive exterior glass and good lighting.
2. Establish the structure quality class by applying the information on page 143.
3. Compute the building floor area. This should include everything within the exterior walls and all insets outside the main walls but under the main roof.
4. Multiply the floor area by the appropriate cost below.
5. Multiply the total cost by the factor listed on page 7.
6. Add the cost of heating and air conditioning systems, booths and counters, kitchen equipment, fire sprinklers, exterior signs, paving and curbing, yard improvements and canopies. See the section beginning on page 201.

Coffee Shop Restaurant, Class 1

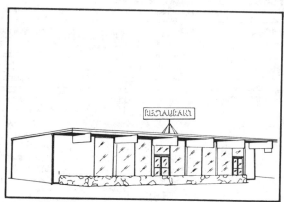

Coffee Shop Restaurant, Class 3

Square Foot Area

Quality Class	1,000	1,200	1,400	1,600	1,800	2,000	2,200	2,400	2,600	2,800	3,000
1, Best	172.78	159.96	150.70	143.65	138.13	133.63	129.93	126.81	124.15	121.86	119.84
1 & 2	149.73	138.64	130.59	124.49	119.68	115.80	112.58	109.87	107.59	105.59	103.84
2, Good	129.31	119.74	112.79	107.52	103.37	100.01	97.24	94.91	92.91	91.19	89.69
2 & 3	112.03	103.72	97.71	93.14	89.55	86.64	84.24	82.22	80.49	79.01	77.70
3, Average	96.78	89.61	84.42	80.47	77.36	74.85	72.78	71.02	69.53	68.24	67.13
3 & 4	83.72	77.52	73.03	69.62	66.94	64.76	62.97	61.46	60.17	59.06	58.07
4, Low	72.41	67.04	63.17	60.20	57.89	56.00	54.44	53.14	52.03	51.07	50.22

Square Foot Area

Quality Class	3,200	3,400	3,600	3,800	4,000	4,400	4,800	5,200	5,600	6,000	8,000
1, Best	116.68	115.38	114.22	113.13	112.13	110.38	108.83	107.47	106.25	105.18	101.05
1 & 2	100.93	99.83	98.80	97.87	97.00	95.48	94.14	92.96	91.92	90.97	87.42
2, Good	87.39	86.43	85.55	84.75	84.01	82.68	81.51	80.49	79.59	78.79	75.69
2 & 3	75.63	74.77	74.02	73.32	72.67	71.53	70.52	69.64	68.86	68.16	65.49
3, Average	65.42	64.69	64.03	63.43	62.87	61.87	61.01	60.25	59.57	58.96	56.65
3 & 4	56.58	55.97	55.39	54.88	54.39	53.54	52.79	52.12	51.53	51.01	49.02
4, Low	48.94	48.39	47.91	47.46	47.04	46.29	45.66	45.08	44.56	44.13	42.39

Coffee Shop Restaurants - Wood Frame

Estimating Procedure

1. Use these figures to estimate the cost of restaurants with counters, or with counter, booth and table seating. A lounge or bar may be part of the building. Higher quality coffee shops usually have cut-up or complex sloping ceiling and roof structures. Coffee shops have extensive exterior glass and good lighting.
2. Establish the structure quality class by applying the information on page 143.
3. Compute the building floor area. This should include everything within the exterior walls and all insets outside the main walls but under the main roof.
4. Multiply the floor area by the appropriate cost below.
5. Multiply the total cost by the factor listed on page 7.
6. Add the cost of heating and air conditioning systems, booths and counters, kitchen equipment, fire sprinklers, exterior signs, paving and curbing, yard improvements and canopies. See the section beginning on page 201.

Coffee Shop Restaurant, Class 3

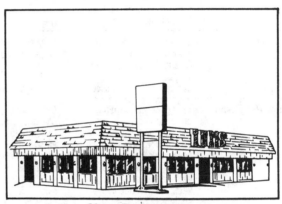

Coffee Shop Restaurant, Class 2

Square Foot Area

Quality Class	1,000	1,200	1,400	1,600	1,800	2,000	2,200	2,400	2,600	2,800	3,000
1, Best	179.44	166.03	156.35	149.01	143.26	138.62	134.77	131.55	128.81	126.44	124.37
1 & 2	155.35	143.74	135.36	129.02	124.04	120.02	116.69	113.91	111.52	109.46	107.70
2, Good	134.35	124.33	117.06	111.57	107.25	103.77	100.91	98.49	96.45	94.67	93.12
2 & 3	116.16	107.49	101.20	96.46	92.73	89.72	87.24	85.17	83.38	81.85	80.51
3, Average	100.45	92.95	87.54	83.43	80.21	77.59	75.45	73.65	72.11	70.78	69.63
3 & 4	86.88	80.38	75.69	72.14	69.35	67.10	65.26	63.69	62.37	61.22	60.21
4, Low	75.10	69.49	65.43	62.37	59.96	58.00	56.40	55.06	53.91	52.92	52.05

Square Foot Area

Quality Class	3,200	3,400	3,600	3,800	4,000	4,400	4,800	5,200	5,600	6,000	8,000
1, Best	121.25	119.91	118.71	117.59	116.56	114.72	113.13	111.71	110.46	109.32	105.01
1 & 2	104.88	103.73	102.69	101.73	100.84	99.25	97.86	96.64	95.55	94.57	90.85
2, Good	90.69	89.69	88.81	87.97	87.19	85.82	84.62	83.56	82.63	81.78	78.57
2 & 3	78.43	77.58	76.80	76.07	75.40	74.22	73.20	72.26	71.46	70.73	67.96
3, Average	67.81	67.08	66.41	65.77	65.21	64.18	63.28	62.48	61.78	61.16	58.75
3 & 4	58.65	58.02	57.42	56.89	56.39	55.48	54.73	54.04	53.43	52.89	50.80
4, Low	50.68	50.13	49.63	49.16	48.74	47.99	47.30	46.72	46.18	45.71	43.92

Conventional Restaurants

Quality Classification

	Class 1 Best Quality	Class 2 Good Quality	Class 3 Average Quality	Class 4 Low Quality
Foundation	Reinforced concrete.	Reinforced concrete.	Reinforced concrete.	Reinforced concrete.
Floor Structure	Reinforced concrete or standard wood frame.	Reinforced concrete or standard wood frame.	Reinforced concrete or standard wood frame.	Reinforced concrete or standard wood frame.
Walls, *Wood Frame Construction*	Standard wood frame.	Standard wood frame.	Standard wood frame.	Standard wood frame.
Walls, *Masonry or Concrete Construction*	Reinforced decorative concrete block or brick.	Reinforced decorative concrete block or brick.	Reinforced concrete block.	Reinforced concrete block.
Roof Structure	Complex angular structure.	Standard wood frame structure, low or medium pitch.	Standard wood frame, low pitch shed or gable type.	Standard wood frame, low pitch shed or gable type.
Roof Cover	Concrete tile, mission tile, or heavy shake.	Composition tar and gravel or medium shake.	Composition tar and gravel.	Composition or composition shingle.
Floor Finish	Terrazzo or good carpeting in dining area. Quarry tile in kitchen. Terrazzo or natural flagstone in entry vestibule.	Vinyl tile or average carpeting in dining area. Terrazzo or quarry tile in kitchen. Terrazzo in entry vestibule.	Good tile in dining area. Grease proof tile in kitchen area. Terrazzo in entry vestibule.	Composition tile in dining area. Grease proof tile in kitchen area.
Exterior Wall Finish	Very good wood siding, ornamental brick or natural stone veneer. 10% of front and side walls in plate glass in heavy metal frames or rabbeted timbers. Entry vestibule with decorative screens of ornamental concrete block or natural stone panels.	Good wood siding or plywood; ornamental brick or concrete block. Good stucco with color finish and with 2' to 4' brick, or natural stone veneer bulkheads. 10% to 20% of front and side walls in plate glass set in good wood or metal frames. Double entry doors, recessed entries with decorative screen of concrete block.	Stucco or average wood siding with 2' to 4' brick or stone veneer bulkheads. 10% to 20% of front and side walls in plate glass set in wood or aluminum extrusions. Double entry doors of metal or wood.	Stucco, low cost wood siding or plywood. 10% to 20% of front & side walls in crystal plate glass set in low-cost wood or metal frames. Double entry doors.
Interior Finish	Select and matched wood paneling, decorative wallpaper or synthetic fabrics.	Plaster with putty coat finish; wood or plastic coated veneer paneling; decorative wallpapers or fabrics.	Plaster putty coat finish; textured gypsum wallboard with portions of wall finished with natural wood or plastic coated veneer.	Painted plaster or gypsum wallboard.
Ceilings	Multi-level structure with acoustical plaster.	Multi-level structure with good plaster or acoustical tile.	Flat or low slope with textured gypsum wallboard or acoustical tile.	Flat or low slope with textured gypsum wallboard or acoustical tile
Plumbing	Good fixtures with metal or marble toilet partitions. Full wall height ceramic tile wainscot. Numerous floor drains.	Commercial fixtures with metal toilet partitions. Ceramic tile wainscot. Numerous floor drains in kitchen and dishwasher area.	4 to 6 standard commercial fixtures with metal toilet partitions. Floor drains in kitchen and dishwashing area.	4 to 6 standard commercial fixtures with wood toilet partitions. Minimum floor drainage in kitchen area.
Lighting	Decorative and ornate incandescent fixtures.	Decorative incandescent fixtures.	Double encased louvered fluorescent fixtures with decorative incandescent fixtures.	Double open strip fluorescent fixtures or low cost incandescent fixtures.

Square foot costs include the following components: Foundations as required for normal soil conditions. Floor, wall and roof structures. Interior floor, wall and ceiling finishes. Exterior wall finish and roof cover. All glass and glazing. Interior partitions. Basic electrical systems and lighting fixtures. Rough and finish plumbing. Permits and fees. Contractors' mark-up.

Conventional Restaurants - Masonry or Wood Frame

Estimating Procedure

1. Use these figures to estimate the cost of restaurants with limited use of glass as exterior walls and subdued lighting. Roof and ceiling structures are simple and conventional. Seating may be any combination of counter, stools, booths or tables. A lounge or bar is often a part of a building of this type.
2. Establish the structure quality class by applying the information on page 146.
3. Compute the building floor area. This should include everything within the exterior walls and all insets outside the main walls but under the main roof.
4. Multiply the floor area by the appropriate cost below.
5. Multiply the total cost by the factor listed on page 7.
6. Add the cost of heating and air conditioning systems, booths and counters, kitchen equipment, fire sprinklers, exterior signs, paving and curbing, yard improvements and canopies. See the section beginning on page 201.

Conventional Restaurant, Class 2

Masonry
Square Foot Area

Quality Class	1,000	1,500	2,000	2,500	3,000	3,500	4,000	5,000	6,000	8,000	10,000
1, Best	168.39	150.59	140.66	134.15	129.52	126.01	123.22	119.07	116.09	111.99	109.28
1 & 2	145.65	130.25	121.65	116.04	112.01	108.98	106.58	102.97	100.40	96.86	94.53
2, Good	125.98	112.66	105.23	100.37	96.90	94.28	92.19	89.10	86.86	83.80	81.76
2 & 3	108.95	97.43	91.01	86.80	83.80	81.52	79.74	77.04	75.10	72.47	70.71
3, Average	94.21	84.26	78.69	75.07	72.47	70.49	68.94	66.62	64.97	62.67	61.15
3 & 4	81.47	72.84	68.05	64.92	62.67	60.96	59.62	57.60	56.16	54.18	52.87
4, Low	70.46	63.01	58.85	56.12	54.19	52.72	51.56	49.80	48.56	46.85	45.74

Wood Frame
Square Foot Area

Quality Class	1,000	1,500	2,000	2,500	3,000	3,500	4,000	5,000	6,000	8,000	10,000
1, Best	173.08	154.78	144.57	137.89	133.13	129.51	126.64	122.39	119.33	115.11	112.33
1 & 2	149.75	133.89	125.03	119.29	115.16	112.02	109.54	105.86	103.21	99.58	97.15
2, Good	129.48	115.79	108.15	103.17	99.58	96.88	94.74	91.57	89.26	86.12	84.04
2 & 3	112.01	100.15	93.54	89.22	86.14	83.81	81.95	79.19	77.21	74.50	72.68
3, Average	96.87	86.61	80.89	77.17	74.50	72.48	70.88	68.47	66.77	64.42	62.87
3 & 4	83.75	74.90	69.96	66.73	64.42	62.68	61.30	59.23	57.76	55.72	54.35
4, Low	72.41	64.76	60.49	57.70	55.71	54.19	52.98	51.22	49.94	48.16	47.00

"A-Frame" Restaurants

Quality Classification

	Class 1 Best Quality	Class 2 Good Quality	Class 3 Average Quality	Class 4 Low Quality
Foundation	Reinforced concrete.	Reinforced concrete.	Reinforced concrete.	Reinforced concrete.
Floor Structure	Reinforced concrete or standard wood frame.	Reinforced concrete or standard wood frame.	Reinforced concrete or standard wood frame.	Reinforced concrete or standard wood frame.
Roof Structure	Wood frame or post and beam, high pitch.	Wood frame or post and beam, high pitch.	Standard wood frame or post and beam, high pitch.	Standard wood frame or post and beam, high pitch.
Roof Cover	Concrete tile, heavy wood shake.	Aluminum shake or medium wood shake.	Wood or composition shingles.	Composition or composition shingles.
Floor Finish	Terrazzo or good carpeting in dining area. Quarry tile in kitchen. Terrazzo or natural flagstone in entry vestibule.	Vinyl tile or average carpeting in dining area Terrazzo or quarry tile in kitchen, terrazzo in entry vestibule.	Good tile in dining area. Grease proof tile in kitchen area. Terrazzo in entry vestibule.	Composition tile in dining area. Grease proof tile in kitchen area.
Interior Finish	Select and matched wood paneling, decorative wallpapers or synthetic fabrics.	Plaster with putty coat finish, wood or plastic coated veneer paneling, decorative wallpapers or fabrics.	Plaster, putty coat finish; textured gypsum wallboard with portions of wall finished with natural wood or plastic coated veneer.	Painted plaster or gypsum wallboard.
End Walls	Very good wood siding; ornamental brick or natural stone veneer. 60% on front in plate glass in heavy metal extrusions or rabbeted timbers. Entry vestibule with decorative screens of ornamental concrete block or natural stone panels.	Good wood siding; ornamental brick or concrete blocks. Good stucco with color finish and with 2' to 4' brick or natural stone veneer bulkheads. 50% of front in plate glass set in good wood or metal frames. Double entry doors, recessed entries with decorative screen of concrete block.	Stucco or average wood siding with 2' to 4' brick or stone veneer bulkheads. 50% to 60% of front in plate glass set in average quality wood or aluminum frames. Double entry doors of metal or wood.	Stucco, low cost wood siding or plywood. 40% to 60% of front in crystal glass set in low cost wood or metal frames.
Ceilings	Decorative wood.	Natural wood, good plaster or acoustical tile.	Painted wood, textured gypsum wallboard or tile.	Painted wood, textured gypsum wallboard or acoustical tile.
Plumbing	Good fixtures with metal or marble toilet partitions. Full wall height ceramic tile wainscot. Numerous floor drains.	Commercial fixtures with metal toilet partitions. Ceramic tile wainscot. Numerous floor drains in kitchen and dishwashing area.	4 to 6 standard commercial fixtures with metal toilet partitions. Floor drains in kitchen and dishwashing area.	4 to 6 standard commercial fixtures with wood toilet partitions. Minimum floor drainage in kitchen area.
Lighting	Triple encased fluorescent fixtures or decorative and ornate incandescent fixtures.	Triple encased fluorescent or decorative incandescent fixtures.	Triple open, double encased louvered fluorescent fixtures or average incandescent fixtures.	Double open strip fluorescent fixtures or low cost incandescent fixtures.

Square foot costs include the following components: Foundations as required for normal soil conditions. Floor, wall and roof structures. Interior floor, wall and ceiling finishes. Exterior wall finish and roof cover. All glass and glazing. Interior partitions. Basic electrical systems and lighting fixtures. Rough and finish plumbing. Permits and fees. Contractors' mark-up.

The in-place cost of these extra components should be added to the basic building cost to arrive at the total structure cost. See the section beginning on page 201: Heating and air conditioning systems. Booths and counters. Kitchen equipment. Fire sprinklers. Exterior signs. Paving and curbing. Yard improvements. Canopies and overhang.

"A-Frame" Restaurants - Wood Frame

Estimating Procedure

1. Use these figures to estimate the cost of restaurants with a sloping roof that forms two or more exterior walls. Either conventional or self service may be used and a lounge or bar may be part of the building.
2. Establish the structure quality class by applying the information on page 148.
3. Compute the building floor area. This should include everything within the exterior walls and all insets outside the main walls but under the main roof.
4. Multiply the floor area by the appropriate cost below.
5. Multiply the total cost by the factor listed on page 7.
6. Add the cost of heating and air conditioning systems, booths and counters, kitchen equipment, fire sprinklers, exterior signs, paving and curbing, yard improvements and canopies. See the section beginning on page 201.

"A-Frame" Restaurant, Class 1

Square Foot Area

Quality Class	400	500	600	800	1,000	1,200	1,400	1,600	1,800	2,000	2,400
1, Best	228.64	203.28	186.16	164.41	151.11	142.08	135.53	130.55	126.60	123.44	118.60
1 & 2	197.74	175.84	161.03	142.22	130.70	122.88	117.22	112.90	109.51	106.76	102.56
2, Good	171.05	152.09	139.29	123.01	113.05	106.30	101.39	97.67	94.73	92.35	88.74
2 & 3	147.97	131.56	120.50	106.41	97.80	91.95	87.72	84.50	81.94	79.89	76.75
3, Average	128.00	113.81	104.23	92.06	84.59	79.55	75.88	73.09	70.89	69.11	66.40
3 & 4	110.58	98.34	90.04	79.53	73.11	68.72	65.55	63.14	61.25	59.71	57.36
4, Low	95.67	85.06	77.91	68.80	63.23	59.45	56.72	54.62	52.98	51.65	49.63

Square Foot Area

Quality Class	2,800	3,200	3,600	4,000	4,500	5,000	6,000	7,000	8,000	9,000	10,000
1, Best	115.11	112.49	110.38	108.63	106.83	217.66	103.00	101.25	99.87	98.74	97.82
1 & 2	99.57	97.29	95.47	93.96	92.39	91.12	89.08	87.57	86.39	85.40	84.61
2, Good	86.14	84.17	82.58	81.28	79.93	78.82	77.07	75.75	74.71	73.88	73.18
2 & 3	74.43	72.87	71.51	70.37	69.20	68.24	66.73	65.59	64.70	63.96	63.36
3, Average	64.45	62.98	61.80	60.82	59.81	58.99	57.68	56.68	55.91	55.30	54.76
3 & 4	55.69	54.42	53.38	52.54	51.68	50.94	49.83	48.97	48.30	47.77	47.30
4, Low	48.22	47.12	46.23	45.51	44.74	44.12	43.15	42.41	41.83	41.37	40.97

Theaters - Masonry or Concrete

Quality Classification

	Class 1 Best w/ Balcony	Class 2 Very Good w/Balcony	Class 3 Good Quality	Class 4 Average Quality	Class 5 Low Quality
Foundation	Reinforced concrete.	Reinforced concrete.	Reinforced concrete.	Reinforced concrete.	Reinforced concrete.
Floor Structure	6" reinforced concrete on 6" rock fill or 2" x 12" joists, 16" o.c.	4" to 6" reinforced concrete on 6" rock fill or 2" x 10" joists, 16" o.c.	4" reinforced concrete on 6" rock fill or 2" x 10" joists, 16" o.c.	4" reinforced concrete on 6" rock fill or 2" x 8" joists, 16" o.c.	4" reinforced concrete on 6" rock fill or 2" x 6" joists, 16" o.c.
Wall Structure	8" reinforced concrete or 12" common brick.	8" reinforced concrete or 12" common brick.	8" reinforced concrete or 12" common brick.	8" reinforced concrete or 12" common brick.	8" reinforced concrete block or 6" reinforced concrete.
Roof Framing	2" x 10" joists 16" o.c. Trusses on heavy pilasters 20' o.c. 2" x 12" rafters or purlins 16" o.c.	2" x 10" joists 16" o.c. Trusses on heavy pilasters 20' o.c. 2" x 12" rafters or purlins 16" o.c.	2" x 10" joists 16" o.c. Trusses on heavy pilasters 20' o.c. 2" x 12" rafters or purlins 16" o.c.	2" x 10" joists 16" o.c. Trusses on heavy pilasters 20' o.c. 2" x 10" rafters or purlins.	2" x 10" joists 16" o.c. Trusses on pilasters 20' o.c. 2" x 10" rafters or purlins 16" o.c.
Roof Covering	5 ply composition roof on 1" sheathing with insulation.	5 ply composition roof on 1" sheathing with insulation.	5 ply composition roof on 1" sheathing with insulation.	5 ply composition roof on 1" sheathing with insulation.	4 ply composition roof on 1" sheathing.
Front	Highly ornamental stucco or plaster finishes, custom select stone veneers or highly ornamental terra cotta	Ornamental stucco, custom select brick, natural stone veneers or marble.	Highly ornamental stucco or custom brick or natural stone or terra cotta veneers.	Ornamental stucco or select brick veneers or partial terra cotta veneers.	Plain or colored stucco or common brick or carrara glass veneers.
Floors, Entry	Custom terrazzo with intricate and ornamental designs, custom select natural stone, marble or very good carpet.	Custom terrazzo with highly ornamental designs, natural stone veneers, marble or very good carpet.	Custom designed terrazzo with ornamental designs with portions natural stone veneers, marble or good carpet.	Colored terrazzo with designs, or average carpet.	Colored concrete with some plain colored terrazzo.
Floors, Interior	Concrete with carpet throughout.	Concrete with carpet throughout.	Concrete with carpet throughout.	Concrete with carpet throughout.	Concrete with carpet runners at aisles.
Restrooms	Per code requirement capacity. Custom select ornamental ceramic tile, terrazzo or natural stone on floors and walls.	Per code requirement capacity. Custom ceramic tile or terrazzo on floor and walls.	Per code requirement capacity. Ceramic tile on floors and walls or terrazzo floors and walls.	Per code requirement capacity. Ceramic tile on walls and floors or terrazzo on floors.	Per code requirement capacity. Ceramic or vinyl tile on floors.
Walls	Painted or custom select canvas backed or custom molded cloth tapestry wall papers or oriental plaster finished with detailed sirocco type moldings and trim, walls with selected matched wood veneers.	Painted or custom canvas backed, custom molded cloth tapestry or ornamental plaster finished with custom moldings and trimmings. Portions of wall select matched wood veneers.	Painted or finished with custom canvas backed wallpapers or molded tapestry finished wallpapers or select wood veneer matched full height at lobby.	Painted or finished with durable canvas or select quality wood veneers on gypsum board or plaster.	Painted and papered with durable canvas materials with portion wood veneers on gypsum board or plaster.
Ceiling	Suspended acoustical with highly ornate moldings and trim with acoustical baffles.	Suspended acoustical with highly ornate moldings and trim with acoustical baffles.	Suspended acoustical with ornate plaster cove moldings and trim with acoustical baffles.	Suspended acoustical with plaster moldings and sound baffles.	Suspended acoustical tile.
Lighting	Incandescent or recessed lights throughout. Interior with ornate chandelier type fixtures throughout. All dimmer controlled.	Incandescent or recessed lights throughout. Interior with ornate chandelier type fixtures throughout. All dimmer controlled.	Incandescent fixtures with chandelier and recessed at theater area, dimmer controlled.	Incandescent fixtures in lobby with fluorescent or chandelier type fixtures in theater area, recessed lighting, dimmer controlled.	Incandescent fixtures with recessed fixtures, dimmer controlled.
Seating	Main floor and balcony seating.	Main floor and balcony seating.	May or may not have balcony.	May or may not have balcony.	May or may not have balcony.

Square foot costs include the following components: Foundations as required for normal soil conditions. Floor, wall, and roof structures. Interior floor, wall and ceiling finishes. Exterior wall finish and roof cover. Display fronts. Interior partitions. All doors. Ticket booth. Basic lighting and electrical systems. Rough and finish plumbing. A mezzanine floor projection booth. A framework for mounting a picture screen. A balcony in auditorium type theaters. Permits and fees. Contractors' mark-up.

Theaters with Balcony - Masonry or Concrete

Estimating Procedure

1. Establish the structure quality class by applying the information on page 150.
2. Compute the building floor area. This should include everything within the main walls and all insets outside the main walls but under the main roof.
3. Add to or subtract from the square foot cost below the appropriate amount from the Wall Height Adjustment Table on page 155 if the wall height is more or less than 28 feet.
4. Multiply the adjusted square foot cost by the building floor area.
5. Deduct, if appropriate, for common walls, using the figures on page 155.
6. Multiply the total cost by the location factor listed on page 7.
7. Add the cost of heating and air conditioning systems, fire extinguishers, exterior signs, paving and curbing. See the section beginning on page 201.

Length less than twice width
Square Foot Area

Quality Class	5,000	6,000	7,000	8,000	9,000	10,000	12,000	15,000	20,000	24,000	30,000
1 Best	133.65	128.76	125.04	122.10	119.72	117.73	114.59	111.19	107.43	105.37	103.12
1 & 2	128.63	123.91	120.34	117.52	115.22	113.30	110.28	107.00	103.39	101.41	99.24
2, Very Good	123.83	119.26	115.82	113.12	110.90	109.05	106.14	102.99	99.52	97.60	95.52
2 & 3	118.81	114.45	111.15	108.55	106.42	104.65	101.85	98.83	95.50	93.66	91.67
3, Good	113.78	109.62	106.44	103.95	101.91	100.23	97.55	94.66	91.45	89.70	87.79
3 & 4	108.42	104.45	101.44	99.06	97.12	95.50	92.96	90.18	87.15	85.48	83.65
4, Average	103.36	99.56	96.67	94.43	92.58	91.04	88.59	85.97	83.08	81.48	79.74
4 & 5	97.73	94.13	91.41	89.26	87.52	86.08	83.78	81.29	78.55	77.04	75.39
5, Low	92.29	88.90	86.35	84.32	82.66	81.29	79.12	76.77	74.18	72.76	71.21

Length between 2 and 4 times width
Square Foot Area

Quality Class	5,000	6,000	7,000	8,000	9,000	10,000	12,000	15,000	20,000	24,000	30,000
1, Best	142.31	137.08	133.11	129.99	127.45	125.33	121.98	118.37	114.36	112.17	109.78
1 & 2	136.96	131.94	128.12	125.11	122.66	120.63	117.40	113.91	110.08	107.96	105.66
2, Very Good	131.87	127.03	123.35	120.47	118.10	116.13	113.03	109.67	105.99	103.96	101.73
2 & 3	126.49	121.83	118.32	115.54	113.27	111.40	108.41	105.20	101.65	99.69	97.58
3, Good	121.17	116.71	113.34	110.68	108.50	106.70	103.84	100.77	97.37	95.49	93.47
3 & 4	114.54	110.32	107.14	104.62	102.58	100.87	98.18	95.26	92.05	90.29	88.37
4, Average	110.12	106.08	103.00	100.59	98.62	96.97	94.39	91.58	88.49	86.80	84.96
4 & 5	104.09	100.26	97.37	95.07	93.21	91.66	89.21	86.57	83.64	82.04	80.30
5, Low	98.28	94.67	91.94	89.78	88.01	86.55	84.25	81.75	78.98	77.47	75.82

Length more than 4 times width
Square Foot Area

Quality Class	5,000	6,000	7,000	8,000	9,000	10,000	12,000	15,000	20,000	24,000	30,000
1, Best	151.55	146.00	141.81	138.49	135.78	133.53	129.95	126.06	121.79	119.42	116.84
1 & 2	145.84	140.52	136.48	133.29	130.68	128.52	125.05	121.34	117.21	114.93	112.45
2, Very Good	140.39	135.27	131.38	128.31	125.78	123.69	120.39	116.79	112.84	110.63	108.24
2 & 3	134.79	129.22	126.13	123.19	120.79	118.76	115.59	112.13	108.32	106.21	103.92
3, Good	128.97	124.27	120.69	117.86	115.57	113.64	110.59	107.29	103.65	101.63	99.46
3 & 4	122.94	118.44	115.04	112.35	110.16	108.33	105.42	102.27	98.79	96.88	94.79
4, Average	117.62	113.32	110.07	107.49	105.39	103.63	100.85	97.86	94.52	92.69	90.69
4 & 5	110.79	106.73	103.65	101.25	99.27	97.62	94.99	92.17	89.03	87.30	85.41
5, Low	104.52	100.80	97.90	95.61	93.75	92.18	89.72	87.04	84.08	82.46	80.68

Length Less Than Twice Width

Estimating Procedure

1. Establish the structure quality class by applying the information on page 150.
2. Compute the building floor area. This should include everything within the main walls and all insets outside the main walls but under the main roof.
3. Add to or subtract from the square foot cost below the appropriate amount from the Wall Height Adjustment Table on page 155 if the wall height is more or less than 20 feet.
4. Multiply the adjusted square foot cost by the building floor area.
5. Deduct, if appropriate, for common walls, using the figures on page 155.
6. Multiply the total cost by the location factor listed on page 7.
7. Add the cost of heating and air conditioning systems, fire extinguishers, exterior signs, paving and curbing. See the section beginning on page 201.

Theater Without Balcony, Class 4

Square Foot Area

Quality Class	3,000	3,500	4,000	5,000	6,000	7,000	8,000	10,000	12,000	15,000	20,000
3, Good	74.29	71.94	70.11	67.41	65.53	64.12	63.01	61.40	60.25	59.02	57.70
3 & 4	71.31	69.05	67.29	64.72	62.90	61.55	60.49	58.95	57.83	56.65	55.36
4, Average	68.49	66.32	64.62	62.14	60.41	59.11	58.10	56.60	55.54	54.40	53.17
4 & 5	66.14	64.03	62.40	60.01	58.31	57.05	56.08	54.64	53.62	52.54	51.35
5, Low	63.67	61.66	60.08	57.78	56.17	54.95	54.02	52.62	51.64	50.58	49.46

Theaters Without Balcony - Masonry or Concrete

Length Between 2 and 4 Times Width

Estimating Procedure

1. Establish the structure quality class by applying the information on page 150.
2. Compute the building floor area. This should include everything within the main walls and all insets outside the main walls but under the main roof.
3. Add to or subtract from the square foot cost below the appropriate amount from the Wall Height Adjustment Table on page 155 if the wall height is more or less than 20 feet.
4. Multiply the adjusted square foot cost by the building floor area.
5. Deduct, if appropriate, for common walls, using the figures on page 155.
6. Multiply the total cost by the location factor listed on page 7.
7. Add the cost of heating and air conditioning systems, fire extinguishers, exterior signs, paving and curbing. See the section beginning on page 201.

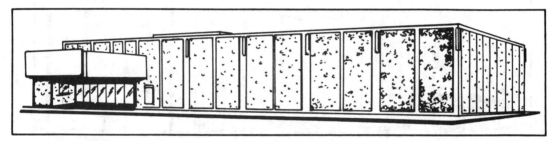

Theater Without Balcony, Class 3

Square Foot Area

Quality Class	3,000	3,500	4,000	5,000	6,000	7,000	8,000	10,000	12,000	15,000	20,000
3, Good	80.11	76.72	74.25	70.93	68.83	67.41	66.39	65.09	64.30	63.61	63.05
3 & 4	76.93	73.67	71.28	68.09	66.09	64.74	63.77	62.51	61.73	61.07	60.52
4, Average	73.96	70.82	68.54	65.48	63.55	62.23	61.31	60.09	59.38	58.73	58.20
4 & 5	71.40	68.36	66.17	63.21	61.35	60.07	59.18	57.99	57.30	56.68	56.16
5, Low	68.82	65.90	63.77	60.92	59.13	57.91	57.04	55.92	55.24	54.64	54.13

Length More Than 4 Times Width

Estimating Procedure

1. Establish the structure quality class by applying the information on page 150.
2. Compute the building floor area. This should include everything within the main walls and all insets outside the main walls but under the main roof.
3. Add to or subtract from the square foot cost below the appropriate amount from the Wall Height Adjustment Table on page 155 if the wall height is more or less than 20 feet.
4. Multiply the adjusted square foot cost by the building floor area.
5. Deduct, if appropriate, for common walls, using the figures on page 155.
6. Multiply the total cost by the location factor listed on page 7.
7. Add the cost of heating and air conditioning systems, fire extinguishers, exterior signs, paving and curbing. See the section beginning on page 201.

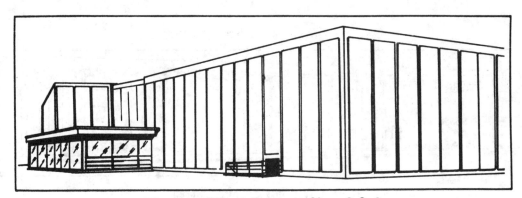

Theater Without Balcony, Class 3 & 4

Square Foot Area

Quality Class	3,000	3,500	4,000	5,000	6,000	7,000	8,000	10,000	12,000	15,000	20,000
3, Good	83.71	80.99	78.86	75.81	73.64	72.07	70.83	69.01	67.74	66.39	64.95
3 & 4	80.00	77.38	75.37	72.43	70.39	68.88	67.69	65.96	64.75	63.45	62.05
4, Average	77.19	74.67	72.71	69.89	67.90	66.43	65.30	63.62	62.45	61.22	59.89
4 & 5	74.50	72.07	70.19	67.46	65.56	64.12	63.05	61.42	60.30	59.10	57.80
5, Low	71.75	69.41	67.60	64.97	63.12	61.77	60.71	59.16	58.06	56.92	55.66

Theaters - Masonry or Concrete

Wall Height Adjustment

Add or subtract the appropriate amount listed in this table to the square foot of floor cost for each foot of wall height more or less than 28 feet, if adjusting for a theater with balcony, or 20 feet, if adjusting for a theater without a balcony.

Square Foot Area

Quality Class	3,000	3,500	4,000	5,000	6,000	7,000	8,000	10,000	12,000	15,000	20,000
1, Best	2.31	2.23	2.18	2.09	2.04	2.00	1.95	1.91	1.88	1.83	1.80
2, Very Good	2.12	2.06	2.02	1.94	1.86	1.82	1.80	1.76	1.73	1.69	1.65
3, Good	1.99	1.93	1.87	1.79	1.75	1.72	1.69	1.64	1.62	1.58	1.53
4, Average	1.82	1.76	1.72	1.66	1.61	1.58	1.54	1.50	1.48	1.45	1.42
5, Low	1.70	1.64	1.60	1.54	1.51	1.46	1.44	1.40	1.37	1.35	1.31

Perimeter Wall Adjustment

A common wall exists when two buildings share one wall. Adjust for common walls by deducting the linear foot costs below from the total structure cost. In some structures one or more walls are not owned at all. In this case, deduct the "No Ownership" cost per linear foot of wall not owned. For Common Wall, deduct $170.00 per linear foot. For no Wall Ownership, deduct $340.00 per linear foot.

Theaters - Wood Frame

Quality Classification

	Class 1 Best Quality	Class 2 Good Quality	Class 3 Average Quality	Class 4 Low Quality
Foundation	Reinforced concrete.	Reinforced concrete.	Reinforced concrete.	Reinforced concrete.
Floor Structure	4" reinforced concrete on 6" rock fill or 2" x 10" joists, 16" o.c.	4" reinforced concrete on 6" rock fill or 2" x 8" joists, 16" o.c.	4" reinforced concrete on 6" rock fill or 2" x 6" joists, 16" o.c.	4" reinforced concrete on 4" rock fill or 2" x 6" joists, 16" o.c.
Wall Structure	2" x 6", 16" o.c.	2" x 4" or 2" x 6", 16" o.c.	2" x 4", 16" o.c. up to 14' high, 2" x 6", 16" o.c. over 14' high.	2" x 4", 16" o.c. up to 14' high, 2" x 6", 16" o.c. over 14' high.
Roof Framing	2" x 10" joists, 16" o.c. Trusses on heavy pilasters, 20' o.c. 2" x 12" rafters or purlins, 16" o.c.	2" x 10" joists, 16" o.c. Trusses on heavy pilasters, 20' o.c. 2" x 10" rafters or purlins.	2" x 10" joists, 16" o.c. Steel trusses on pilasters, 20' o.c. 2" x 10" rafters or purlins, 16" o.c.	2" x 10" joists, 16" o.c. Wood trusses, 2" x 8" purlins, 16" o.c.
Roof Covering	5 ply composition roof on 1" x 6" sheathing with insulation.	5 ply composition roof on 1" x 6" sheathing with insulation.	4 ply composition roof on 1" x 6" sheathing.	4 ply composition roof on 1" x 6" sheathing.
Front	Highly ornamental stucco or custom brick or natural stone or terra cotta veneers.	Ornamental stucco or select brick veneers or partial terra cotta veneers.	Plain or colored stucco or common brick or ornamental wood.	Plain stucco.
Floors, Entry & Lobby	Custom designed terrazzo with ornamental designs with portions natural stone veneers, marble or good carpet.	Colored terrazzo with designs or average carpet.	Colored concrete and portions terrazzo, plain colored.	Plain or colored concrete.
Floors, Interior	Concrete with carpet throughout.	Concrete with carpet throughout.	Concrete with carpet runners at aisles.	Plain or colored concrete.
Restrooms	As per code requirement capacity. Ceramic tile on floors and walls or terrazzo floors and walls.	As per code requirement capacity. Ceramic tile on walls and floors or terrazzo on floors.	As per code requirement capacity. Ceramic or vinyl tile on floors.	As per code requirement capacity. Plain concrete floors and walls, painted.
Walls	Painted and finished with custom backed wallpapers or molded tapestry finished wallpapers or select wood veneer, matched full height at lobby.	Painted and finished with durable canvas or wood veneers, select quality on gypsum wallboard or plaster.	Painted and papered with durable canvas materials with portion wood veneers on gypsum wallboard or plaster.	Painted with or without stencil type painted molded designs on gypsum wallboard taped and textured.
Ceilings	Suspended acoustical, ornate cove moldings and trim with acoustical baffles.	Suspended acoustical with plaster moldings and sound baffles.	Suspended acoustical tile.	Gypsum wallboard taped, textured and painted.
Lighting	Incandescent fixtures with chandelier fixtures, recessed at theater area, dimmer controlled.	Incandescent fixtures in lobby with fluorescent or chandelier type fixtures in theater area, recessed lighting, dimmer controlled.	Incandescent recessed fixtures, dimmer controlled.	Plain incandescent fixtures, with dimmers.

Square foot costs include the following components: Foundations as required for normal soil conditions. Floor, wall, and roof structures. Interior floor, wall and ceiling finishes. Exterior wall finish and roof cover. Display fronts. Interior partitions. All doors. Ticket booth. Basic lighting and electrical systems. Rough and finish plumbing. A mezzanine floor projection booth. A framework for mounting a picture screen. A balcony in auditorium type theaters. Permits and fees. Contractors' mark-up.

Theaters - Wood Frame
Length Less Than Twice Width

Estimating Procedure
1. Establish the structure quality class by applying the information on page 156.
2. Compute the building floor area. This should include everything within the main walls and all insets outside the main walls but under the main roof.
3. Add to or subtract from the square foot cost below the appropriate amount from the Wall Height Adjustment Table on page 160 if the wall height is more or less than 20 feet.
4. Multiply the adjusted square foot cost by the building floor area.
5. Deduct, if appropriate, for common walls, using the figures on page 160.
6. Multiply the total cost by the location factor listed on page 7.
7. Add the cost of heating and air conditioning systems, fire extinguishers, exterior signs, paving and curbing. See the section beginning on page 201.

Theater, Class 4 Front, Class 3 Rear

Square Foot Area

Quality Class	3,000	3,500	4,000	5,000	6,000	7,000	8,000	10,000	12,000	15,000	20,000
1, Best	65.97	63.88	62.25	59.86	58.21	56.95	55.98	54.53	53.52	52.43	51.24
1 & 2	63.91	61.87	60.29	58.00	56.38	55.16	54.20	52.81	51.82	50.77	49.62
2, Good	61.84	59.87	58.34	56.13	54.54	53.38	52.46	51.11	50.14	49.13	48.02
2 & 3	59.86	57.96	56.48	54.34	52.80	51.67	50.79	49.48	48.54	47.56	46.48
3, Average	57.79	55.97	54.54	52.46	50.98	49.89	49.04	47.78	46.88	45.93	44.89
3 & 4	56.06	54.29	52.91	50.89	49.46	48.40	47.56	46.35	45.47	44.55	43.54
4, Low	54.19	52.46	51.14	49.17	47.81	46.78	45.97	44.78	43.95	43.06	42.10

Theaters - Wood Frame

Length Between 2 and 4 Times Width

Estimating Procedure

1. Establish the structure quality class by applying the information on page 156.
2. Compute the building floor area. This should include everything within the main walls and all insets outside the main walls but under the main roof.
3. Add to or subtract from the square foot cost below the appropriate amount from the Wall Height Adjustment Table on page 160 if the wall height is more or less than 20 feet.
4. Multiply the adjusted square foot cost by the building floor area.
5. Deduct, if appropriate, for common walls, using the figures on page 160.
6. Multiply the total cost by the location factor listed on page 7.
7. Add the cost of heating and air conditioning systems, fire extinguishers, exterior signs, paving and curbing. See the section beginning on page 201.

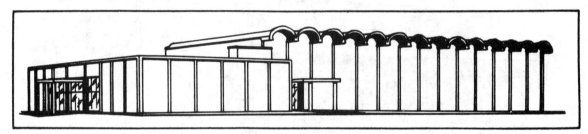

Theater, Class 3

Square Foot Area

Quality Class	3,000	3,500	4,000	5,000	6,000	7,000	8,000	10,000	12,000	15,000	20,000
1, Best	70.31	68.09	66.34	63.79	62.02	60.67	59.63	58.10	57.00	55.83	54.56
1 & 2	68.06	65.90	64.23	61.77	60.03	58.74	57.72	56.23	55.18	54.05	52.82
2, Good	65.83	63.74	62.10	59.73	58.07	56.81	55.83	54.38	53.36	52.27	51.08
2 & 3	63.69	61.67	60.10	57.79	56.17	54.97	54.03	52.63	51.63	50.56	49.42
3, Average	61.60	59.64	58.12	55.89	54.32	53.15	52.24	50.89	49.93	48.90	47.80
3 & 4	59.72	57.82	57.51	56.51	52.67	51.53	50.64	49.33	48.42	47.40	46.35
4, Low	57.61	55.78	54.35	52.27	50.80	49.72	48.85	47.59	46.68	45.73	44.70

Theaters - Wood Frame
Length More Than 4 Times Width

Estimating Procedure

1. Establish the structure quality class by applying the information on page 156.
2. Compute the building floor area. This should include everything within the main walls and all insets outside the main walls but under the main roof.
3. Add to or subtract from the square foot cost below the appropriate amount from the Wall Height Adjustment Table on page 160 if the wall height is more or less than 20 feet.
4. Multiply the adjusted square foot cost by the building floor area.
5. Deduct, if appropriate, for common walls, using the figures on page 160.
6. Multiply the total cost by the location factor listed on page 7.
7. Add the cost of heating and air conditioning systems, fire extinguishers, exterior signs, paving and curbing. See the section beginning on page 201.

Theater, Class 3

Square Foot Area

Quality Class	3,000	3,500	4,000	5,000	6,000	7,000	8,000	10,000	12,000	15,000	20,000
1, Best	74.77	72.38	70.54	67.85	65.95	64.53	63.41	61.80	60.63	59.40	58.04
1 & 2	72.41	70.12	68.33	65.71	63.89	62.52	61.44	59.86	58.73	57.53	56.23
2, Good	70.08	67.85	66.13	63.60	61.82	60.49	59.46	57.92	56.83	55.68	54.43
2 & 3	67.84	65.68	64.03	61.56	59.84	58.55	57.56	56.07	55.03	53.90	52.69
3, Average	65.62	63.52	61.91	59.54	57.89	56.63	55.66	54.24	53.21	52.13	50.95
3 & 4	63.57	61.56	59.98	57.69	56.09	54.88	53.93	52.53	51.55	50.49	49.37
4, Low	61.52	59.56	58.04	55.83	54.28	53.10	52.19	50.85	49.90	48.87	47.77

Theaters - Wood Frame

Wall Height Adjustment

Add or subtract the amount listed in this table to the square foot of floor cost for each foot of wall height more or less than 20 feet.

Square Foot Area

Quality Class	3,000	3,500	4,000	5,000	6,000	7,000	8,000	10,000	12,000	15,000	20,000
1, Best	1.75	1.71	1.67	1.59	1.56	1.53	1.50	1.47	1.44	1.40	1.37
2, Good	1.65	8.07	1.58	1.49	1.45	1.43	1.40	1.37	1.35	1.32	1.29
3, Average	1.55	1.50	1.47	1.41	1.38	1.35	1.33	1.29	1.27	1.24	1.21
4, Low	1.43	1.38	1.34	1.30	1.27	1.25	1.23	1.20	1.18	1.16	1.11

Perimeter Wall Adjustment

A common wall exists when two buildings share one wall. Adjust for common walls by deducting the linear foot costs below from the total structure cost. In some structures one or more walls are not owned at all. In this case, deduct the "No Ownership" cost per linear foot of wall not owned.

For common wall, deduct $96.00 per linear foot.

For no wall ownership, deduct $192.00 per linear foot.

Mobile Home Parks

Quality Classification

	Class 1 Best Quality	Class 2 Good Quality	Class 3 Average Quality	Class 4 Low Quality
Engineering, Plans, Permits, Surveying	Good planning, necessary permits, good engineering; designed by architect.	Good planning, necessary permits, good engineering; designed by architect.	Average planning, necessary permits, engineered and designed.	Fair planning, necessary permits, minimum surveying.
Grading	Fully graded.	Fully graded.	Fully graded.	Minimum site leveling; grades not engineered; road grading.
Street Paving	2" thick asphalt surface on good base, concrete curbs, 30' width.	2" thick asphalt surface on good base, concrete curbs, 25' width.	20' roads, 2" asphalt on rock base; concrete or wood edging.	Narrow streets, 2" asphalt on ground; no curbs or edging.
Patios & Walks	Patios, 300 to 500 S.F. of good concrete. Walks to utility rooms, pools and recreation areas.	Patios, 200 to 300 S.F. of good concrete. Walks to utility rooms, pools and recreation areas.	Patios, approximately 150 S.F. average concrete or average grade asphalt. Walks to utility buildings.	Some patios, concrete or asphalt paving. No walks.
Trailer Pad & Parking	Concrete or good asphalt pad and driveway.	Asphalt under trailer and extended to one side for driveway.	Gravel under trailer and small asphalt driveway.	Gravel under trailer.

	Class 1 Best Quality	Class 2 Good Quality	Class 3 Average Quality	Class 4 Low Quality
Sewer	8" lines, 10" mains. Meets all code requirements. Storm drain system.	8" lines, 10" mains. Meets all code requirements	4" to 6" and 8" lines. Meets code requirements in most areas.	3" to 6" lines, inadequate. Below good code requirements.
Water	Engineered system for equalized pressure throughout park. Sprinkler system in common areas.	Adequate line size, designed and properly sized for equalized pressure.	Adequate line size; has required valves at each space.	Small lines; has required valves at each space.
Gas	Supplied to each space, sized to code requirements.	Supplied to each space, sized to code requirements.	None except in utility buildings and recreation buildings.	None except in utility buildings.
Electric	Underground service, designed for larger modern trailers with adequate size to enlarge to take care of future needs. Approximately 100 amp or more. Speaker system, underground television system to each space.	Underground service, designed for larger modern trailers with facilities to enlarge capacity to 100 amp. Approximately 70 amp service or more. Speaker system, underground television system to each space.	Underground services, not designed for more capacity. Approximately 30 amp service or more. Speaker system.	Overhead system wired for 15 amp service at each space.
Outdoor Lighting	Lamp post each five spaces, ornate type.	Lamp post each five spaces, inexpensive type.	Overhead street lights at each corner.	Few overhead street lights.
Telephone	Underground to each space.	Underground to each space.	None.	None.
Sign	Large expensive sign.	Good sign.	Average sign.	Inexpensive sign.
Garbage	Built-in ground.	Built-in ground.	None.	None.
Mail Boxes	Good mail box and post each space.	Inexpensive mail box and post each space.	None.	None.
Fences and Gates	Good wood or cyclone. Ornamental fence or wall in front.	Good wood or cyclone. Block wall in front.	Inexpensive wood or wire.	None or inexpensive wire.
Pools	Good quality.	Good quality, adequate size for park.	Small with few extras.	None or inexpensive.
Utility Building	Wood frame and good stucco. Board batt redwood siding or concrete block exterior. Best composition shingle or tar and rock roofing. Good interior plaster or gypsum wallboard. Well finished concrete floors with vinyl tile. Good lighting. Good heating. Showers ceramic tile or fiberglass walls with ceramic tile floor. Glass shower doors. Good quality plumbing fixtures. Good workmanship throughout.	Wood frame and good stucco or concrete block exterior. Thick-butt composition shingles or tar and gravel roof. Good exterior or plaster or gypsum wallboard. Well finished concrete floors. Good lighting. Good heating. Showers ceramic tile walls with ceramic tile base. Good quality plumbing fixtures. Good workmanship throughout.	Wood frame, average stucco exterior. Composition shingle or roll roofing. Gypsum wallboard taped and textured or plaster interior. Average concrete floors. Average heating. Fair heating. Metal stall with showers or showers with enameled cement plaster walls and tile floor with tile base. Average plumbing fixtures. Average workmanship throughout.	Wood frame, fair stucco or fair siding exterior. Plastic interior. Composition roll roofing. Fair concrete finish. Fair lighting. Inexpensive heating. Showers enameled cement plaster walls and tile floors. Fair plumbing fixtures and fair workmanship throughout.
Recreation Building	Wood frame and stucco. Board and batt redwood siding or concrete block exterior. Best composition shingles or tar and rock roofing. Good interior plaster or gypsum wallboard taped, textured and painted. Well finished concrete floors with vinyl tile. Good heating. Good lighting. Rest room for each sex containing at least one each of the following fixtures: Shower, water closet & lavatory. Showers ceramic tile floors and walls or fiberglass walls and tile floor with glass shower doors. Good quality plumbing fixtures. Kitchen sink, range, refrigerator, cabinets and drainboard of formica or equal material. Large glass area in community room.	Wood frame and good stucco or concrete block exterior. Thick butt composition shingles or tar and gravel roof. Good exterior or plaster or gypsum wallboard. Well finished concrete floors. Good heating. Good lighting. Ceramic tile stall showers with ceramic tile base. Good quality plumbing fixtures. Kitchen with tile drain board and some hardwood cabinets. Small office area. Large glass windows in community room.	Wood frame, average stucco or siding exterior. Composition shingle or roll roofing. Gypsum wallboard taped and textured or plaster interior. Average concrete floors. Average lighting. Average heating. Showers with enameled cement plaster walls and ceramic tile base. Water closets and lavatories. One rest room for each sex with at least 1 each shower, water closet and lavatory. Average grade of plumbing fixtures. Ceiling of gypsum wallboard. Average workmanship throughout.	None.

Mobile Home Parks

Estimating Procedure

1. Establish the park quality class by applying the information on pages 160 and 161.
2. Compute the square foot area per home space. This should include the mobile home space, streets, recreation and other community use areas but exclude excess land not improved or not in use as a part of the park operation. Divide this total area by the number of home spaces. The result is the average area per home space.
3. Multiply the appropriate cost below by the number of home spaces.
4. Determine the quality class and area of recreation and utility buildings. Compute the total cost of these buildings and add this amount to or subtract it from the total from step 3 to adjust for more or fewer buildings than included in the quality specification.
5. Multiply the total building costs by the location factor listed on page 7.
6. Add the cost of septic tank systems, wells, and covered areas built at the individual spaces.

Space costs with community facilities include the cost of the following components: Grading associated with a level site under normal soil conditions. Street paving and curbs. Patios and walks. Pads and parking paving. Sewer, electrical, gas and water systems including normal hook-up costs. Outdoor lighting. Signs. Mail boxes. Fences and gates. Contractors' mark-up.

Space costs with community facilities and buildings include the cost of all the above components plus these components in amounts proportionate to the size of the park: Recreation, administrative and utility buildings adequate for the size of the park. Recreation facilities such as pools, shuffle board courts, playground equipment, fire pits, etc. Telephones. Restrooms.

The cost of the following components are not included in the basic building cost: Septic tank systems. Wells. Structures or covered areas on individual spaces. The cost of grading beyond that associated with a level site.

Parks Without Community Facilities or Buildings
Square Foot Area

Quality Class	1,500	2,000	2,500	3,000	3,500	4,000	4,500	5,000	5,500	6,000	6,500
1, Best	--	--	--	--	--	7,690	7,835	8,136	8,970	8,615	8,615
2, Good	--	--	--	6,760	7,106	7,252	7,386	7,564	7,700	7,700	7,700
3, Average	--	--	4,993	5,428	5,293	5,920	6,042	6,042	6,042	6,042	--
4, Low	3,028	3,331	3,644	3,878	4,092	4,270	4,270	4,270	4,270	--	--

Parks With Community Facilities and Buildings
Square Foot Area

Quality Class	1,500	2,000	2,500	3,000	3,500	4,000	4,500	5,000	5,500	6,000	6,500
1, Best	--	--	--	--	--	11,524	11,793	12,152	12,498	12,845	13,050
2, Good	--	--	--	10,328	10,578	10,822	11,030	11,219	11,644	11,644	11,644
3, Average	--	--	8,143	8,143	8,525	8,660	8,762	8,762	8,762	8,762	--
4, Low	5,492	5,570	5,729	5,977	6,190	6,315	6,302	6,302	6,302	--	--

Square Foot Costs for Building Alone

	Recreational Buildings	Utility Buildings
1, Best	$39.00 to $56.30	$37.00 to $44.70
2, Good	34.00 to 44.80	27.30 to 37.00
3, Average	27.50 to 39.25	25.40 to 27.50
4, Low	--	23.30 to 26.60

Quality Classification

	Wood Frame	Masonry or Concrete	Painted Steel, Good	Painted Steel, Average	Painted Steel, Low
Foundation & Floor	Reinforced concrete.	Reinforced concrete.	Reinforced concrete.	Reinforced concrete.	Reinforced concrete.
Walls	Wood frame 2 x 4, 16" o.c.	8" concrete block.	Steel frame.	Steel frame.	Steel frame.
Roof Structure	Light wood frame, flat or shed type.	Light wood frame, flat or shed type.	Steel frame, flat or shed type.	Steel frame, flat or shed type.	Steel frame, flat or shed type.
Exterior Finish	Painted wood siding or stucco.	Painted concrete block.	Painted steel.	Painted steel.	Painted steel.
Roof Cover	Composition.	Composition.	Steel deck.	Steel deck.	Steel deck.
Glass Area	Small area, painted wood frames.	Small area, painted steel frames.	Large area, painted steel frames.	Average area, painted steel frames.	Small area, painted steel frames.
Lube Room Doors	Folding steel gate.	Folding steel gate.	Painted steel sectional roll up.	Painted steel sectional roll up.	Folding steel gate.
Floor Finish	Concrete.	Concrete.	Concrete, colored concrete in office.	Concrete.	Concrete.
Interior Wall Finish	Exposed studs, painted.	Concrete block, painted.	Exposed structure painted. Painted steel panels in office.	Exposed structure painted. Painted steel panels in office.	Exposed structure painted. Painted steel panels in office.
Ceiling Finish	Exposed structure painted.	Exposed structure painted.	Exposed structure painted. Painted steel panels in office.	Exposed structure painted. Painted steel panels in office.	Exposed structure painted.
Rest Room Finish	Wallboard and paint walls and ceilings.	Concrete block and paint walls, wallboard and paint ceilings.	Ceramic tile floors, 4' ceramic tile wainscot, painted steel ceilings.	Ceramic tile floors, 4' ceramic tile wainscot, painted steel ceilings.	Concrete floors, painted steel walls, painted steel ceilings.
Rest Room Fixtures	4 low cost fixtures.	4 low cost fixtures.	5 average cost fixtures.	5 average cost fixtures.	4 low cost fixtures.
Exterior Appointments	None.	None.	2' overhang on 3 sides, 3' raised walk on 3 sides, fluorescent soffit lights on 3 sides.	1' overhang on 2 sides, 3' raised walk on 2 sides.	None.

Square foot costs include the cost of the following components: Foundations as required for normal soil conditions. Floor, wall and roof structure. Interior floor, wall and ceiling finishes as described above. Interior partitions. Exterior finish and roof cover. A built-in work bench, tire rack and shelving. Electrical services and fixtures contained within the building. Air and water lines within the building. That portion of rough plumbing serving the building and plumbing fixtures within the building. Roof overhangs and raised walks as described above. Lube room doors. Permits and fees. Contractor's mark-up.

The in-place cost of these extra components should be added to the basic building cost to arrive at the total structure cost. See the section "Additional Costs for Service Stations" on Page 169. Canopies. Pumps, dispensers and turbines. Air and water services outside the building. Island lighters. Gasoline storage tanks. Hoists. Compressors. Yard lights. Signs. Paving. Curbs and fences. Miscellaneous equipment and accessories. Island office and storage buildings. Site improvements. Heating and cooling systems

Land improvement costs: Most service stations sites require an expenditure of $10,000 or more for items such as leveling, excavation, curbs, driveways, relocation of power poles, replacement of sidewalks with reinforced walks and street paving.

Estimating Procedure

1. Establish the structure quality class by applying the information on page 163.
2. Compute the building floor area.
3. Multiply the square foot cost by the building floor area.
4. Multiply the total cost by the location factor listed on page 7.
5. Add the cost of appropriate equipment and fixtures from the section "Additional Costs for Service Stations" beginning on page 169.

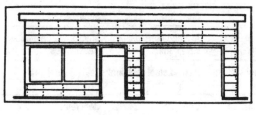

Wood Frame

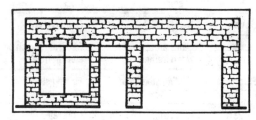

Masonry

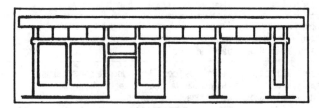

Painted Steel, Good

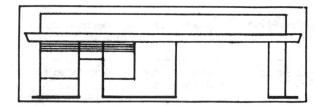

Painted Steel, Average

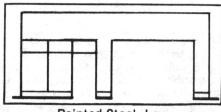

Painted Steel, Low

Square Foot Area

Quality Class	500	600	700	800	900	1,000	1,100	1,200	1,300	1,400	1,800
Wood Frame	60.78	55.81	52.21	49.59	47.51	45.85	44.50	43.37	42.42	41.59	39.22
Masonry or Concrete	68.43	62.69	58.70	55.70	53.37	51.52	50.00	48.73	47.65	46.73	44.06
Painted Steel, Good	105.55	96.89	90.72	86.08	82.48	79.62	77.25	75.30	73.63	72.22	68.11
Painted Steel, Average	91.87	84.93	79.92	75.47	72.31	65.12	67.71	65.99	64.53	63.29	59.71
Painted Steel, Low	80.09	73.53	68.85	65.34	62.60	60.42	58.62	57.14	55.87	54.50	51.70

Services Stations - Porcelain Finished Steel
Quality Classification

	Good Quality	Average Quality	Low Quality
Foundation & Floor	Reinforced concrete.	Reinforced concrete.	Reinorced concrete.
Walls	Steel frame.	Steel frame.	Steel frame.
Roof Structure	Steel frame, flat or shed type.	Steel frame, flat or shed type.	Steel frame, flat or shed type.
Exterior Finish	Porcelain and steel.	Porcelain and steel.	Porcelain and steel.
Roof Cover	Steel deck.	Steel deck.	Steel deck.
Glass Area	Large area, aluminum frames.	Large area, aluminum frames.	Average area, painted steel frames.
Lube Room Doors	Aluminum and glass sectional roll up.	Aluminum and glass sectional roll up.	Painted steel and glass sectional roll up.
Floor Finish	Concrete floors, ceramic tile in office.	Concrete floors, colored concrete in office.	Concrete floors, colored concrete in office.
Interior Wall Finish	Porcelain steel panels. Painted steel panels in office.	Exposed structure painted. Painted steel panels in office.	Exposed structure painted.
Ceiling Finish	Exposed structure painted. Porcelain steel panels in office.	Exposed structure painted. Painted steel panels in office.	Exposed structure painted. Painted steel panels in office.
Rest Room Finish	Ceramic tile floors, 8' ceramic tile or porcelain panel. Painted steel ceiling.	Ceramic tile floors, 5' ceramic tile wainscot. Painted steel ceiling.	Ceramic tile floors, 5' ceramic tile wainscot. Painted steel ceiling.
Rest Room Fixtures	5 good fixtures.	5 good fixtures.	5 good fixtures.
Exterior Appointments	3' to 4' overhang on 3 sides, 6' x 8' sign pylon, 3' raised walk on 3 sides, fluorescent soffit lights on 3 sides.	3' to 4' overhang on 3 sides, 6' x 8' sign pylon, 3' raised walk on 3 sides, fluorescent soffit lights on 3 sides.	3' raised walk on 3 sides, fluorescent soffit lights on 3 sides.

Square foot costs include the cost of the following components: Foundations as required for normal soil conditions. Floor, wall and roof structure. Interior floor, wall and ceiling finishes as described above. Interior partitions. Exterior finish and roof cover. A built-in work bench, tire rack and shelving. Electrical services and fixtures contained within the building. Air and water lines within the building. That portion of rough plumbing serving the building and plumbing fixtures within the building. Roof overhangs and raised walks as described above. Lube room doors. Permits and fees. Contractor's mark-up.

The in-place cost of these extra components should be added to the basic building cost to arrive at the total structure cost. See the section "Additional Costs for Service Stations" on page 169. Canopies. Pumps, dispensers and turbines. Air and water services outside the building. Island lighters. Gasoline storage tanks. Hoists. Compressors. Yard lights. Signs. Paving. Curbs and fences. Miscellaneous equipment and accessories. Island office and storage buildings. Site improvements. Heating and cooling systems

Land improvement costs: Most service station sites require an expenditure of $10,000 or more for items such as leveling, excavation, curbs, driveways, relocation of power poles, replacement of sidewalks with reinforced walks and street paving.

Estimating Procedure

1. Establish the structure quality class by applying the information on page 165.
2. Compute the building floor area.
3. Multiply the square foot cost by the building floor area.
4. Multiply the total cost by the location factor listed on page 7.
5. Add the cost of appropriate equipment and fixtures from the section "Additional Costs for Service Stations" beginning on page 169.

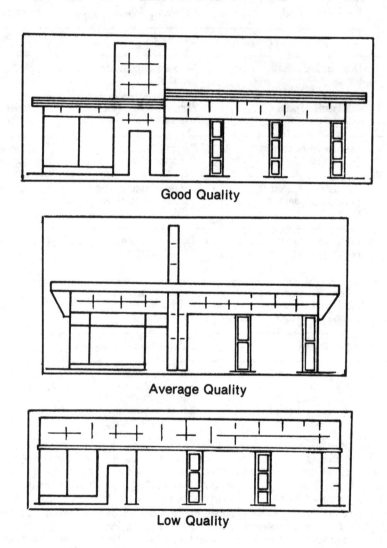

Good Quality

Average Quality

Low Quality

Square Foot Area

Quality Class	1,000	1,100	1,200	1,300	1,400	1,500	1,600	1,700	1,800	2,000	2,400
Good	98.01	94.60	91.91	89.73	87.95	86.53	85.30	84.29	83.45	82.13	80.45
Average	91.97	88.77	86.24	84.21	82.55	81.18	80.05	79.10	78.30	77.05	75.49
Low	83.16	80.26	77.96	76.13	74.65	73.39	72.38	71.53	70.80	69.68	68.26

Services Stations - Ranch or Rustic Type

Quality Classification

	Best Quality	Good Quality	Average Quality	Low Quality
Foundation & Floor	Reinforced concrete.	Reinforced concrete.	Reinforced concrete.	Reinforced concrete.
Walls	Steel frame.	Steel frame.	Steel frame, wood frame or masonry.	Steel frame, wood frame or masonry.
Roof Structure	Steel frame, hip or gable type.	Steel frame, hip or gable type.	Steel or wood frame, hip or gable type.	Steel or wood frame, hip or gable type.
Exterior Finish	Natural stone veneer.	Used brick veneer.	Painted steel and masonry veneer.	Painted steel or wood siding.
Roof Cover	Shingle tile or mission tile.	Heavy wood shakes or shingle tile.	Wood shakes or tar and rock.	Composition shingle or tar and gravel.
Glass Area	Large area plate glass in heavy aluminum frame.	Large area plate glass in heavy aluminum frame.	Large area, painted steel frame.	Average area, painted steel frame.
Lube Room Doors	Painted steel or aluminum and glass sectional roll up.	Painted steel or aluminum and glass sectional roll up.	Painted steel sectional roll up.	Painted steel sectional roll up.
Floor Finish	Concrete floors, ceramic tile in office.	Concrete floors, ceramic tile in office.	Concrete floors.	Concrete floors.
Interior Wall Finish	Painted steel panels or gypsum wallboard and paint.	Painted steel panels or gypsum wallboard and paint.	Painted steel panels or gypsum wallboard and paint.	Painted steel panels or gypsum wallboard and paint
Ceiling Finish	Painted steel panels.	Painted steel panels.	Painted steel panels, gypsum wallboard, or "V" rustic and paint.	Painted steel panels, gypsum wallboard or "V" rustic and paint.
Restroom Finish	Ceramic tile floors, ceramic tile walls, painted steel ceiling.	Ceramic tile floors, ceramic tile walls, painted steel ceiling.	Ceramic tile floors, 5' ceramic tile wainscot, painted steel ceiling.	Ceramic tile floors, 5' ceramic tile wainscot, painted steel ceiling.
Restroom Fixtures	5 good fixtures.	5 good fixtures.	5 good fixtures.	5 good fixtures.
Exterior Appointments	3' to 6' overhang on all sides, 3' raised walk on 3 sides, fluorescent soffit lights on all sides.	3' to 6' overhang on all sides, 3' raised walk on 3 sides, fluorescent soffit lights on all sides.	3' to 6' overhang on 3 sides, 6' x 8' sign pylon, 3' raised walk on 3 sides, fluorescent soffit lights on 3 sides.	2' to 3' overhang on 3 sides, 3' raised walk on 3 sides, fluorescent soffit lights on 3 sides.

Square foot costs include the cost of the following components: Foundations as required for normal soil conditions. Floor, wall and roof structure. Interior floor, wall and ceiling finishes as described above. Interior partitions. Exterior finish and roof cover. A built-in work bench, tire rack and shelving. Electrical services and fixtures contained within the building. Air and water lines within the building. That portion of rough plumbing serving the building and plumbing fixtures within the building. Roof overhangs and raised walks as described above. Lube room doors. Permits and fees. Contractor's mark-up.

The in-place cost of these extra components should be added to the basic building cost to arrive at the total structure cost. See the section "Additional Costs for Service Stations" on page 169. Canopies. Pumps, dispensers and turbines. Air and water services outside the building. Island lighters. Gasoline storage tanks. Hoists. Compressors. Yard lights. Signs. Paving. Curbs and fences. Miscellaneous equipment and accessories. Island office and storage buildings. Site improvements. Heating and cooling systems

Land improvement costs: Most service stations sites require an expenditure of $10,000 or more for items such as leveling, excavation, curbs, driveways, relocation of power poles, replacement of sidewalks with reinforced walks and street paving.

Service Stations - Ranch or Rustic Type

Estimating Procedure

1. Establish the structure quality class by applying the information on page 167.
2. Compute the building floor area.
3. Multiply the square foot cost by the building floor area.
4. Multiply the total cost by the location factor listed on page 7.
5. Add the cost of appropriate equipment and fixtures from the section "Additional Costs for Service Stations" beginning on page 169.

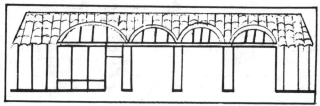

Best Quality

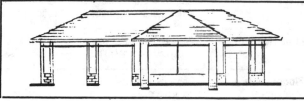

Good Quality

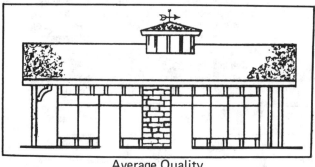

Average Quality

Low Quality

Square Foot Area

Quality Class	1,000	1,100	1,200	1,300	1,400	1,500	1,600	1,700	1,800	2,000	2,400
Best	107.98	104.16	101.13	98.72	96.76	95.14	93.82	92.71	91.79	90.35	88.58
Good	101.11	97.52	94.70	92.43	90.60	89.10	87.83	86.82	85.93	84.62	82.95
Average	94.72	91.36	88.72	86.59	84.86	83.45	82.31	81.32	80.52	79.25	77.71
Low	89.03	85.87	83.39	81.41	79.78	78.46	77.35	76.45	75.67	74.50	73.04

Additional Costs for Service Stations

A portion of the typical plumbing or electrical cost has been added to each item of equipment requiring these services. It will not be necessary except in rare instances to add extra cost for these items.

Canopies, cost per square foot

Type	Less than 500 S.F.	500 to 1,000 S.F.	Over 1,000 S.F.
Painted steel	$19.60 to $22.00	$17.90 to $20.60	$15.40 to $17.50
Porcelain and steel	22.50 to 23.50	19.60 to 21.80	17.80 to 19.30
Ranch style or gable roof type	25.50 to 27.80	20.10 to 23.40	18.20 to 20.50
Deluxe steel with illuminated plastic signs on sides or in gables. Also includes illuminated plastic island lighters.			
Complete,	$39.30 to $46.20		
Round type, good steel,	37.50 to 44.00		

Costs include cost of foundation, steel support column or columns, complete canopy, painting or porcelaining, light fixtures, and electrical service. Ranch or gable roof types include the cost of a rock or shake roof cover. Concrete pads under canopies or masonry trim on support columns are not included in these costs.

Island Office and Storage Buildings, cost per square foot

Type	Under 30	31 - 40	41 - 50	Area 51 - 60	61 - 80	81 - 100	101 - 120
Steel and glass or concrete block	234.12	223.83	199.85	177.01	158.75	148.46	118.77
Wood frame with stucco and glass	211.05	199.64	164.28	152.87	117.51	106.10	104.95

These buildings are usually found at self-service stations. Add $1,300 per unit for any plumbing fixtures in these buildings. Steel island offices cost about $1,250.

Pumps, Dispensers and Turbines, cost each

Type	Installed Cost	Type	Installed Cost
Single pump	$3,900	Blendomatic pump	$6,330
Twin pump	5,175	Blendomatic dispenser	5,150
Single dispenser	2,900	Turbine pump, 1/3 HP	1,050
Twin disposer	4,900	Turbine pump, 3/4 HP	1,450

Installed cost includes the cost of the pump or dispenser, installation cost, electrical hookup cost, a portion of the piping cost and a portion of the island block cost. Concrete islands 4" to 6" thick cost from $8.15 to $9.25 per square foot.

All of the above pump and dispenser costs are for the computing type. Add for electronic remote control totalizer, per hose, $1,300. Add for vapor control system, per hose/dispenser, $1,300.

Dispenser cost does not include the cost of the pump. Turbine pump costs must be added. 1/3 HP turbines will serve a single product up to four dispensers. 3/4 HP turbines will serve a single product up to eight dispensers.

Additional Costs for Service Stations

Air and Water Services

Type	Air Only		Air and Water	
	Equipment Cost	Installed Cost	Equipment Cost	Installed Cost
Underground disappearing hose type	$320.00	$480.00	$348.00	$723.00
Post type with auto inflater	$513.00	$757.00	$649.00	$1,028.00
Post type with auto inflator and disappearing hoses	$857.00	$1,168.00	$1,310.00	$1,616.00

Costs include cost of installation and a portion of the cost of air and water lines.

Island Lighters

Width	Length	4 Tubes	6 Tubes
42"	9'-5"	$1,370 ea.	$1,670 ea.
42"	11'-5"	1,720 ea.	2,170 ea.
42"	15'-6"	2,240 ea.	2,470 ea.
42"	19'-6"	2,730 ea.	2,920 ea.
36"	30'-0"	--	3,450 ea.

Cost includes foundation, davit poles or steel support columns and electrical service.

Cash Boxes complete, with pedestal $295 each

Gasoline Storage Tanks (Fiberglass):

Capacity in Gallons	Tank Cost	Installed Cost	Capacity in Gallons	Tank Cost	Installed Cost
110	$502.40	$882.00	5,300	$5,211.00	$6,766.00
150	622.40	1,049.00	6,300	5,463.00	6,976.00
280	743.40	1,211.00	7,400	6,755.00	7,544.00
550	983.40	1,542.00	8,400	6,966.00	8,531.00
1,000	1,910.00	3,278.00	10,500	7,141.00	9,262.00
2,000	3,575.00	4,371.00	12,600	9,064.00	11,500.00
4,000	4,785.00	7,214.00	---	---	---

Installed cost includes cost of tank, excavation (4' bury and soil disposal), placing backfill, fill box (concrete slab over tank), tank piping and vent piping.

Miscellaneous Lube Room Equipment

Air hose reel	$785.00	Pneumatic tube changer	$8,820.00
Water hose reel	800.00	Automatic lube equipment	6,500.00
Grease pit for trucks	$285.00 to $340.00 per L.F.	5 hose reel assembly	8,230.00

Yard Lights

High pressure sodium luminaires. Costs include electrical connection and mounting on a building soffit. For pole mounted yard lights, add pole mounting costs from page 172. Cost per light fixture.

70 Watt	100 Watt	200 Watt	300 Watt	400 Watt
$370.00	$380.00	$480.00	$500.00	$620.00

Additional Costs for Service Stations

Vehicle Hoist

Type	Equipment Cost	Installed Cost
One post 8,000 lb. semi hydraulic hoist	$ 4,550.00	$ 6,900.00
One post 8,000 lb. fully hydraulic hoist	4,900.00	9,100.00
Two post 11,000 lb. semi hydraulic hoist	5,600.00	9,700.00
Two post 11,000 fully hydraulic hoist	6,700.00	10,900.00
Two post 11,000 lb. pneumatic hoist	6,300.00	9,200.00
Two post 24,000 lb. pneumatic hoist	12,300.00	14,100.00

Air Compressors

Horsepower	Equipment Cost	Installed Cost	Horsepower	Equipment Cost	Installed Cost
1/2	$ 880.00	$1,140.00	2	$1,760.00	$2,070.00
3/4	1,090.00	1,345.00	3	1,970.00	2,430.00
1	1,325.00	1,450.00	5	2,590.00	3,000.00
1-1/2	1,530.00	1,760.00	7-1/2	3,420.00	3,830.00

Costs include compressor and tank only.

Paving, cost per S.F.

Asphalt, 2" with 4" base	$1.70 to $2.25
Concrete 4", with base	2.15 to 2.95
Concrete 6", with base	2.65 to 3.50
Oil macadam	1.60
Pea gravel	.80

Site Improvement

Vertical curb and gutter	$4.90 to $15.40 LF
Concrete apron	6.15 to 13.50 SF
6" reinforced concrete sidewalks	3.70 to 4.90 SF
Standard 4" sidewalk	3.10 to 3.40 SF

The above costs are normally included in land value.

Fencing and Curbing, cost per L.F.

Heavy 2 rail fence, 2" x 6"	$7.90 to $9.60
Rails on 4" x 4" posts 6' to 8' o.c.	8.50 to 10.00
Chain link 3' to 4' high	7.30 to 10.75
Solid board 3' to 4' high	8.35 to 9.90
Log barrier	7.25 to 13.00
Metal guard rail on wood posts	27.00 to 49.00
6" x 6" doweled wood bumper strip	7.90 to 10.20
6" x 6" concrete bumper strip	6.00 to 8.50
Cable railing on wood posts	7.60 to 9.00
6" x 12" concrete curb and gutter	13.00 to 15.00
6" concrete block walls, per S.F.	5.30 to 7.60

plus $6.50/LF for foundation

Service Station Signs, cost per square foot of sign area on one side

Painted sheet metal with floodlights	$42.50 to $57.00
Porcelain enamel with floodlights	47.00 to 63.00
Plastic with interior lights	53.50 to 81.50
Simple rectangular neon with painted sheet metal faces and a moderate amount of plain letters	56.50 to 100.00
Round or irregular neon with porcelain enamel faces and more elaborate lettering	81.50 to 123.00

All of the above sign costs are for single faced signs. Add 50% to these costs for double faced signs. Sign costs include costs of installation and normal electrical hookup. They do not include the post cost. See page 172. These costs are intended for use on **service station signs only** and are based on volume production. Costs of custom-built signs will be higher.

Rotators, cost per sign for rotating mount

Small signs	Less than 50 S.F.	$1,700.00 to $1,850.00
Medium signs	50 to 100 S.F.	1,800.00 to 3,625.00
Large signs	100 to 200 S.F.	3,575.00 to 6,600.00
Extra large signs	Over 200 S.F.	$37.50 per S.F. of sign area

Additional Costs for Service Stations

Post Mounting Costs

Post Height	4"	6"	Pole Diameter at Base 8"	10"	12"	14"
15	$ 810	$ 960	$ 1,410	$ 1,910	$ 2,970	$ 3,560
20	960	1,130	1,640	1,980	3,540	3,980
25	1,050	1,300	1,700	2,360	3,940	4,350
30	1,150	1,540	1,760	2,580	4,160	4,820
35	----	1,680	1,980	2,850	4,710	5,180
40	----	1,780	2,410	3,040	4,890	5,760
45	----	----	2,830	3,460	5,340	5,970
50	----	----	3,040	3,940	5,660	6,540
55	----	----	----	4,160	5,970	7,120
60	----	----	----	4,390	6,390	7,440
65	----	----	----	----	6,860	7,910

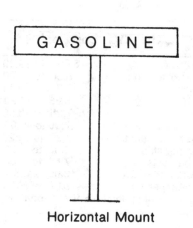

Horizontal Mount

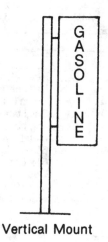

Vertical Mount

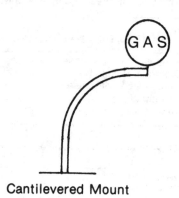

Cantilevered Mount

If signs are mounted on separate posts, post mounting costs must be added. Post mounting costs include the installed cost of a galvanized steel post and foundation. On horizontally mounted signs, post height is the distance from the ground to the bottom of the sign. On vertically mounted signs, post height is the distance to the top of the post.

For cantilevered posts, use one and one-half to two times the conventional post cost.

All of the above post costs are for single posts. Use 90% of the single post costs for each additional post.

If signs are mounted on buildings or canopies and if, because of the extra weight of the sign, extra heavy support posts or foundations are required, 125% of the post mounting cost should be used.

For example, the cost of a 4' x 25' plastic sign mounted on a 15' by 6" post shared by an adjacent canopy might be estimated as follows:

Sign Cost (100 x $67.25)	$6,725.00
Post Cost ($960.00 x 1/2)	480.00
Total Cost	$7,205.00

If this sign were mounted on an 8" post 20' above the canopy with extra supports not needed, the cost might be estimated as follows:

Sign Cost (100 x $67.25)	$6,725.00
Post Cost ($1,640 x 1)	1,640.00
Total Cost	$8,365.00

Service Garages - Masonry or Concrete

Quality Classification

	Class 1 Best Quality	Class 2 Good Quality	Class 3 Average Quality	Class 4 Low Quality
Foundation	Reinforced concrete or masonry.	Reinforced concrete or masonry.	Reinforced concrete or masonry.	Unreinforced concrete or masonry.
Floor Structure	6" rock fill, 4" concrete with reinforcing mesh.	6" rock fill, 4" concrete with reinforcing mesh.	4" rock fill, 4" concrete with reinforcing mesh.	Unreinforced 4" concrete.
Walls	8" reinforced concrete block, 12" common brick.	8" reinforced concrete block, 6" reinforced concrete.	8" reinforced concrete block, 6" reinforced concrete or 8" common brick.	8" unreinforced concrete block or 8" clay tile.
Roof Structure	Glu-lams or steel trusses on heavy pilasters 20' o.c. 2" x 10" purlins 16" o.c..	Glu-lams or steel trusses on pilasters 20' o.c., 2" x 10" purlins 16" o.c.	Glu-lams or wood trusses with 2" x 8" purlins 16" o.c.	Glu-lams or light wood trusses, 2" x 8" rafters 24" o.c.
Roof Cover	5 ply built-up roof on wood sheathing, with small rock.	4 ply built-up roof on wood sheathing, with small rock.	4 ply built-up roof on wood sheathing.	4 ply built-up roof on wood sheathing.
Restrooms	Two rest rooms with three average fixtures each.	Two rest rooms with two average fixtures each.	One rest room with two low cost fixtures.	One rest room with two low cost fixtures.
Lighting	One incandescent fixture per 300 square feet of floor area.	One incandescent fixture per 300 square feet of floor area.	One incandescent fixture per 300 square feet of floor area.	One incandescent fixture per 300 square feet of floor area.
Windows	3% to 5% of wall area.	3% to 5% of wall area.	3% to 5% of wall area.	3% to 5% of wall area.

Square foot costs include the cost of the following components: Foundations as required for normal soil conditions. Floor, wall and roof structures. Exterior wall finish and roof cover. Entry doors. Basic lighting and electrical systems. Rough and finish plumbing. Permits and fees. Contractors' mark-up.

The in-place cost of these extra components should be added to the basic building cost to arrive at the total structure cost. See page 201. Heating and air conditioning systems. Fire sprinklers. Interior finish costs. Interior partitions. Drive-through doors. Canopies and walks. Exterior signs. Paving and curbing. Miscellaneous yard improvements. Hoists, gas pump and compressor costs are listed in the section "Additional Costs for Service Stations" on page 169.

Service Garages - Masonry or Concrete

Length Less Than Twice Width

Estimating Procedure

1. Use these figures to estimate buildings designed primarily for motor vehicle repair. Sales area should be figured separately. Use the costs for urban stores beginning on page 42.
2. Establish the building quality class by applying the information on page 173.
3. Compute the floor area.
4. If the wall height is more or less than 18 feet, add to or subtract from the square foot costs below the appropriate amount from the Wall Height Adjustment Table on page 177.
5. Multiply the adjusted square foot cost by the floor area.
6. Deduct for common walls or no wall ownership. Use the figures on page 177.
7. Multiply the total cost by the location factor on page 7.
8. Add the cost of heating and air conditioning systems, fire sprinklers, interior finish and partitions, drive-thru doors, canopies and walks, exterior signs, paving, curbing, and yard improvements. See page 201. Add the cost of hoists, pumps and compressors from pages 169 and 171.

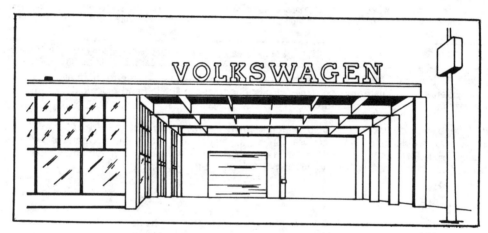

Service Garage (rear portion), Class 2

Square Foot Area

Quality Class	2,000	2,500	3,000	4,000	5,000	6,000	7,500	10,000	15,000	20,000	30,000
1, Best	36.27	32.93	30.57	27.41	25.32	23.85	22.25	20.52	18.53	17.40	16.10
1 & 2	34.80	31.60	29.33	26.30	24.31	22.88	21.35	19.69	17.78	16.71	15.43
2, Good	33.26	30.21	28.05	25.14	23.22	21.86	20.41	18.82	17.00	15.96	14.75
2 & 3	31.66	28.76	26.69	23.93	22.12	20.82	19.44	17.91	16.19	15.19	14.03
3, Average	30.15	27.37	25.41	22.76	21.05	19.81	18.50	17.04	15.40	14.47	13.38
3 & 4	28.51	25.89	24.02	21.53	19.91	18.75	17.50	16.12	14.56	13.67	12.64
4, Low	26.88	24.43	22.67	20.33	18.79	17.68	16.49	15.23	13.75	12.90	11.94

Service Garages - Masonry or Concrete

Length Between 2 and 4 Times Width

Estimating Procedure

1. Use these figures to estimate buildings designed primarily for motor vehicle repair. Sales area should be figured separately. Use the costs for urban stores beginning on page 42.
2. Establish the building quality class by applying the information on page 173.
3. Compute the floor area.
4. If the wall height is more or less than 18 feet, add to or subtract from the square foot costs below the appropriate amount from the Wall Height Adjustment Table on page 177.
5. Multiply the adjusted square foot cost by the floor area.
6. Deduct for common walls or no wall ownership. Use the figures on page 177.
7. Multiply the total cost by the location factor on page 7.
8. Add the cost of heating and air conditioning systems, fire sprinklers, interior finish and partitions, drive-thru doors, canopies and walks, exterior signs, paving, curbing, and yard improvements. See page 201. Add the cost of hoists, pumps and compressors from pages 169 and 171.

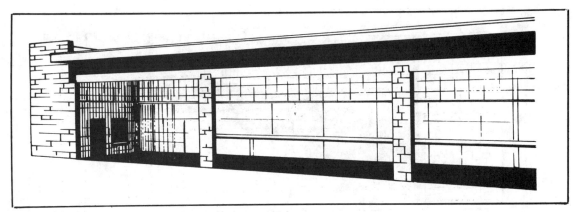

Service Garage, Class 3

Square Foot Area

Quality Class	2,000	2,500	3,000	4,000	5,000	6,000	7,500	10,000	15,000	20,000	30,000
1, Best	38.63	35.07	32.56	29.18	26.97	25.40	23.70	21.84	19.74	18.51	17.12
1 & 2	36.99	33.60	31.19	27.94	25.82	24.32	22.70	20.92	18.89	17.73	16.39
2, Good	35.34	32.09	29.81	26.71	24.67	23.23	21.68	19.98	18.05	16.95	15.66
2 & 3	33.61	30.52	28.34	25.39	23.48	22.10	20.61	19.00	17.15	16.10	14.89
3, Average	30.87	28.92	26.85	24.07	22.25	20.96	19.55	18.01	16.27	15.27	14.12
3 & 4	30.24	27.44	25.47	22.82	21.11	19.88	18.54	17.09	15.44	14.14	13.41
4, Low	28.51	25.89	24.03	21.55	19.92	18.75	17.49	16.13	14.56	13.66	12.63

Service Garages - Masonry or Concrete

Length More Than 4 Times Width

Estimating Procedure

1. Use these figures to estimate buildings designed primarily for motor vehicle repair. Sales area should be figured separately. Use the costs for urban stores beginning on page 42.
2. Establish the building quality class by applying the information on page 173.
3. Compute the floor area.
4. If the wall height is more or less than 18 feet, add to or subtract from the square foot costs below the appropriate amount from the Wall Height Adjustment Table on page 177.
5. Multiply the adjusted square foot cost by the floor area.
6. Deduct for common walls or no wall ownership. Use the figures on page 177.
7. Multiply the total cost by the location factor on page 7.
8. Add the cost of heating and air conditioning systems, fire sprinklers, interior finish and partitions, drive-thru doors, canopies and walks, exterior signs, paving, curbing, and yard improvements. See page 201. Add the cost of hoists, pumps and compressors from pages 169 and 171.

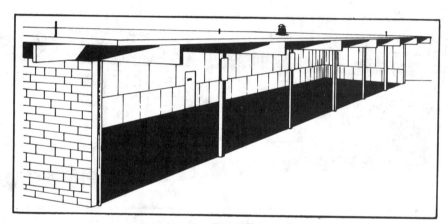

Service Garage, Class 2 & 3

Square Foot Area

Quality Class	2,000	2,500	3,000	4,000	5,000	6,000	7,500	10,000	15,000	20,000	30,000
1, Best	41.16	37.36	34.68	31.07	28.71	27.04	25.23	23.26	21.01	19.74	18.25
1 & 2	39.60	35.96	33.36	29.89	27.64	26.01	24.27	22.37	20.21	18.98	17.55
2, Good	37.67	34.19	31.72	28.42	26.27	24.71	23.09	21.28	19.22	18.05	16.69
2 & 3	35.88	32.57	30.23	27.08	25.03	23.57	21.99	20.26	18.33	17.21	15.92
3, Average	34.12	30.95	28.74	25.75	23.78	22.40	20.91	19.27	17.40	16.35	15.12
3 & 4	32.21	29.23	27.14	24.30	22.46	21.15	19.74	18.21	16.44	15.44	14.29
4, Low	32.00	27.53	25.55	22.90	21.16	19.93	18.59	17.13	15.48	14.53	13.45

Service Garages - Masonry or Concrete

Wall Height Adjustment

Add or subtract the amount listed in this table to the square foot of floor cost for each foot of wall height more or less than 18 feet.

Area	2,000	2,500	3,000	4,000	5,000	6,000	7,500	10,000	15,000	20,000	30,000
Cost	.38	.34	.31	.27	.24	.22	.20	.14	.11	.09	.06

Perimeter Wall Adjustment

A common wall exists when two buildings share one wall. Adjust for common walls by deducting the linear foot costs below from the total structure cost. In some structures, one or more walls are not owned at all. In this case, deduct the "No Ownership" cost per linear foot of wall not owned.

For common wall, deduct $106.00 per linear foot.

For no wall ownership, deduct $212.00 per linear foot.

Service Garages - Wood Frame

Quality Classification

	Class 1 Best Quality	Class 2 Good Quality	Class 3 Average Quality	Class 4 Low Quality
Foundation	Concrete, heavily reinforced.	Reinforced concrete.	Masonry or reinforced concrete.	Masonry or concrete.
Floor Structure	4" reinforced concrete on 6" rock fill.	4" reinforced concrete on 6" rock fill.	4" concrete on 6" rock fill.	4" concrete on 4" rock file.
Walls	2" x 4" studs 16" o.c. in walls 14' high; 2" x 6" studs 16" o.c. in walls over 14' high; 3" sill, double plate, adequate blocking and bracing.	2" x 4" studs 16" o.c. in walls 14' high; 2" x 6" studs 16" o.c. in walls over 14' high; 2" sill, double plate, adequate blocking and bracing.	2" x 4" studs 16" o.c. in walls to 14' high; 2" x 6" studs 16" o.c. in walls over 14' high; 2" sill, double plate, minimum blocking and bracing.	2" x 4" studs 24" o.c.; 2" x 4" sill, double 2" x 4" plate, minimum diagonal bracing.
Exterior	Good corrugated iron or board and batt.	Good corrugated iron or board and batt.	Average corrugated iron or board and batt.	Light corrugated iron or board and batt.
Roof Structures	Glu-lams, trusses or tapered steel girders on steel intermediate columns; 2" x 10" rafters 16" o.c.	Glu-lams, average wood trusses, tapered steel girders on steel intermediate columns; 2" x 8" purlins or rafters 16" o.c.	Glu-lams or light wood trusses, on wood posts 18' o.c.; 2" x 8" rafters on purlins 24" o.c.	Light trussed rafters, clear span in small buildings, post and beam support in large buildings.
Roof Cover	Good quality 4 ply composition roofing on wood sheathing.	Average quality 4 ply composition roofing on wood sheathing.	Average quality 3 ply composition roofing on wood sheathing, or good corrugated aluminum.	Light weight 3 ply composition roofing on wood sheathing, or heavy corrugated iron.
Rest Rooms	Two restrooms with three average fixtures each.	Two restrooms with three average fixtures each.	One restroom with two low cost fixtures.	One restroom with two low cost fixtures.
Lighting	One incandescent fixture per 300 square feet of floor area.	One incandescent fixture per 300 square feet of floor area.	One incandescent fixture per 300 square feet of floor area.	One incandescent fixture per 300 square feet of floor area.
Windows	3% to 5% of wall area.	3% to 5% of wall area.	3% to 5% of wall area.	3% to 5% of wall area.

Square foot costs include the cost of the following components: Foundations as required for normal soil conditions. Floor, wall and roof structure. Exterior wall finish and roof cover. Entry doors. Basic lighting and electrical systems. Rough and finish plumbing. Permits and fees. Contractors' mark-up.

The in-place cost of these extra components should be added to the basic building cost to arrive at the total structure cost. See page 201. Heating and air conditioning systems. Fire sprinklers. Interior finish costs. Interior partitions. Drive-through doors. Canopies and walks. Exterior signs. Paving and curbing. Miscellaneous yard improvements. Hoists, gas pump and compressor costs are listed in the section "Additional Costs for Service Stations" on page 169.

Service Garages - Wood Frame

Length Less Than Twice Width

Estimating Procedure

1. Use these figures to estimate buildings designed primarily for motor vehicle repair. Sales area should be figured separately. Use the costs for urban stores beginning on page 42.
2. Establish the building quality class by applying the information on page 178.
3. Compute the floor area.
4. If the wall height is more or less than 16 feet, add to or subtract from the square foot costs below the appropriate amount from the Wall Height Adjustment Table on page 182.
5. Multiply the adjusted square foot cost by the floor area.
6. Deduct for common walls or no wall ownership. Use the figures on page 182.
7. Multiply the total cost by the location factor on page 7.
8. Add the cost of heating and air conditioning systems, fire sprinklers, interior finish and partitions, drive-thru doors, canopies and walks, exterior signs, paving, curbing, and yard improvements. See page 201. Add the cost of hoists, pumps and compressors from pages 169 and 171.

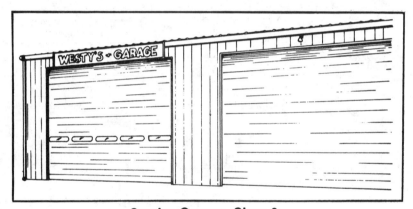

Service Garage, Class 3

Square Foot Area

Quality Class	2,000	2,500	3,000	4,000	5,000	6,000	7,500	10,000	15,000	20,000	30,000
1, Best	24.40	22.15	20.55	18.43	17.03	16.03	14.96	13.80	12.48	11.72	10.83
1 & 2	23.15	21.01	19.52	17.48	16.15	15.21	14.21	13.09	11.84	11.12	10.27
2, Good	21.91	19.89	18.46	16.55	15.29	14.41	13.44	12.39	11.21	10.53	9.74
2 & 3	20.61	18.70	17.35	15.55	14.39	13.54	12.64	11.65	10.54	9.90	9.16
3, Average	19.31	17.52	16.27	14.58	13.47	12.69	11.84	10.93	9.86	9.28	8.59
3 & 4	18.12	16.45	15.28	13.69	12.67	11.90	11.12	10.25	9.27	8.71	8.07
4, Low	16.97	15.39	14.31	12.81	11.84	11.15	10.41	9.59	8.66	8.15	7.55

Service Garages - Wood Frame

Length Between 2 and 4 Times Width

Estimating Procedure

1. Use these figures to estimate buildings designed primarily for motor vehicle repair. Sales area should be figured separately. Use the costs for urban stores beginning on page 42.
2. Establish the building quality class by applying the information on page 178.
3. Compute the floor area.
4. If the wall height is more or less than 16 feet, add to or subtract from the square foot costs below the appropriate amount from the Wall Height Adjustment Table on page 182.
5. Multiply the adjusted square foot cost by the floor area.
6. Deduct for common walls or no wall ownership. Use the figures on page 182.
7. Multiply the total cost by the location factor on page 7.
8. Add the cost of heating and air conditioning systems, fire sprinklers, interior finish and partitions, drive-thru doors, canopies and walks, exterior signs, paving, curbing, and yard improvements. See page 201. Add the cost of hoists, pumps and compressors from pages 169 and 171.

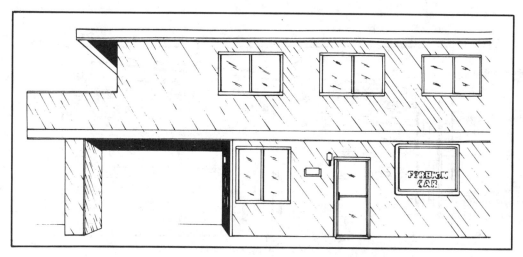

Service Garage (rear portion) Class 2 & 3

Square Foot Area

Quality Class	2,000	2,500	3,000	4,000	5,000	6,000	7,500	10,000	15,000	20,000	30,000
1, Best	25.95	23.56	21.87	19.60	18.13	17.08	15.93	14.70	13.27	12.46	11.53
1 & 2	24.64	22.39	20.77	18.61	17.21	16.22	15.13	13.95	12.61	11.84	10.95
2, Good	23.31	21.18	19.66	17.61	16.28	15.34	14.32	13.18	11.94	11.21	10.36
2 & 3	21.94	19.92	18.50	16.58	15.33	14.44	13.48	12.42	11.23	10.55	9.76
3, Average	20.65	18.75	17.40	15.61	14.43	13.58	12.68	11.69	10.56	9.92	9.17
3 & 4	19.24	17.46	16.22	14.55	13.44	12.67	11.81	10.89	9.83	9.24	8.55
4, Low	17.85	16.21	15.05	13.49	12.47	11.76	10.95	10.11	9.14	8.57	7.94

Service Garages - Wood Frame

Length More Than 4 Times Width

Estimating Procedure

1. Use these figures to estimate buildings designed primarily for motor vehicle repair. Sales area should be figured separately. Use the costs for urban stores beginning on page 42.
2. Establish the building quality class by applying the information on page 178.
3. Compute the floor area.
4. If the wall height is more or less than 16 feet, add to or subtract from the square foot costs below the appropriate amount from the Wall Height Adjustment Table on page 182.
5. Multiply the adjusted square foot cost by the floor area.
6. Deduct for common walls or no wall ownership. Use the figures on page 182.
7. Multiply the total cost by the location factor on page 7.
8. Add the cost of heating and air conditioning systems, fire sprinklers, interior finish and partitions, drive-thru doors, canopies and walks, exterior signs, paving, curbing, and yard improvements. See page 201. Add the cost of hoists, pumps and compressors from pages 169 and 171.

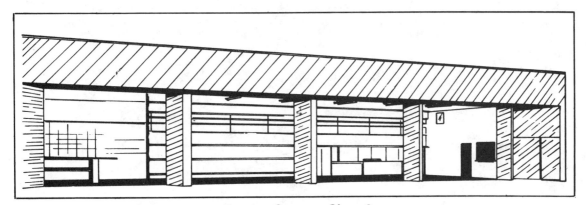

Service Garage, Class 2

Square Foot Area

Quality Class	2,000	2,500	3,000	4,000	5,000	6,000	7,500	10,000	15,000	20,000	30,000
1, Best	27.67	25.13	23.34	20.91	19.33	18.21	16.98	15.66	14.15	13.27	12.27
1 & 2	26.29	23.88	22.16	19.88	18.37	17.29	16.14	14.88	13.44	12.62	11.65
2, Good	24.93	22.62	21.01	18.82	17.40	16.38	15.29	14.10	12.73	11.96	11.04
2 & 3	23.45	21.30	19.79	17.74	16.39	15.43	14.40	13.28	12.00	11.26	10.40
3, Average	21.99	19.99	18.53	16.63	15.36	14.47	13.49	12.45	11.25	10.56	9.76
3 & 4	20.52	18.64	17.30	15.50	14.34	13.49	12.59	11.59	10.49	9.84	9.10
4, Low	19.06	17.30	16.07	14.40	13.30	12.53	11.69	10.78	9.74	9.15	8.45

Service Garages - Wood Frame

Wall Height Adjustment

Add or subtract the amount listed in this table to the square foot cost for each foot of wall height more or less than 16 feet.

Area	2,000	2,500	3,000	4,000	5,000	6,000	7,500	10,000	15,000	20,000	30,000
Cost	.30	.26	.23	.20	.17	.12	.11	.09	.07	.06	.05

Perimeter Wall Adjustment

A common wall exists when two buildings share one wall. Adjust for common walls by deducting the linear foot costs below from the total structure cost. In some structures one or more walls are not owned at all. In this case, deduct the "No Ownership" cost per linear foot of wall not owned.

For common wall, deduct $37.50 per linear foot.

For no wall ownership, deduct $75.00 per linear foot.

Auto Service Centers - Masonry or Concrete

Quality Classification

	Class 1 Best Quality	Class 2 Good Quality	Class 3 Average Quality	Class 4 Low Quality
Foundation	Reinforced concrete.	Reinforced concrete.	Reinforced concrete.	Reinforced concrete.
Floor Structure	4" reinforced concrete on 6" rock fill.	4" reinforced concrete on 6" rock fill.	4" reinforced concrete on 6" rock fill.	4" reinforced concrete on 6" rock file.
Walls	8" reinforced decorative colored concrete block.	8" reinforced detailed concrete block.	8" reinforced concrete block.	8" reinforced concrete block.
Roof Structures	Steel open web joists, steel deck.	Glu-lams, 3" x 12" purlins 3' o.c., or wood open web joists (truss joists) 4' o.c.	Glu-lams, 3" x 12" purlins 3' o.c., or wood open web joists (truss joists) 4' o.c., 1/2" plywood sheathing.	Glu-lams, 3" x 12" purlins 3' o.c., or wood open web joists (truss joists) 4' o.c., 1/2" plywood sheathing.
Floor Finish	Concrete in wood area, resilient tile in sales area.	Concrete in wood area, resilient tile in sales area.	Concrete in wood area, minimum grade tile in sales area.	Concrete.
Interior Wall Finish	Painted concrete block in work area; gypsum wallboard, texture and paint in sales area.	Painted concrete block in work area and sales area.	Unfinished in work area, painted concrete block in sales area	Unfinished.
Ceiling Finish	Open in work area, acoustical tile suspended in exposed grid in sales area.	Open in work area, acoustical tile suspended in exposed grid in sales area.	Open in work area. Celotex tile in sales area.	Open.
Exterior Finish	Decorative colored concrete block.	Colored or painted detailed block.	Painted concrete block.	Unpainted concrete block.
Display Front *(Covers about 25% of the exterior wall)*	1/4" plate glass in good aluminum frame. Good aluminum and glass doors.	1/4" plate glass in average aluminum frame. Aluminum and glass doors.	1/4" plate glass in light aluminum frame. Wood and glass door.	Crystal glass in wood frame, wood door. door.
Roof and Cover	5 ply built-up roofing with insulation.	4 ply built-up roofing.	4 ply built-up roofing.	4 ply built-up roofing.
Plumbing	Two rest rooms with three good fixtures. Metal toilet partitions.	Two rest rooms with two average fixtures. Wood toilet partitions.	One rest room with two fixtures.	One rest room with two fixtures.
Electrical & Wiring	Conduit wiring with triple tube fluorescent strips, 8' o.c.	Conduit wiring with triple tube fluorescent strips, 8' o.c.	Conduit wiring, double tube fluorescent strips, 8' o.c.	Conduit wiring, incandescent fixtures, 10' o.c. or single tube fluorescent strips, 8' o.c.

Square foot costs include the cost of the following components: Foundations as required for normal soil conditions. Floor, wall and roof structures. Interior floor, wall and ceiling finishes (in sales area). Exterior wall finish and roof cover. Display windows. Interior partitions. Entry doors. Basic lighting and electrical systems. Rough and finish plumbing. Permits and fees. Contractors' mark-up.

Wall Height Adjustment: Add or subtract the amount listed in this table to the square foot of floor cost for each foot of wall height more or less than 16 feet.

Area	1,500	2,000	2,500	3,000	3,500	4,000	5,000	6,000	7,500	10,000	15,000
Cost	1.06	.91	.78	.72	.67	.63	.57	.52	.45	.41	.36

Perimeter Wall Adjustment: For common wall, deduct $85.00 per linear foot. For no wall ownership, deduct $169.00 per linear foot.

Auto Service Centers - Masonry or Concrete

Length Less Than Twice Width

Estimating Procedure

1. Use these figures to estimate buildings designed for selling and installing automobile accessories. The square foot costs below allow for a sales area occupying 25% of the building space. The sales area has finished floors, walls and ceiling as described in the quality classification. The remaining 75% of the building is service area and has no interior finish.
2. Establish the building quality class by applying the information on page 183.
3. Compute the floor area. This should include everything within the exterior walls.
4. If the wall height is more or less than 16 feet, add to or subtract from the square foot costs below the appropriate amount from the Wall Height Adjustment Table on page 183.
5. Deduct for common walls or no wall ownership. See page 183.
6. Multiply the total cost by the location factor on page 7.
7. Add the cost of heating and air conditioning systems, fire sprinklers, canopies, walks, exterior signs, paving, curbing, loading docks, ramps, and yard improvements. See page 201. Add the cost of service station equipment from page 169.

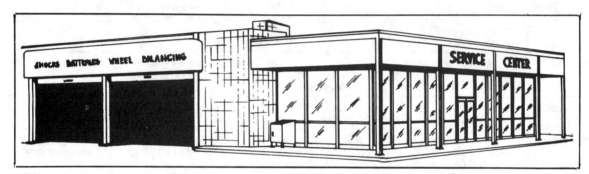

Auto Service Center, Class 2

Square Foot Area

Quality Class	1,500	2,000	2,500	3,000	3,500	4,000	5,000	6,000	7,500	10,000	15,000
Exceptional	71.32	65.25	61.26	58.41	56.20	54.48	51.89	50.01	47.96	45.70	43.08
1, Best	67.58	61.84	58.05	55.33	53.27	51.63	49.17	47.38	45.45	43.31	40.82
1 & 2	64.61	59.11	55.49	52.90	50.92	49.35	47.00	45.29	43.44	41.39	39.02
2, Good	61.47	56.25	52.80	50.34	48.45	46.97	44.72	43.11	41.35	39.40	37.12
2 & 3	58.63	53.62	50.35	47.99	46.20	44.78	42.65	41.09	39.41	37.56	35.41
3, Average	55.84	51.09	47.96	45.73	44.01	42.66	40.63	39.14	37.56	35.78	33.73
3 & 4	53.12	48.61	45.64	43.50	41.87	40.60	38.66	37.25	35.73	34.05	32.09
4, Low	50.67	46.36	43.52	41.50	39.94	38.72	36.87	35.53	34.09	32.48	30.60

Auto Service Centers - Masonry or Concrete

Length Between 2 and 4 Times Width

Estimating Procedure

1. Use these figures to estimate buildings designed for selling and installing automobile accessories. The square foot costs below allow for a sales area occupying 25% of the building space. The sales area has finished floors, walls and ceiling as described in the quality classification. The remaining 75% of the building is service area and has no interior finish.
2. Establish the building quality class by applying the information on page 183.
3. Compute the floor area. This should include everything within the exterior walls.
4. If the wall height is more or less than 16 feet, add to or subtract from the square foot costs below the appropriate amount from the Wall Height Adjustment Table on page 183.
5. Deduct for common walls or no wall ownership. See page 183.
6. Multiply the total cost by the location factor on page 7.
7. Add the cost of heating and air conditioning systems, fire sprinklers, canopies, walks, exterior signs, paving, curbing, loading docks, ramps, and yard improvements. See page 201. Add the cost of service station equipment from page 169.

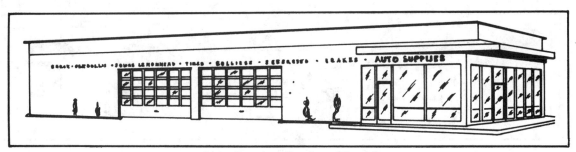

Auto Service Center, Class 2

Square Foot Area

Quality Class	1,500	2,000	2,500	3,000	3,500	4,000	5,000	6,000	7,500	10,000	15,000
Exceptional	75.79	68.99	64.52	61.34	58.89	56.97	54.09	52.01	49.75	47.25	44.38
1, Best	71.83	65.38	61.16	58.12	55.82	54.00	51.26	49.29	47.15	44.77	42.06
1 & 2	68.59	62.42	58.38	55.49	53.29	51.54	48.94	47.06	45.02	42.76	40.15
2, Good	65.18	59.32	55.48	52.73	50.65	48.98	46.51	44.73	42.79	40.64	38.17
2 & 3	62.26	56.66	53.01	50.38	48.37	46.81	44.43	42.72	40.88	38.82	36.46
3, Average	59.12	53.81	50.32	47.83	45.93	44.43	42.18	40.56	38.80	36.86	34.61
3 & 4	56.46	51.40	48.07	45.68	43.87	42.44	40.29	38.74	37.06	35.21	33.05
4, Low	53.58	48.78	45.62	43.37	41.64	40.27	38.25	36.76	35.18	33.41	31.38

Length More Than 4 Times Width

Estimating Procedure

1. Use these figures to estimate buildings designed for selling and installing automobile accessories. The square foot costs below allow for a sales area occupying 25% of the building space. The sales area has finished floors, walls and ceiling as described in the quality classification. The remaining 75% of the building is service area and has no interior finish.
2. Establish the building quality class by applying the information on page 183.
3. Compute the floor area. This should include everything within the exterior walls.
4. If the wall height is more or less than 16 feet, add to or subtract from the square foot costs below the appropriate amount from the Wall Height Adjustment Table on page 183.
5. Deduct for common walls or no wall ownership. See page 183.
6. Multiply the total cost by the location factor on page 7.
7. Add the cost of heating and air conditioning systems, fire sprinklers, canopies, walks, exterior signs, paving, curbing, loading docks, ramps, and yard improvements. See page 201. Add the cost of service station equipment from page 169.

Auto Service Center, Class 3

Square Foot Area

Quality Class	1,500	2,000	2,500	3,000	3,500	4,000	5,000	6,000	7,500	10,000	15,000
Exceptional	81.74	74.32	69.39	65.81	63.05	60.87	57.55	55.14	52.49	49.53	46.05
1, Best	77.42	70.40	65.72	62.33	59.72	57.64	54.51	52.22	49.70	46.92	43.62
1 & 2	73.87	67.18	62.73	59.48	57.00	55.02	52.02	49.83	47.44	44.76	41.64
2, Good	70.30	63.95	59.70	56.62	54.25	52.35	49.50	47.42	45.17	42.60	39.62
2 & 3	67.05	60.97	56.93	53.97	51.73	49.93	47.21	45.22	43.05	40.62	37.79
3, Average	63.84	58.04	54.20	51.40	49.24	47.53	44.95	43.06	40.99	38.67	35.97
3 & 4	60.90	55.39	51.72	49.05	46.98	45.34	42.88	41.07	39.11	36.89	34.33
4, Low	57.79	52.56	49.06	46.53	44.59	43.05	40.69	38.98	37.11	35.02	32.57

Industrial Structures Section

Section Contents

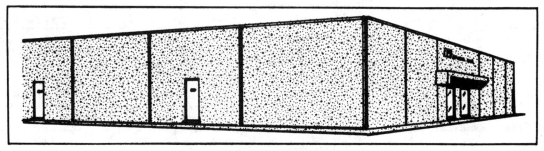

Warehouse, Class 3

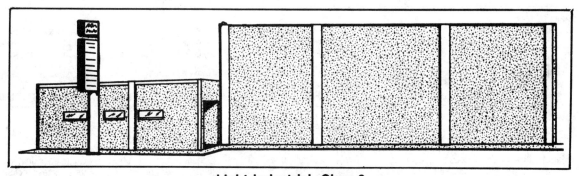

Light Industrial, Class 2

Quality Classification

	Class 1 Best Quality	Class 2 Good Quality	Class 3 Average Quality	Class 4 Low Quality
Foundations	Continuous reinforced concrete.	Continuous reinforced concrete.	Continuous reinforced concrete.	Reinforced concrete pads under pilasters.
Floor Structure	6" rock base, 6" concrete with reinforcing mesh or bars.	6" rock base, 6" concrete with reinforcing mesh or bars.	6" rock base, 5" concrete with reinforcing mesh or bars.	6" rock base, 4" concrete with reinforcing mesh.
Wall Structure	8" reinforced concrete block or brick with pilasters 20' o.c., painted sides and rear exterior, front wall brick veneer.	8" reinforced concrete block or brick with pilasters 20' o.c., painted sides and rear exterior, stucco and some brick veneer on front.	8" reinforced concrete block or brick, unpainted.	8" reinforced concrete block or brick, unpainted.
Alternate Wall Structure	6" concrete tilt-up, ornamental rock on 1/4 of exterior walls, remainder painted and patterned panels.	6" concrete tilt-up exposed aggregate on 1/4 of exterior walls or patterned and painted panels on 1/2 of exterior, remainder painted.	6" concrete tilt-up, unpainted.	6" concrete tilt-up, unpainted.
Roof Structure	Glu-lams, wood or steel trusses on steel intermediate columns, span exceeds 70'.	Glu-lams, wood or steel trusses on steel intermediate columns, span exceeds 70'.	Glu-lams or steel beams on steel intermediate columns, short span.	Glu-lams on steel intermediate columns, short span.
Roof Cover	Panelized roof system, 1/2" plywood sheathing, 5 ply built-up roof.	Panelized roof system, 1/2" plywood sheathing, 4 ply built-up roof.	Panelized roof system, 1/2" plywood sheathing, 4 ply built-up roof.	Panelized roof system, 1/2" plywood sheathing, 4 ply built-up roof.
Skylights	48 S.F. of skylight per 2500 S.F. of floor area (1-6' x 8' skylight 40' to 50' o.c.).	32 S.F. of skylight per 2500 S.F. of floor area (1-4' x 8' skylight 40' to 50' o.c.).	24 S.F. of skylight per 2500 S.F. of floor area (1-4' x 6' skylight 40' to 50' o.c.).	10 S.F. of skylight per 2500 S.F. of floor area (1-2' x 4' skylight 40' to 50' o.c.).
Ventilators	1 large rotary vent per 2500 S.F. of floor area	1 medium rotary vent per 2500 S.F. of floor area.	1 medium rotary vent per 2500 S.F. of floor area.	1 small rotary vent per 2500 S.F. of floor area.
Rest Rooms, Finish	Good vinyl asbestos tile floors enameled gypsum wallboard partitions.	Vinyl asbestos tile floors, enameled gypsum wallboard partitions.	Concrete floors, painted gypsum wallboard partitions.	Concrete floors, unfinished sheet rock partitions.
Fixtures	2 rest rooms, 3 good fixtures in each.	2 rest rooms, 3 average fixtures in each.	2 rest rooms, 2 average fixtures in each.	1 rest room, 2 low cost fixtures in each.
Lighting	4' single tube fluorescent fixtures 10' x 12' spacing.	Low cost single tube fluorescent fixtures 12' x 20' spacing.	Low cost 4' single tube fluorescent fixtures 20' x 20' spacing.	Low cost incandescent fixtures 20' x 30' spacing.
Warehouse Fixtures **Light Industrial**	Same as above plus 1 additional fixture per 10,000 S.F. of floor area.	Same as above plus 1 additional fixture per 10,000 S.F. of floor area.	Same as above plus 1 additional fixture per 10,000 S.F. of floor area.	Same as above plus 1 additional fixture per 20,000 S.F. of floor area.
Light Manufacturing	Same as above plus 2 additional fixtures per 10,000 S.F. of floor area.	Same as above plus 2 additional fixtures per 10,000 S.F. of floor area.	Same as above plus 2 additional fixtures per 10,000 S.F. of floor area.	Same as above plus 2 additional fixtures per 20,000 S.F. of floor.
Warehouse Lighting **Light Industrial**	Continuous single tube fluorescent strips 8' o.c.	Continuous single tube fluorescent strips 12' o.c.	Continuous single tube fluorescent strips 16' o.c.	Continuous single tube fluorescent strips 20' oc
Light Manufacturing	Continuous triple tube fluorescent strip fixtures 8' o.c.	Continuous triple tube fluorescent strip fixtures 12' o.c.	Continuous double tube fluorescent strips 8' o.c.	Continuous double tube fluorescent strips 12' o.c.

Square foot costs include the following components: Foundations as required for normal soil conditions. Floor, wall and roof structures. Exterior wall finish and roof cover. Basic lighting and electrical systems. Rough and finish plumbing. A usual or normal parapet wall. Walk-through doors. Contractors' mark-up.

Warehouses

Estimating Procedure

1. Establish the structure quality class by applying the information on page 188.
2. Compute the building floor area.
3. Add to or subtract from the square foot cost below the appropriate amount from the Wall Height Adjustment Table (at the bottom of page 191) if the wall height is more or less than 20 feet.
4. Multiply the adjusted square foot cost by the building floor area.
5. Deduct, if appropriate, for common walls, using the figures at the bottom of page 191.
6. Multiply the total cost by the location factor on page 7.
7. Add the cost of heating and air conditioning equipment, fire sprinklers, interior offices, drive-through or delivery doors, canopies, interior partitions, docks and ramps, paving and curbing, and miscellaneous yard improvements. See the section beginning on page 201.

Length less than twice width
Square Foot Area

Quality Class	3,000	4,000	5,000	7,500	10,000	15,000	20,000	30,000	50,000	100,000	200,000
1, Best	63.39	58.02	54.42	48.89	45.66	41.88	39.66	37.04	34.45	31.88	30.07
1 & 2	59.03	54.03	50.68	45.53	42.52	39.01	36.94	34.50	32.09	29.69	28.01
2, Good	54.68	50.05	46.94	42.18	39.39	36.12	34.21	31.95	29.72	27.49	25.94
2 & 3	51.15	46.79	43.89	39.46	36.84	33.80	32.00	29.89	27.80	25.71	24.25
3, Average	47.43	43.41	40.70	36.58	34.16	31.34	29.67	27.72	25.77	23.86	22.51
3 & 4	44.37	40.61	38.07	34.23	31.95	29.32	27.76	25.93	24.11	22.31	21.04
4, Low	41.15	37.66	35.31	31.73	29.64	27.18	25.74	24.04	22.38	20.69	19.52

Length between 2 and 4 times width
Square Foot Area

Quality Class	3,000	4,000	5,000	7,500	10,000	15,000	20,000	30,000	50,000	100,000	200,000
1, Best	67.62	61.57	57.54	51.39	47.82	43.63	41.19	38.32	35.48	32.68	30.72
1 & 2	63.15	57.50	53.73	47.99	44.64	40.74	38.45	35.77	33.14	30.51	28.68
2, Good	58.47	53.25	49.75	44.44	41.34	37.73	35.60	33.13	30.68	28.25	26.55
2 & 3	54.68	49.81	46.53	41.55	38.66	35.28	33.30	30.98	28.69	26.42	24.84
3, Average	50.72	46.19	43.15	38.54	35.84	32.72	30.88	28.74	26.62	24.52	23.02
3 & 4	47.45	43.21	40.37	36.07	33.55	30.61	28.90	26.88	24.90	22.93	21.57
4, Low	43.89	39.96	37.35	33.36	31.02	28.33	26.73	24.86	23.02	21.20	19.93

Length more than 4 times width
Square Foot Area

Quality Class	3,000	4,000	5,000	7,500	10,000	15,000	20,000	30,000	50,000	100,000	200,000
1, Best	73.28	66.57	62.06	55.14	51.06	46.28	43.44	40.13	36.82	33.53	31.22
1 & 2	70.28	62.17	57.94	51.48	47.67	43.21	40.56	37.47	34.39	31.31	29.14
2, Good	65.18	57.60	53.69	47.69	44.17	40.04	37.59	34.72	31.86	29.00	27.01
2 & 3	59.32	53.88	50.23	44.62	41.33	37.45	35.17	32.47	29.81	27.13	25.26
3, Average	55.01	49.98	46.60	41.38	38.32	34.75	32.63	30.13	27.66	25.15	23.42
3 & 4	51.42	46.70	43.53	38.67	35.83	32.47	30.48	28.15	25.83	23.51	21.91
4, Low	47.63	43.29	40.34	35.85	33.19	30.09	28.25	26.09	23.94	21.80	20.29

Light Industrial Buildings

Estimating Procedure

1. Establish the structure quality class by applying the information on page 188.
2. Compute the building floor area.
3. Add to or subtract from the square foot cost below the appropriate amount from the Wall Height Adjustment Table (at the bottom of page 191) if the wall height is more or less than 20 feet.
4. Multiply the adjusted square foot cost by the building floor area.
5. Deduct, if appropriate, for common walls, using the figures at the bottom of page 191.
6. Multiply the total cost by the location factor on page 7.
7. Add the cost of heating and air conditioning equipment, fire sprinklers, interior offices, drive-through or delivery doors, canopies, interior partitions, docks and ramps, paving and curbing, and miscellaneous yard improvements. See the section beginning on page 201.

Length less than twice width
Square Foot Area

Quality Class	3,000	4,000	5,000	7,500	10,000	15,000	20,000	30,000	50,000	100,000	200,000
1, Best	64.67	59.71	56.38	51.26	48.25	44.73	42.64	40.21	37.79	35.38	33.67
1 & 2	60.35	55.69	52.58	47.81	45.01	41.73	39.79	37.51	35.25	32.99	31.42
2, Good	55.90	51.62	48.72	44.30	41.71	38.67	36.87	34.76	32.67	30.57	29.10
2 & 3	52.83	48.78	46.04	41.87	39.42	36.54	34.85	32.85	30.88	28.90	27.50
3, Average	48.72	45.00	42.47	38.61	36.37	33.70	32.14	30.30	28.47	26.66	25.37
3 & 4	45.65	42.14	39.80	36.18	34.05	31.57	30.11	28.39	26.68	24.97	23.77
4, Low	42.49	39.23	37.03	33.68	31.70	29.39	28.03	26.43	24.82	23.23	22.12

Length between 2 and 4 times width
Square Foot Area

Quality Class	3,000	4,000	5,000	7,500	10,000	15,000	20,000	30,000	50,000	100,000	200,000
1, Best	68.78	63.16	59.40	53.69	50.33	46.42	44.13	41.43	38.79	36.14	34.30
1 & 2	64.33	59.08	55.57	50.21	47.08	43.43	41.27	38.77	36.29	33.81	32.07
2, Good	59.79	54.92	51.65	46.67	43.76	40.36	38.38	36.03	33.72	31.43	29.81
2 & 3	55.87	51.30	48.25	43.60	40.88	37.70	35.84	33.65	31.48	29.35	27.85
3, Average	51.85	47.62	44.80	40.48	37.94	35.02	33.27	31.25	29.24	27.24	25.85
3 & 4	48.58	44.62	41.96	37.91	35.54	32.79	31.16	29.26	27.40	25.52	24.22
4, Low	45.19	41.50	39.02	35.27	33.07	30.50	29.00	27.24	25.48	23.75	22.54

Length more than 4 times width
Square Foot Area

Quality Class	3,000	4,000	5,000	7,500	10,000	15,000	20,000	30,000	50,000	100,000	200,000
1, Best	74.09	67.77	63.54	57.06	53.26	48.79	46.16	43.09	40.02	36.96	34.82
1 & 2	69.39	63.49	59.50	53.44	49.87	45.71	43.24	40.35	37.48	34.62	32.61
2, Good	64.35	58.87	55.19	49.57	46.27	42.37	40.10	37.42	34.76	32.11	30.24
2 & 3	60.28	55.14	51.70	46.42	43.33	39.70	37.55	35.04	32.55	30.07	28.33
3, Average	55.99	51.24	48.02	43.13	40.24	36.87	34.88	32.56	30.23	27.92	26.32
3 & 4	52.28	47.84	44.84	40.27	37.58	34.44	32.58	30.41	28.24	26.08	24.57
4, Low	48.69	44.55	41.77	37.51	34.99	32.06	30.35	28.32	26.29	24.30	22.88

Factory Buildings

Estimating Procedure

1. Establish the structure quality class by applying the information on page 188.
2. Compute the building floor area.
3. Add to or subtract from the square foot cost below the appropriate amount from the Wall Height Adjustment Table (at the bottom of this page) if the wall height is more or less than 20 feet.
4. Multiply the adjusted square foot cost by the building floor area.
5. Deduct, if appropriate, for common walls, using the figures at the bottom of this page.
6. Multiply the total cost by the location factor on page 7.
7. Add the cost of heating and air conditioning equipment, fire sprinklers, interior offices, drive-through or delivery doors, canopies, interior partitions, docks and ramps, paving and curbing, and miscellaneous yard improvements. See the section beginning on page 201.

Length less than twice width
Square Foot Area

Quality Class	3,000	4,000	5,000	7,500	10,000	15,000	20,000	30,000	50,000	100,000	200,000
1, Best	64.04	59.48	56.43	51.74	49.02	45.82	43.93	41.71	39.51	37.34	35.81
1 & 2	60.13	55.86	52.98	48.59	46.03	43.02	41.24	39.16	37.11	35.06	33.63
2, Good	56.12	52.11	49.43	45.34	42.93	40.13	38.48	36.55	34.62	32.72	31.37
2 & 3	52.55	48.80	46.30	42.45	40.22	37.59	36.04	34.23	32.43	30.64	29.39
3, Average	48.91	45.42	43.08	39.51	37.43	34.98	33.55	31.85	30.18	28.51	27.36
3 & 4	45.68	42.42	40.25	36.91	34.96	32.68	31.32	29.76	28.19	26.64	25.54
4, Low	42.45	39.43	37.40	34.30	32.47	30.37	29.12	27.65	26.20	24.75	23.74

Length between 2 and 4 times width
Square Foot Area

Quality Class	3,000	4,000	5,000	7,500	10,000	15,000	20,000	30,000	50,000	100,000	200,000
1, Best	67.64	62.57	59.15	53.96	50.92	47.38	45.29	42.85	40.44	38.03	36.34
1 & 2	63.45	58.68	55.50	50.61	47.77	44.43	42.48	40.20	37.93	35.67	34.09
2, Good	59.34	54.87	51.87	47.34	44.67	41.56	39.73	37.58	35.46	33.36	31.88
2 & 3	55.56	51.38	48.58	44.31	41.81	38.91	37.19	35.19	33.20	31.24	29.85
3, Average	51.67	47.78	45.18	41.22	38.88	36.17	34.59	32.72	30.87	29.04	27.76
3 & 4	48.28	44.66	42.23	38.53	36.34	33.82	32.33	30.58	28.84	27.14	25.95
4, Low	44.83	41.47	39.20	35.75	33.75	31.40	30.03	28.40	26.80	25.19	24.09

Length more than 4 times width
Square Foot Area

Quality Class	3,000	4,000	5,000	7,500	10,000	15,000	20,000	30,000	50,000	100,000	200,000
1, Best	72.42	66.75	62.93	57.06	53.61	49.56	47.17	44.36	41.57	38.76	36.81
1 & 2	68.05	62.72	59.12	53.62	50.37	46.58	44.32	41.68	39.05	36.42	34.58
2, Good	63.54	58.57	55.20	50.06	47.03	43.48	41.38	38.91	36.46	34.02	32.28
2 & 3	59.54	54.88	51.72	46.91	44.07	40.74	38.78	36.46	34.17	31.88	30.26
3, Average	55.38	51.08	48.16	43.68	41.03	37.93	36.07	33.92	31.79	29.64	28.14
3 & 4	51.74	47.68	44.95	40.74	38.28	35.40	33.70	31.68	29.69	27.70	26.30
4, Low	48.01	44.26	41.72	37.83	35.54	32.85	31.27	29.41	27.55	25.71	24.40

Wall Height Adjustment: Industrial building costs are based on a 20' wall height, measured from the bottom of the floor slab to the top of the roof cover. Add or subtract the amount listed to the square foot cost for each foot more or less than 20 feet.

Area	3,000	4,000	5,000	7,500	10,000	15,000	20,000	30,000	50,000	100,000	200,000
Cost	.48	.44	.40	.34	.30	.25	.22	.18	.10	.06	.03

Perimeter Wall Adjustment: A common wall exists when two buildings share one wall. Adjust for common walls by deducting $64.00 per linear foot from the total structure cost. In some structures one or more walls are not owned at all. In this case, deduct $127.00 per liner foot of wall not owned.

Internal Offices

Internal offices are office areas built into the interior area of an industrial building. Add the square foot costs in this section to the basic building cost. The costs include floor finish, partition framing, wall finish, trim and doors, counter, ceiling structure, ceiling finish, and the difference between the cost of lighting and windows in an industrial building and the cost of lighting and windows in an office building.

Plumbing costs are not included in these internal office costs. if a building has fixtures in excess of those listed in the quality classification for the main building, add $1,195 to $1,475 for each extra fixture. For two-story internal offices, apply a square foot cost based on the total area of both floor levels.

	Best	Good	Average	Low Cost
Floor Finish	Carpet, some vinyl tile or terrazzo.	Resilient tile, some carpet.	Composition tile.	Colored concrete.
Wall Finish	Hardwood veneer, some textured cloth wall cover.	Gypsum board, texture and paint, some hardwood veneer.	Gypsum board, texture texture and paint.	Plywood and paint.
Ceiling Finish	Illuminated plastic ceilings.	Suspended "T" bar and acoustical tile.	Gypsum board, texture and paint.	Exposed or plywood and paint.
Doors - Interior	Good grade hardwood.	Good grade hardwood.	Low cost hardwood.	Standard paint grade.
Exterior	Pair aluminum entry doors with extensive sidelights.	Aluminum entry door with sidelights.	Wood store door.	Standard paint grade.
Windows	Average amount in good aluminum frame, fixed plate glass in good frame in front.	Average amount in good aluminum frame, some fixed plate glass.	Average amount of average cost aluminum sliding type.	Average amount of low cost aluminum sliding type.
Counters	Good grade hardwood counter, good shelving below.	Good grade plastic counter, average amount of shelving below.	Low cost plastic counter, average amount of shelving.	Paint grade counter, small amount of shelving.
Lighting	Illuminated ceilings.	Recessed fluorescent fixtures.	Average amount of fluorescent fixtures.	One incandescent per each 150 S.F.
S.F. Costs	$35.00 to $49.00.	$27.50 to $35.60.	$22.25 to $29.00.	$11.00 to $16.00.

The range of costs is primarily due to the density of partitions in the office area. Use the lower costs for offices with larger rooms and fewer partitions.

	Best	Good	Average	Low Cost
Wall Height Adjustment*	$0.46 to $0.67.	$0.39 to $0.51.	$0.41 to $0.55.	$0.35 to $0.55.

Wall Height Adjustment: Add or subtract the amount listed in this column to the square foot cost for each foot when the office wall height is more or less than 8 feet.

External Offices

External offices outside the main building walls have one side that is common to the main building. Square foot costs for external office areas should be estimated with the figures for general office buildings with interior suite entrances. See pages 106 to 109 or 114 to 117.

In selecting a square foot cost for the external office area as well as the main building area, the area of each portion should be used separately for area classification. The area and the perimeter formed by these individual areas is used for shape classification of the main area and for calculating the area and perimeter relationship of the office area. A "no wall ownership" deduction based on the office wall cost should be made for the length of the common side. Wall height adjustments should be made for each area individually.

Steel Buildings

Engineered steel buildings are constructed to serve as warehouses, factories, airplane hangars, garages, and stores, among other uses. They are generally either high-profile (4 in 12 rise roof) or low-profile (1 in 12 rise roof).

The square foot costs for the basic building include: Foundations as required for normal soil conditions. A 4 inch concrete floor with reinforcing mesh and a 2 inch sand fill. A steel building made up of steel frames or bents set 20 or 24 feet on centers, steel roof purlins 4-1/2 to 5-1/2 feet on centers, steel wall girts 3-1/2 to 4-1/2 feet on centers, post and beam type end wall frames, 26-gauge galvanized steel on ends, sides, and roof, window area equal to 2% of the floor area. Basic wiring and minimum lighting fixtures. One small or medium gravity vent per 2,500 square feet of floor area.

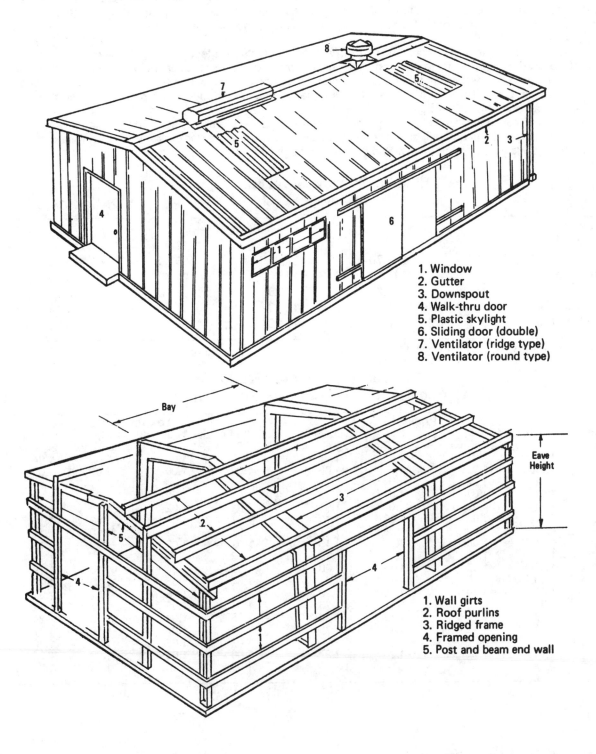

1. Window
2. Gutter
3. Downspout
4. Walk-thru door
5. Plastic skylight
6. Sliding door (double)
7. Ventilator (ridge type)
8. Ventilator (round type)

1. Wall girts
2. Roof purlins
3. Ridged frame
4. Framed opening
5. Post and beam end wall

Steel Buildings

Estimating Procedure

1. Compute the building area.
2. Add to or subtract from the square foot cost below $.21 for each foot of wall height at the eave more or less than 14 feet. Subtract $.35 from the square foot cost below for low profile buildings.
3. Multiply the adjusted square foot cost by the building area.
4. Multiply the total by the appropriate live load adjustment. The basic building costs are for a 12 pound live load usually built where snow load is not a design factor. A 20 pound live load building is usually found in light snow areas, and a 30 pound live load building is usually found in heavy snow areas.
5. Add or subtract alternate costs from pages 195 to 197. Alternate costs reflect the difference between unit costs of components that are included in the basic building square foot costs and the unit costs of alternate components. Total alternate cost is found by multiplying the alternate cost by the area or number of alternate components.
6. Add the appropriate costs from pages 198 to 199, "Additional Costs for Steel Buildings." These costs reflect the in-place cost of components that are not included in the basic square foot cost but are often found as part of steel buildings. The cost of items that alter or replace a portion of the building reflect the net added cost of the component in-place. The cost of the item that is replaced has been deducted from the total cost of the additive components. No further deduction is necessary.
7. Add the cost of heating and air conditioning systems, fire sprinklers, plumbing, additional electrical work, alternate interior partitions and offices, dock height additive costs, loading docks or ramps, paving and curbing, and miscellaneous yard improvements. See the section beginning on page 201.

High Profile Rigid Steel Buildings
Square Foot Cost

Width	Length			
	60'	80'	120'	Over 140'
20'	$17.70	$17.00	$16.50	$16.10
30'	16.30	15.70	14.60	13.50
40'	15.20	14.20	13.20	12.70
50' to 70'	13.40	12.10	11.20	11.10
80' to 140'	---	11.30	10.70	10.30

Deduct $.35 per square foot for low profile buildings.

Live Load Adjustment Factors
(multiply base cost by factor shown)

Building Width	High Profile				Low Profile			
	20 lb. Live Load		30 Lb. Live Load		20 Lb. Live Load		30 Lb. Live Load	
	20' Bays	24' Bays	20' Bays	24' Bays	20' Bays	24' Bays	20' Bays	24' Bays
20'	1.00	1.01	1.05	1.07	1.00	1.00	1.03	1.04
24'	1.01	1.02	1.06	1.08	1.00	1.00	1.07	1.09
28'	1.01	1.02	1.06	1.08	--	--	--	--
30'	1.01	1.02	1.06	1.08	1.01	1.02	1.09	1.10
32'	1.01	1.02	1.07	1.09	1.01	1.02	1.10	1.11
36'	1.01	1.02	1.08	1.10	1.01	1.02	1.10	1.11
40'	1.02	1.03	1.08	1.11	1.01	1.02	1.10	1.11
50'	1.04	1.06	1.10	1.15	1.02	1.03	1.10	1.12
60'	1.05	1.07	1.12	1.18	1.03	1.04	1.15	1.18
70'	1.07	1.09	1.14	1.20	1.03	1.04	1.15	1.19
80'	1.08	1.10	1.15	1.21	1.04	1.05	1.16	1.20
90'	1.09	1.10	1.16	1.22	1.04	1.05	1.17	1.21
100'	1.09	1.10	1.17	1.23	1.04	1.05	1.17	1.22
110'	1.10	1.11	1.18	1.24	1.05	1.06	1.18	1.24
120'	1.11	1.12	1.19	1.25	1.06	1.07	1.19	1.27
130'	--	--	--	--	1.07	1.08	1.20	1.30
140'	1.13	1.14	1.20	1.27	1.08	1.09	1.22	1.33
150'	--	--	--	--	1.09	1.10	1.24	1.35

Alternate Costs for Steel Buildings

Floors, cost difference per square foot

4" concrete with reinforced mesh and 2" sand fill	0
5" concrete with reinforced mesh and 2" sand fill	+.27
6" concrete with reinforced mesh and 2" sand fill	+.54
6" concrete with 1/2" rebars, 12" x 12" o.c. and 2" sand fill	+1.08
3" plant mix asphalt on 4" untreated rock base	-.41
2" plant mix asphalt on 4" untreated rock base	-.50
Gravel (compacted)	-.92
Dirt (compacted)	-1.38

Roof Covering, cost difference per square foot

26 gauge galvanized steel	0
24 gauge galvanized steel	+.13
22 gauge galvanized steel	+.44
26 gauge colored steel	+.35
24 gauge colored steel	+.56
.024 aluminum	+.31
.032 aluminum	+.65
.032 colored aluminum	+.96
No roof cover (just purlins)	-1.30

Wall Covering, cost difference per square foot

26 gauge colored steel	+.35
24 gauge colored steel	+.54
26 gauge galvanized steel	0
24 gauge galvanized steel	+.14
22 gauge galvanized steel	+.45
26 gauge aluminum and steel	+.29
24 gauge aluminum and steel	+.49
.024 aluminum	+.31
.032 aluminum	+.64
.032 colored aluminum	+.98
26 gauge galvanized steel insulated panel	+.44
26 gauge colored steel insulated panel	+.86
26 gauge aluminum steel insulated panel	+.69
26 gauge galvanized steel mono panel	+.87
26 gauge colored steel mono panel	+1.33
No wall cover (just girts)	-1.41

These costs are to be added to or subtracted from the building square foot costs on page 194.

Open Endwall (includes ridged frame in lieu of post and beam), low profile building

Cost Deduction Per Each Endwall
Eave Heights

Width	10'	12'	14'	16'	18'	20'	24'
20'	$414	$495	$600	700	--	---	---
24'	600	640	800	820	---	---	---
30'	715	840	1,000	1,170	---	---	---
32'	765	895	1,075	1,240	$1,375	$1,600	$2,270
36'	840	950	---	---	---	---	---
40'	900	1,055	1,275	1,440	1,575	1,860	2,270
50'	990	1,170	1,450	1,645	1,920	2,225	2,670
60'	1,110	1,300	1,575	1,840	2,280	2,670	3,360
70'	---	1,430	1,580	2,100	2,525	2,930	3,725
80'	---	1,500	1,840	2,280	1,350	3,125	4,340
90'	---	---	1,975	2,225	2,920	3,230	4,575
100'	---	---	1,660	2,220	2,740	3,250	4,720
110'	---	---	1,540	2,050	2,620	3,210	4,890
120'	---	---	1,375	1,780	2,475	3,210	4,890
130'	---	---	1,180	1,720	2,370	3,100	4,890
140'	---	---	860	1,430	2,290	2,900	4,630
150'	---	---	685	1,300	1,970	2,670	4,530

These costs are to be subtracted from the basic building cost.

Open Sidewalls, per linear foot of wall

Eave Height	10'	12'	14'	16'	18'	20'	24'
	$2.50	$3.00	$3.50	$4.00	$4.50	$5.00	$6.00

These costs are to be subtracted from the basic building cost.

Alternate Costs for Steel Buildings

Open Endwall (includes ridged frame in lieu of post and beam), high profile building

Cost Deduction Per Each Endwall

Width	Eave Heights						
	10'	12'	14'	16'	18'	20'	24'
20'	$510	$540	$765	---	---	---	---
24'	590	775	960	---	---	---	---
28'	870	975	1,130	---	---	---	---
30'	950	1,045	1,240	---	---	---	---
32'	1,000	1,090	1,300	$1,480	$1,720	$1,940	$2,220
36'	1,130	1,220	1,470	---	---	---	---
40'	1,345	1,440	1,670	1,950	2,210	2,410	2,815
50'	1,690	1,940	2,190	2,410	2,620	2,815	3,640
60'	2,225	2,410	2,815	3,010	3,390	3,830	4,625
70'	2,600	3,200	3,390	3,860	4,220	4,625	5,710
80'	3,080	3,620	3,830	4,220	4,625	5,220	6,220
90'	---	4,450	5,030	5,370	5,340	6,350	7,740
100'	---	4,620	5,390	5,800	6,420	7,110	8,500
110'	---	---	6,990	7,740	8,520	9,075	10,660

Ridged Frame Endwall in lieu of post and beam, low profile building

Cost Deduction Per Each Endwall

Width	Eave Heights						
	10'	12'	14'	16'	18'	20'	24'
20'	$490	$525	$560	---	---	---	---
24'	525	550	570	---	---	---	---
28'	560	620	630	---	---	---	---
30'	590	630	660	---	---	---	---
32'	630	680	690	$790	$880	$930	$1,080
36'	745	790	830	---	---	---	---
40'	810	880	930	990	1,090	1,120	1,365
50'	1,110	1,175	1,260	1,325	1,410	1,480	1,710
60'	1,400	1,490	1,560	1,680	1,760	1,780	2,050
70'	1,930	1,975	2,280	2,175	2,260	2,290	2,490
80'	2,180	2,290	2,490	2,490	2,670	2,900	3,050
90'	---	2,800	3,100	3,290	3,290	3,520	3,970
100'	---	3,500	3,760	3,970	4,210	4,350	4,770
110'	---	---	4,360	4,770	4,770	5,010	5,540

Ridged Frame Endwall in lieu of post and beam, high profile building

Cost Deduction Per Each Endwall

Width	Eave Heights						
	10'	12'	14'	16'	18'	20'	24'
20'	$570	$600	$630	$640	---	---	---
24'	610	620	630	680	---	---	---
30'	630	650	820	---	---	---	---
32'	790	790	830	850	$900	$910	$1,010
36'	850	880	---	---	---	---	---
40'	870	930	980	1,030	1,080	1,110	1,260
50'	1,200	1,280	1,300	1,375	1,470	1,510	1,730
60'	1,540	1,600	1,680	1,375	1,760	1,830	2,060
70	---	2,060	2,220	2,280	2,280	2,425	2,680
80	---	2,490	2,710	2,800	2,900	2,990	3,100
90	---	2,490	2,710	2,800	2,900	3,800	3,970
100	---	---	4,170	4,275	4,360	4,430	4,700
110	---	---	5,160	5,160	5,320	5,370	5,775
120'	---	---	6,020	6,160	6,390	6,550	6,780
130'	---	---	6,780	7,180	7,310	6,975	8,020
140'	---	---	8,180	8,380	8,720	8,750	9,170

Alternate Costs for Steel Buildings

Canopies (sidewall location), cost per linear foot

Type and Size	5'	6'	8'	10'	12'	15'
High profile 20' bays	$25.64	$29.20	$36.56	$42.40	$46.61	$52.78
High profile 24' bays	24.98	28.12	35.15	40.24	44.02	50.40
Low profile 20' bays	20.11	23.15	29.74	35.26	41.42	47.69
Low profile 24' bays	19.36	22.71	28.44	33.75	38.94	45.32
			Cost Per Each Canopy			
Basic end cost	$192.52	$233.64	$323.40	$431.56	$581.90	$744.14

The above costs are for 26 gauge galvanized metal canopy.

Canopies (endwall location), add the following to sidewall location costs above.

Eave Height	10'	12'	14'	16'	18'	20'	24'
Cost	$224	$291	$291	$328	$364	$385	$520

Sliding Sidewall Doors, cost per opening, including framed opening.

Single Doors	10'	12'	14'	16'	18'	20'	24'
8' x 9'	$895	$895	$900	$1,000	$1,010	$1,080	$1,080
10' x 9'	945	1,010	975	1,080	1,110	1,120	1,140
10' x 11'	---	975	990	1,110	1,140	1,140	1,165
10' x 13'	---	---	930	1,120	1,165	1,165	1,175
12' x 9'	1,280	1,330	1,280	1,500	1,450	1,470	1,440
12' x 11'	---	---	1,310	1,460	1,500	1,520	1,520
12' x 13'	---	---	1,280	1,500	1,520	1,530	1,550
12' x 15'	---	---	---	1,490	1,550	1,620	1,620
14' x 9'	1,470	1,530	1,410	1,585	1,620	1,640	1,640
14' x 11'	---	1,560	1,310	1,560	1,640	1,600	1,710
14' x 13'	---	---	1,390	1,670	1,750	1,620	1,730
14' x 15'	---	---	---	1,585	1,740	1,800	1,810
16' x 9'	1,585	1,710	1,690	1,800	1,830	1,830	1,810
16' x 11'	---	1,640	1,710	1,830	1,860	1,880	1,880
16' x 13'	---	---	1,610	1,890	---	---	---

The above costs are for 26 gauge colored sliding. Doors in eave heights listed at the top of each column. If door is in an endwall, use the lowest cost of size desired and add $220.00.

Hollow Metal Walk-Thru Doors, cost per opening including framed opening.

	1-3/8" Thick				1-3/4" Thick			
	Flush		Half Glass		Flush		Half Glass	
Door Size	Galvanized	Colored	Galvanized	Colored	Galvanized	Colored	Galvanized	Colored
2'0" x 6'8" single	$455	$485	$550	$595	---	---	---	---
2'6" x 6'8" single	465	490	550	595	---	---	---	---
3'0" x 6'8" single	455	530	570	660	$660	$730	$750	$800
3'0" x 7'0" single	455	530	570	660	660	730	750	835
3'4" x 7'0" single	---	---	---	---	680	770	790	870
4'0" x 7'0" single	---	---	---	---	750	870	900	975
6'0" x 6'8" pair	760	900	930	1,050	1,050	1,190	1,290	1,380
6'0" x 7'0" pair	810	960	970	1,110	1,120	1,250	1,285	1,440
8'0" x 7'0" pair	---	---	---	---	1,200	1,330	1,440	1,620

Half glass doors include glazing.

Additives: When walk-thru door is used as a pilot in a sliding door, **add** $77.50 to the costs above.

Heavy duty lockset	Add $150 each	Panic hardware, pair	Add $1,000 pair
Door closer	Add $145 each	Weatherstripping, bronze	
Panic hardware, single	Add $500 each	and neoprene	Add $4.25/LF

Additional Costs for Steel Buildings

These costs are to be added to the basic building cost.

Framed Openings, cost per opening

				Eave Height			
Location	10'	12'	14'	16'	18'	20'	24'
Sidewalls	$300	$290	$320	$330	$430	$460	$700
Endwalls	430	410	440	450	520	570	650

These costs are for the standard widths of 8', 10', 12', 14', 16', 18', and 20'. If the width is other than standard, add $75.00 per opening. If there is only one opening per building, add $100.00.

Gutters and Downspouts, cost per linear foot

Eave gutter (4")	
24 gauge galvanized steel	$7.00
24 gauge colored steel	7.30
.032 aluminum	6.50
Valley gutter (6")	
12 gauge galvanized steel	14.10
24 gauge galvanized steel	11.90
Downspouts	
4" x 4" x 26 gauge galvanized steel	4.90
4" x 4" x 26 gauge colored steel	5.15
6" x 6" x 24 gauge galvanized steel	5.40
6" x 6" x 24 gauge colored steel	6.20

Door Hoods, cost each

Width of Door	Galvanized	Colored
12'	$140.00	$160.00
16'	180.00	200.00
20'	210.00	230.00

Louvers, cost each, including screens

	Fixed		Adjustable	
Size	Galvanized	Color	Galvanized	Color
3' x 2'	$240	$250	$315	$330
3' x 3'	250	325	330	380
5' x 5'	305	305	475	535

Insulation, cost per square foot of surface area

	Roof Thickness			Wall Thickness		
Size and Type	1"	1-1/2"	2"	1"	1-1/2"	2"
.6 pound density						
White vinyl faced	$.80	$.85	$.90	$.85	$.85	$1.05
Colored vinyl faced	.95	1.00	1.10	1.05	1.10	1.25
Aluminum faced	1.05	1.10	1.25	1.15	1.25	1.35
.75 pound density						
White vinyl faced	.85	.90	1.00	.90	1.00	1.10
Colored vinyl faced	1.00	1.05	1.15	1.10	1.15	1.35
Aluminum faced	1.15	1.15	1.35	1.25	1.35	1.45

Overhangs (sidewall location), cost per linear foot of overhang, 26 gauge galvanized

	20' Bays			24' Bays		
Size and Type	3'	4'	5'	3'	4'	5'
No soffit	16.31	19.77	22.82	14.99	17.91	20.00
Soffit	35.17	41.39	45.86	34.07	38.99	46.34
Color, soffit	39.36	41.39	47.15	35.40	41.39	45.86
Color, no soffit	17.42	21.30	24.77	16.32	19.55	22.93
Color, galvanized soffit	36.82	42.48	48.43	35.40	41.39	47.15
Color, color soffit	37.82	43.90	49.54	54.36	41.39	48.45

Overhangs (endwall location), cost per linear foot of overhang, 26 gauge galvanized

Size and Type	3'	4'	5'	6'
No soffit	10.48	17.36	16.98	19.92
Galvanized soffit	25.36	31.34	37.10	41.67
Color, soffit	31.65	30.80	38.62	43.07
Color, no soffit	11.47	14.74	18.71	22.38
Color, galvanized soffit	26.02	33.74	38.62	44.25
Color, color soffit	27.52	33.74	39.70	43.07

Additional Costs for Steel Buildings

These costs are to be added to the basic building cost

Skylights, Polycarbonate, with curb

	2' x 2'	4' x 4'	4' x 8
Single dome	$101.00	$350.00	$425.00
Double dome	120.00	390.00	480.00
Triple dome	150.00	450.00	670.00
Double, ventilating	295.00	590.00	750.00

Partitions, Interior, 26 gauge steel, cost per square foot of partition with two sides finished

Painted drywall finish	$1.40
Painted plywood, fire retardant	5.55

Ventilators, round type, includes screen (gravity type), cost each

Diameter	Galvanized	Stationary Colored	Aluminum	Galvanized	Rotary Colored	Aluminum
12"	$135	$145	$235	$210	$235	$300
16"	190	200	275	270	290	380
20"	225	235	320	340	355	435
24"	245	250	365	380	390	500

Ventilators, ridge type, includes screen and damper

Throat Size	4"	9"	12"	14"
Cost per linear foot, galvanized	$35.50	$54.50	$68.50	$78.50
Cost per linear foot, colored	37.50	59.00	74.00	84.50

Ventilator-Dampers, cost each

Diameter	12"	16"	20"	24"
Damper only	$53.30	$73.00	$76.60	$88.00
Dampers with cords and pulleys	109.00	135.00	165.00	190.00

Continuous Ridge Ventilator, includes screen and damper, cost per 10 foot unit

Size & Type	First 10' Galvanized	Color	Each Additional 10' Galvanized	Color
9" throat	$535	$605	$500	$555
10" throat	605	665	540	605
12" throat	390	870	720	770

Steel Sliding Windows, includes glass and screens, cost per window

3' x 2'6"	$280
6' x 2'6"	345
6' x 3'8"	400

Smoke and Heat Vents, automatic control, cost per 10 foot unit

Size & Type	First 10' Galvanized	Color	Each Additional 10' Galvanized	Color
9" ridge mounted	$2,100	$2,200	$1,710	$1,760
9" slope mounted	2,150	2,310	1,760	1,860

Add for operators:	
One or two 10 foot sections	Add $70.00
Two to seven 10 foot sections	Add $145.00

Aluminum Industrial Windows, includes glass and screens, cost per window

Size and Type	Project Out	Fixed
3' x 2'6"	$260	$185
2' x 2'8"	280	205
6' x 2'6"	370	255
6' x 3'8"	450	320

Aluminum Sliding Windows, includes glass and screens, cost per window

			Height		
Width	2'	2'6"	3'	3'6"	4'
2'	$205.00	$225.00	$230.00	$245.00	$260.00
3'	225.00	240.00	245.00	260.00	270.00
4'	---	255.00	270.00	270.00	300.00
5'	---	270.00	280.00	340.00	350.00
6'	---	300.00	340.00	370.00	395.00

If window is fixed, deduct $4.10 per window. For mullions add $8.25 each.

Commercial and Industrial Structures

Typical Physical Lives in Years by Quality Class

Building Type	Masonry or Concrete 1	2	3	4	5	6	Wood or Wood and Steel Frame 1	2	3	4	5
Urban Stores	70	70	70	60	60	60	70	60	60	60	--
Suburban Stores	70	60	60	60	60	--	60	50	50	45	--
Supermarkets	70	60	60	60	--	--	60	50	50	45	--
Small Food Stores	60	60	60	60	--	--	50	50	45	45	--
Discount Houses	70	60	60	60	--	--	60	50	50	45	--
Banks and Savings Offices	70	70	70	60	60	--	60	60	60	50	50
Department Stores	70	60	60	60	--	--	60	50	50	45	--
General Office Buildings	60	60	60	60	--	--	60	50	50	45	--
Medical-Dental Buildings	60	60	60	50	--	--	60	50	50	45	--
Convalescent Hospitals	60	60	60	50	--	--	55	50	50	45	--
Funeral Homes	70	70	70	60	60	--	60	60	60	50	50
Restaurants	70	60	60	60	--	--	60	50	50	45	--
Theaters	50	50	50	50	--	--	50	45	45	40	--
Service Garages	60	60	50	50	--	--	45	45	40	40	--
Auto Service Centers	50	50	45	45	--	--	--	--	--	--	--
Warehouses	55	55	50	50	--	--	--	--	--	--	--
Light Industrial Buildings	55	55	50	50	--	--	--	--	--	--	--
Factory Buildings	40	40	35	35	--	--	--	--	--	--	--

Service Stations located on main highways or in high land value areas can be expected to become obsolete in 20 years. Other service stations can be expected to become obsolete in 25 years. Reinforced Concrete Department Stores have a typical physical life of 80 years for class one structures and 70 years for lower quality class structures. Steel Buildings have a typical physical life of 50 years.

Normal Percent Good Table

Average Life in Years — 20, 25, 30, 35 Years

Age	20 Yr Rem. Life	% Good	25 Yr Rem. Life	% Good	30 Yr Rem. Life	% Good	35 Yr Rem. Life	% Good
0	20	100	25	100	30	100	35	100
1	19	95	24	97	29	98	34	99
2	18	90	23	93	28	96	33	97
3	17	85	22	90	27	93	32	95
4	16	79	21	86	26	90	31	93
5	15	73	20	82	25	88	30	91
6	14	67	19	78	24	85	29	89
7	13	61	18	74	23	82	28	87
8	12	56	17	70	22	79	27	85
9	11	51	16	65	21	75	26	83
10	10	49	15	60	20	72	25	80
11	9	48	14	56	19	68	24	78
12	9	46	13	52	18	65	23	75
13	8	44	12	50	17	61	22	72
14	7	43	11	48	16	58	21	69
15	6	43	10	47	15	54	20	66
16	6	41	9	46	14	50	19	63
17	5	39	8	45	13	49	18	60
18	5	38	8	44	12	48	17	57
19	5	37	7	43	12	47	16	54
20	4	35	7	42	11	47	15	51
21	4	34	6	41	11	46	14	50
22	4	33	6	40	10	45	13	49
23	3	32	5	39	10	44	13	48
24	3	30	5	38	9	43	12	47
25	3	29	5	37	9	43	12	47
26	3	28	4	36	8	42	11	46
27	2	27	4	35	8	41	11	45
28	2	25	4	34	7	40	10	44
29	2	24	4	33	7	39	10	43
30	2	23	3	32	6	38	9	43
31	2	21	3	31	6	37	9	42
32	1	20	3	30	5	36	8	42
33	--	--	3	29	5	35	8	41
34	--	--	3	28	5	35	7	40
35	--	--	2	27	5	34	7	39
36	--	--	2	26	4	33	6	38
38	--	--	2	24	4	32	6	37
	--	--	2	22	3	30	5	36
	--	--	--	--	2	26	4	32
	--	--	--	--	2	23	4	30
	--	--	--	--	1	20	3	27
	--	--	--	--	--	--	2	24
	--	--	--	--	--	--	1	20

Average Life in Years — 40, 45, 50, 55 Years

Age	40 Yr Rem. Life	% Good	45 Yr Rem. Life	% Good	50 Yr Rem. Life	% Good	55 Yr Rem. Life	% Good
0	40	100	45	100	50	100	55	100
2	38	98	43	99	48	99	53	99
4	36	96	41	97	46	98	51	98
6	34	93	39	95	44	97	49	97
8	32	90	37	93	42	95	47	96
10	30	86	35	90	40	93	45	95
12	28	82	33	87	38	91	43	94
14	26	78	31	84	36	88	41	92
16	24	73	29	81	34	85	39	90
18	22	68	27	77	32	82	37	88
20	20	63	25	73	30	80	35	86
22	18	58	23	69	28	77	33	83
24	17	53	21	65	26	73	31	80
26	15	50	20	60	24	69	29	77
28	14	48	18	55	23	65	27	74
30	13	47	17	50	21	61	26	71
32	11	45	15	49	20	57	24	67
34	10	44	14	48	18	53	22	63
36	9	43	13	47	17	50	21	59
38	8	42	12	46	16	48	19	55
40	8	40	11	44	14	47	18	52
42	7	39	10	43	13	46	17	50
44	6	38	9	42	12	45	16	49
46	6	36	8	41	11	44	15	48
49	5	35	7	40	10	43	14	47
50	5	34	7	38	10	42	13	45
52	4	32	6	37	9	41	12	44
54	4	31	6	36	8	40	11	43
56	3	30	5	35	8	39	10	42
58	3	29	5	34	7	38	9	41
60	3	27	4	32	7	37	9	40
62	2	26	4	31	6	36	8	39
64	2	25	4	30	6	35	8	38
66	2	24	3	29	5	34	7	37
68	2	22	3	28	5	33	7	36
70	2	21	3	27	4	32	6	36
72	1	20	3	25	4	31	6	35
74	--	--	2	24	4	30	5	34
76	--	--	2	23	3	28	5	32
82	--	--	1	20	3	26	4	30
84	--	--	--	--	2	24	4	29
88	--	--	--	--	2	22	3	27
92	--	--	--	--	1	20	2	25
96	--	--	--	--	--	--	2	23
102	--	--	--	--	--	--	1	20

Average Life in Years — 60, 70 Years

Age	60 Yr Rem. Life	% Good	70 Yr Rem. Life	% Good
0	60	100	70	100
2	58	99	68	99
4	56	99	66	99
6	54	98	64	99
8	52	97	62	98
10	50	96	60	98
12	48	95	58	97
14	46	94	56	96
16	44	93	54	96
18	42	92	52	95
20	40	89	50	94
22	38	87	48	93
24	36	85	46	92
26	34	83	45	91
28	32	81	42	89
30	30	78	40	87
32	29	75	39	85
34	27	72	37	83
36	25	69	35	81
38	24	66	33	79
40	22	63	31	76
42	21	60	30	73
44	20	56	29	70
46	18	52	27	67
48	17	49	26	64
50	16	48	25	61
52	15	47	23	58
54	14	46	22	56
56	13	46	21	54
58	12	45	20	52
60	11	44	19	50
64	10	42	17	48
68	9	40	15	46
72	8	38	13	44
76	7	36	12	43
80	6	35	11	41
86	5	32	9	39
92	4	29	8	36
100	3	25	6	33
108	2	22	4	29
112	1	20	3	27
122	--	--	2	24
130	--	--	1	20

...nal Costs for Commercial and Industrial Structures

Additional Costs
for Commercial and Industrial Structures

Section Contents

Additional Structure Costs

Basements

Cost includes concrete floor and walls, open ceiling, minimum lighting, no plumbing, and no wall finish. Cost per square foot of floor at 12' wall height.

Area	500	1,000	1,500	2,000	3,000	4,000	5,000	7,500	10,000	15,000	20,000
Cost	29.92	26.76	23.36	21.12	20.54	17.95	17.27	16.70	14.69	14.01	13.19

Add or subtract the amount listed in the table below to the square foot of floor cost for each foot of wall height more or less than 12 feet.

Wall Height Adjustment
Square Foot Area

Area	500	1,000	1,500	2,000	3,000	4,000	5,000	7,500	10,000	15,000	20,000
Cost	1.99	1.47	1.30	1.05	.85	.79	.74	.58	.47	.37	.35

Canopies, per S.F. of canopy area

Light frame, flat roof underside, plywood and paint or cheap stucco supported by wood or light steel posts, 4" to 6" wood fascia.	$12.90 to $13.40
Average frame, underside of good stucco, flat roof, cantilevered from building or supported by steel posts, 6" to 12" metal fascia.	$14.00 to $18.30
Same as above but with sloping shake or tile roof.	$15.60 to $20.40
Corrugated metal on steel frame.	$12.50 to $17.00

Canopy Lights, per S.F. based on one row of lights for 5' canopy

Recessed spots (1 each 6 linear feet)	$2.00
Single tube fluorescent	3.70
Double tube fluorescent	5.10

Public Address Systems, speakers attached to building. No conduit included.

Base cost, master control	$650 to $1,250
Per indoor speaker	130
Per outdoor speaker	260
Sound Systems, cost per unit	
Voice only, per unit	$75 to $110
Music (add to above), small units	60 to 90
Music (add to above), large units	90 to 110
Larger installations cost the least per unit.	

Docks for unloading trucks. Cost per S.F. of dock at 4' height

L x W	10'	20'	30'	50'	100'	200'
5'	22.75	19.90	17.50	15.90	15.50	15.10
10'	18.70	16.00	12.60	11.80	11.10	11.00
15'	17.70	12.90	12.00	12.50	10.00	9.60
20'	16.00	12.20	10.70	11.00	9.60	8.90

Cost includes compacted fill, three concrete walls, concrete floor, and rock base.

Intercommunication Systems

Master control, base cost	$1,200 to $3,600
Cost per station	90 to 130
Nurses call system, per station	130 to 245

Security Systems

Control panel	$660 to $990
Each door or window secured	66 to 112
Heat detectors, each	89 to 150
Smoke detectors, each	165 to 221
Motion detectors, each	221 to 282

Loading Ramps, cost per S.F. of ramp

Size	
Under 300 S.F.	$7.10
Over 300 S.F.	6.60

Dock Levelers and Lifts, cost each

Dock leveler, manual	$510
Dock leveler, mechanical	2,650
Powered platform dock leveler	
6' x 6' recessed	2,755
6' x 8' recessed	2,820
Electro-hydraulic, pit recessed scissor lift	
5,000 lb. capacity, 6' x 8'	8,225
10,000 lb. capacity, 8' x 10'	15,200
20,000 lb. capacity, 8' x 12'	25,200

Doors, with hardware

Exterior, commercial, cost per door	
Glass in wood (3' x 7')	$645 to $1,000
1/4" plate in aluminum (3' x 7')	1,050 to 1,650
Automatic, tempered glass (3' x 7')	4,990 to 7,535
Residential type (3' x 7')	240 to 445
Interior, commercial and industrial, cost per S.F.	
Hollow core wood	$10.75 to $11.60
Solid wood	11.10 to 14.30
Hollow core metal	24.20 to 28.60
Fire, cost per S.F.	
Hollow metal, 1-3/4"	$37.40 to $49.50
Metal clad, rolling	31.25 to 50.00
Metal clad, swinging	43.90 to 62.60

Elevators, Freight, Electric, car and equipment, per shaft, car speed in feet per minute, 2 stop

Capacity	50 to 75	100 to 150	200
2,500 lbs	$44,900	---	---
3,000	47,300	$54,200	$62,700
3,500	50,700	56,800	63,800
4,000	53,200	59,100	68,400
5,000	56,800	64,800	73,200
6,000	62,600	70,900	79,000
8,000	70,900	79,000	89,700
10,000	81,500	89,700	102,700

For manual doors, **add** $4,300 for each stop. For power operated doors, **add** $6,500 for each additional stop. **Add** $6,500 per car for self-leveling cars. **Add** for double center opening doors, per stop $275. **Add** for deluxe cab (raised panel, interior, drop ceiling) $3,250.

Elevators, Freight, Hydraulic, 100 F.P.M.

Shaft, car, machinery	
2,000 lb. capacity	$31,600
6,000 lb. capacity	39,000
Cost per stop	
Manual doors	7,100
Automatic doors	12,900

Roll-Up Metal Warehouse Door
with chain operator, cost each

10' x 10'	$1,660
12' x 12'	2,220
14' x 14'	2,440
Fusible link (add to above)	445
Motor controlled (add to above)	225

Draperies, cost per square yard of opening

	44" high	68" high	96" high
Minimum	$32.00	$71.90	$18.10
Good quality	91.70	71.90	52.80
Better quality	109.30	80.00	63.20

Escalators, cost per flight up or down

Total Rise	32" W	40" W	48" W
10' to 13'	$93,200	$103,600	$103,300
14'	96,100	99,900	107,500
15'	99,500	103,700	112,300
16'	102,400	110,200	112,600
17'	105,100	112,600	113,600
18'	107,500	115,600	114,000
19'	111,400	116,500	114,800
20'	115,600	120,300	118,800
21'	118,900	121,700	121,500

Add for glass side enclosure: $11,100 - $13,100.

Dumbwaiters, includes door, traction type

	1st Two Stops	Add. Stops
Hand operated, 25 fpm (no doors)		
25 lb.	$2,800 to $3,000	$1,400
75 lb.	3,600 to 4,600	1,400
Electrical, with machinery above, floor loading		
100 lb., 50 fpm	$8,800 to $10,325	$2,400
300 lb., 50 fpm	9,600 to 10,325	2,400
500 lb., 50 fpm	11,000 to 13,900	2,400
500 lb., 100 fpm	15,050 to 17,000	(5 stop)

Elevators, Passenger, Electric, car and machinery cost, per shaft

Capacity	200 F.P.M., 5 Stops	350 F.P.M., 5 Stops	500 F.P.M., 10 Stops
2,000 lbs.	$71,400	$76,200	$164,900
2,500 lbs.	87,200	81,500	170,300
3,000 lbs.	88,300	86,200	172,300
3,500 lbs.	88,800	90,900	173,300
4,000 lbs.	89,300	96,700	174,500
4,000 lbs (Hospital)	90,400	98,800	179,700

Add for each additional stop: 200 or 350 F.P.M. units, $5,600; 500 F.P.M. units, $9,250. Deduct for multi-shaft applications, $2,600 to $5,600 per additional shaft. Add for rear-opening door: $7,950 to base cost, plus $5,700 per door.

Fire Extinguishers, cost each

Fire hose and cabinet	$400 to $600
Extinguisher cabinets	150 to 200
Extinguishers	125 to 300

Fill, compacted under raised floor, includes perimeter retaining wall but not slab, per C.F.

Up to 10,000 S.F.	$.80 to $1.20
Over 10,000 to 50,000 S.F	.60 to .90

Fire Escapes

Type	Unit	Cost
Second story	Each	$2,900 to $4,050
Additional floors	Per story	1,745 to 2,600

Fire Sprinklers, cost per S.F. of area served

Area	Wet Pipe System Normal	Special*
to 2,000	$3.00	$3.50
2,001 to 4,000	2.00	3.00
4,001 to 10,000	1.75	2.50
Over 10,000	1.65	2.45
	Dry Pipe System Normal	Special*
to 2,000	$3.50	$4.00
2,001 to 4,000	2.20	2.75
4,001 to 10,000	2.00	2.60
Over 10,000	1.90	2.50

Costs include normal installation, service lines, permit and valves. *Special hazard systems are custom engineered to meet code or insurance requirements and are usually so identified by a metal plate attached to the riser.

Overhead Suspended Heaters, per unit

25 M.B.T.U.	$ 760 to $ 910
50	880 to 970
75	980 to 1,090
100	1,125 to 1,270
150	1,360 to 1,470
200	1,590 to 1,660
250	1,760 to 1,770

Fireplace

	1 Story	2 Story
Freestanding wood burning heat circulating prefab fireplace, with interior flue, base and cap.	$1,200	---
Zero-clearance, insulated prefab metal fireplace, brick face.	1,725	$2,300
5' base, common brick, on interior face.	2,300	2,550
6' base, common brick, used brick, face brick or natural stone on interior face with average wood mantle.	3,625	3,800
8' base, common brick, used brick or natural stone on interior face, raised hearth.	5,050	5,525

Electric Heating Units

Baseboard, per linear foot	$53.00 to $64.00
Add for thermostat	75.00
Cable in ceiling, per S.F.	1.60 to 2.15
Wall heaters, per K.W.	155.00 to 180.00

Heating and Cooling Systems

	Cost per S.F. of Floor Area	
Type and Use	Heating Only	Heating & Cooling
Urban stores	$2.30 to $3.35	$4.60 to $8.60
Suburban stores	2.30 to 3.35	3.25 to 7.10
Small food stores	1.70 to 2.85	3.90 to 7.50
Supermarkets	1.60 to 2.80	3.60 to 8.80
Discount houses	1.60 to 2.80	3.70 to 6.50
Department stores	2.35 to 3.35	3.70 to 9.00
Reinforced concrete	2.35 to 3.45	5.25 to 10.50
General offices		
Forced air	3.25 to 4.50	7.90 to 11.30
Hot & chilled water	---	9.20 to 15.50
Bank and savings	4.00 to 5.90	4.80 to 12.90
Medical-Dental		
Forced air	3.60 to 5.70	9.15 to 14.60
Hot & chilled water	---	12.35 to 17.00
Convalescent hospitals		
Forced air	3.60 to 5.25	6.75 to 10.50
Hot & chilled water	---	9.10 to 15.00
Funeral homes	3.25 to 4.50	9.10 to 15.00
Theaters	3.25 to 4.50	6.50 to 10.40
Restaurants	2.70 to 4.00	9.00 to 15.60
Industrial buildings	1.90 to 3.90	4.80 to 8.00
Interior offices	3.25 to 5.65	5.25 to 9.75

Use the higher figures where more heating and cooling density is required.

Additional Structure Costs

Kitchen Equipment, cost per linear foot of stainless steel fixture

Work tables	$245 to $310
Serving fixtures	255 to 330

Mezzanines, cost per S.F. of floor

Unfinished (min. lighting and plumbing)	16.25 to 19.70
Store mezzanines	27.40 to 32.90
Office mezzanines (without partitions)	30.10 to 38.40
Office mezzanines (with partitions)	38.40 to 60.20

Costs include floor system, floor finish, stairways, lighting, and partitions where applicable.

Seating, cost per seat space

Theater, economy	$125.00
Theater, lodge	240.00
Pews, bench type	80.00
Pews, seat type	105.00

Partitions, cost per S.F. of surface

Gypsum on wood frame, (finished both sides) 2" x 4" wood studs, 24" on center with 1/2" gypsum board, taped, textures and painted.	$3.85
Plaster on wood frame (finished both sides) 2" x 4" wood studs, 24" on center with 2 coats plaster over gypsum lath, painted with primer and 1 coat enamel.	$7.65

Pneumatic Tube Systems

Twin tube, two station system

2-1/4" round, 500 to 1,500 feet	$15,400 to $27,400
3" round, 500 to 1,500 feet	16,000 to 30,700
4" round, 500 to 1,500 feet	16,500 to 35,100
4" x 7" oval, 500 to 1,500 feet	26,400 to 43,900

Automatic System, twin tube, cost per station

4" round, 500 to 1,500 feet	$19,700 to $26,200
4" x 7" oval, 500 to 1,500 feet	26,400 to 28,500

Skylights. Plastic Rectangular Domes, cost per unit

Size	Single Plastic Panel		Double Plastic Panel	
	Skylight Only	With 4" or 9" Insulated Curb	Skylight Only	With 4" or 9" Insulated Curb
16" x 16"	$130	$225	$160	$240
16" x 24"	150	250	180	250
16" x 48"	170	290	240	350
24" x 24"	170	270	250	270
24" x 32"	180	290	275	350
24" x 48"	210	330	290	400
28" x 92"	360	535	570	660
32" x 32"	180	290	250	350
32" x 48"	240	340	325	400
32" x 72"	310	480	530	610
39" x 39"	250	340	340	400
39" x 77"	380	535	660	710
40" x 61"	350	480	530	720
48" x 48"	270	400	400	490
48" x 64"	380	590	590	710
48" x 72"	435	680	680	800
48" x 92"	570	735	880	1,120
48" x 122"	810	970	1,110	1,290
58" x 58"	435	600	710	820
60" x 72"	550	735	855	980
60" x 92"	700	890	1,070	910
64" x 64"	560	710	770	1,220
77" x 77"	800	1,010	1,290	1,420
94" x 94"	1,420	1,660	2,325	2,580

Triple dome skylights cost about 30% more than double dome skylights.

Plastic Circular Dome Skylights, cost each

Size	Single Plastic Panel		Double Plastic Panel		Additives	
	4" Curb	9" Curb	4" Curb	9" Curb	Ceiling Dome	Wall Liner
30"	$ 510	$ 520	$ 600	$ 630	$ 240	$240
36"	520	540	630	680	310	260
48"	690	760	880	920	415	310
60"	870	910	1,170	1,180	590	320
72"	1,310	1,290	1,740	1,820	780	390
84"	1,800	1,910	2,510	2,650	1,120	460
96"	2,640	2,760	3,675	3,930	1,210	550

The above costs are for single skylights. For three or more, deduct 20%.

Plastic Pyramid Skylights, cost each

Size	Height	Installed Cost	
		2 or Less	3 or More
39" x 39"	34"	$ 830	$ 690
48" x 48"	42"	1,170	980
58" x 58"	49"	1,300	1,420

Plastic Continuous Vaulted Skylights, cost per L.F.

Width	Single Panel	Double Panel
16"	$90	$128
20"	95	130
24"	105	150
30"	130	180
36"	140	190
42"	150	200
48"	180	220
54"	190	240
60"	200	250
72	220	270
84"	260	350

Ventilators, Roof, Power Type, cost each

Throat Dia.	2 or Less	3 or More	Add for Insulated Curb
6"	$360	$350	$70
8"	580	530	95
10"	700	650	115
12"	785	755	120
18"	860	820	130
24"	950	910	160
30"	1,720	1,590	170
36"	1,840	1,620	175
48"	4,040	3,660	210

Above costs are for a single-speed motor installation. Dampers and bird screens are included. Add: Explosive-proof units, add $300 each. Two-speed motors, add $550 to $800 each. Plastic coating, depending on size of unit, add $100 to $175 each.

Plastic Ridge Type Skylights, cost per linear foot

Width*	Single Panel	Double Panel
18"	$190	$280
24"	210	340
30"	265	390
36"	330	450
42"	340	540
48"	360	700

*Width is from ridge to curb following slope of roof.

Wire Glass Skylights, Exterior Aluminum Frame, cost each

24" x 48"	$300
24" x 72"	365
24" x 96"	485
48" x 48"	485
48" x 72"	600
48" x 96"	730

Ventilators, Roof, Gravity Type, cost each

Throat Dia.	2 or Less	3 or More	Add for Insulated Curb
8"	$155	$150	$80
12"	170	155	90
18"	265	260	125
24"	320	305	135

Heat and Smoke Vents, cost each

Size	Plastic Dome Lid	Aluminum Covered Lid
32" x 32"	$ 760	$ 860
32" x 48"	900	925
50" x 50"	1,035	1,070
50" x 62"	1,170	1,330
50" x 74"	1,075	1,390
50" x 92"	1,410	1,550
50" x 98"	1,640	1,670
62" x 104"	2,080	2,130
74" x 104"	2,340	2,575

Walk-In Boxes, cost per S.F. of floor area

Temperature Range	50	100	200	300	400	500	600
Over 45°	$107.00	$68.55	$51.64	$45.63	$45.63	$39.74	$39.74
25° to 45°	119.98	74.99	57.42	51.42	51.42	45.42	45.42
0° to 25°	143.54	91.06	63.09	57.10	57.10	51.10	51.10
-25° to 0°	155.33	102.84	91.81	68.56	68.56	59.13	62.56

Cost Includes: Painted wood exterior facing, insulation as required for temperature, interior plaster, one 4 x 7 door per 300 S.F. of floor area. Costs are based upon 8' exterior wall height. Costs do not include machinery and wiring. Figure refrigeration machinery at $1,500.00 per ton capacity.

Material Handling Systems

Belt type conveyors, 24" wide.	
Horizontal sections, per linear foot	$170.00
Elevating, descending sections, per flight	270.00
Mail conveyors, automatic, electronic.	
Horizontal, per linear foot	1,590.00
Vertical, per 12' floor	19,400.00
Mail chutes, cost per floor, 5" x 14", aluminum	1,170.00
Linen chutes, 18 gauge steel, 30" diameter, per 10' floor	910.00
Disinfecting and sanitizing unit, each	260.00

Display Fronts

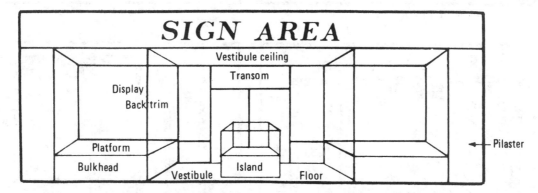

Display front costs may be estimated by calculating the in-place cost of each component or by estimating a cost per linear foot of bulkhead and multiplying by the bulkhead length. This section contains data for both methods. For most fronts, the cost per linear foot method is best suited for rapid preliminary estimates.

Bulkhead length is the distance from the inside of the pilaster and following along the bulkhead or glass to the inside of the opposite pilaster. This measurement includes the distance across entryways.

The cost per linear foot of bulkhead is estimated using the storefront specifications and costs in this section. This manual suggests linear foot costs for four quality types: low cost, average, good, and very good. Costs are related for each quality type in terms of flat or recessed type fronts.

Recessed type fronts include all components described in the specifications. Flat front costs do not include the following components: vestibule floor, vestibule and display area ceiling framing, back trim, display platform, lighting. The cost of automatic door openers is not included in front costs.

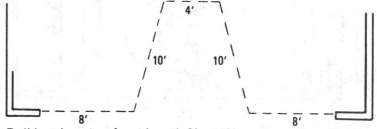

Bulkhead or storefront length 8' + 10' + 4' + 10' + 8' = 40'

Display Front, Best

Display Front, Best

Display Front, Good

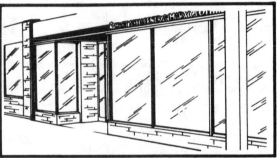

Display Front, Good

Display Front, Average

Display Front, Average

Display Front, Low Cost

Display Front, Low Cost

Display Fronts

All front costs are based upon a height of 10 feet from the floor level to the top of the display window. Variations from this standard should be adjusted by using the display window height adjustment costs shown with the front foot costs. These amounts are added or deducted for each foot of variation from the standard.

Bulkhead height variations will not require adjustment. Cost differentials, due to variations in bulkhead height, will be compensated for by equal variations in display window height if overall heights are equal.

Sign areas are based upon 4 foot heights. Cost adjustments are given for flat type and for recessed type fronts. A cost range is given for recessed type fronts because deeply recessed fronts will have lower linear foot costs for sign area components than will moderately recessed fronts because this cost is spread over a longer distance when recesses are deep.

Display island costs are estimated by applying 60 to 80 percent of the applicable linear foot cost to the island bulkhead length. Window height adjustments should be made, but sign height adjustments will not be necessary.

Components	Best	Good	Average	Low Cost
Bulkhead (0 to 4' high)	Vitrolite domestic marble or stainless steel.	Black Carrara flagstone, terrazzo or good ceramic tile.	Average ceramic tile, Roman brick or imitation flagstone.	Stucco, wood or common brick.
Window Area	Bronze or stainless steel. 1/4" plate glass with mitered joints.	Heavy aluminum. 1/4" plate glass, some mitered joints.	Aluminum. 1/4" plate glass.	Light aluminum with wood stops. Crystal or 1/4" plate glass.
Sign Area (4' high)	Vitrolite, domestic marble or stainless steel.	Black Carrara flagstone, terrazzo or good ceramic tile.	Average ceramic tile, Roman brick or imitation flagstone.	Stucco.
Pilasters	Vitrolite, domestic marble.	Black Carrara flagstone, terrazzo or good ceramic tile.	Average ceramic tile, Roman brick or imitation flagstone.	Stucco.
Vestibule Floor*	Decorative terrazzo.	Decorative terrazzo.	Plain terrazzo.	Concrete.
Vestibule and* Display Area Ceilings	Stucco or gypsum wallboard and texture.	Stucco or gypsum wallboard and texture.	Stucco or gypsum wallboard and texture.	Gypsum wallboard and texture.
Back Trim*	Hardwood veneer on average frame.	Gypsum wallboard and texture or light frame.	None.	None.
Display Platform Cover*	Excellent carpet.	Good carpet.	Average carpet.	Plywood with tile.
Lighting*	1 recessed spot per linear foot of bulkhead.	1 recessed spot per linear foot of bulkhead.	1 exposed spot per 2 linear feet of bulkhead.	1 exposed spot per 4 linear feet of bulkhead.
Doors	3/4" glass double doors.	Good aluminum and glass double door or single 3/4" glass door.	Average aluminum and glass double door.	Wood and glass.
Costs, Flat Fronts	$685.00/linear foot.	$455.00/linear foot.	$290.00/linear foot.	$250.00/linear foot.
Costs, Recessed Fronts	$730.00/linear foot.	$525.00/linear foot.	$350.00/linear foot.	$280.00/linear foot.
Display Window Adjustment per Foot of Height	$31.00/linear foot.	$28.00/linear foot.	$26.00/linear foot.	$26.00/linear foot.
Flat Front, Sign Area Adjustment per Foot of Height	$32.00/linear foot.	$18.90/linear foot.	$7.25/linear foot.	$2.60/linear foot.
Recessed Front, Sign Area Adjustment per Foot of Height	$11.00 to $16.00/ linear foot.	$6.50 to $9.50/ linear foot.	$2.90 to $4.40/ linear foot.	$1.40 to $1.65/ linear foot.

*Not included in flat front costs.

Lighting, cost per fixture

Open incandescent	$38.00 to $56.00
Recessed incandescent	56.00 to 98.00
Fluorescent exposed, 4' single	135.00 to 198.00
Fluorescent recessed, 4' single	145.00 to 250.00

Bulkhead Walls, cost per S.F. of wall

Up to 5' high, nominal 6" thick	
Concrete	$11.00 to $16.30
Concrete block	7.75 to 11.40
Wood frame	4.90 to 8.60

Ceiling, cost per S.F. of floor

Dropped ceiling framing	$1.55 to $2.25
Acoustical tile on wood strips	2.50 to 3.00
Acoustical plaster including lath	2.30 to 3.35
Gypsum board, texture and paint	1.75 to 2.70
Plaster and paint including lath	2.60 to 3.50

Entrances, cost per entrance

Aluminum and 1/4" float glass	
Single door, 3' x 7'	$1,180 to $1,960
Double door, 6' x 7'	2,090 to 2,890
Stainless steel and 1/4" float glass	
Single door, 3' x 7'	2,330 to 3,650
Double door, 6' x 7'	3,190 to 5,220
3/4" tempered glass	
Single door, 3' x 7'	2,440 to 3,480
Double door, 6' x 7'	4,060 to 5,120

Includes door, glass, lock handles, hinges, sill and frame.

Exterior Wall Finish, cost per S.F. of wall

Aluminum sheet baked enamel finish	$3.20 to $6.00
Brick veneer	
Common brick	$6.50 to $8.50
Roman	8.30 to 12.30
Norman	8.30 to 12.30
Glazed	10.60 to 15.90
Carrara glass	
Black	16.25 to 25.30
Red	17.60 to 26.70
Flagstone veneer	
Imitation	11.10 to 14.30
Natural	17.20 to 27.30
Marble	
Plain colors	25.60 to 33.00
With color variations	45.50 to 53.50
Stucco	2.50 to 3.30
Terrazzo	16.50 to 22.70
Tile, ceramic	10.10 to 15.30

Display Platforms, cost per S.F. of platform area

Framing up to 5' high	$5.20 to $8.00
Hardwood cover	3.10 to 5.00
Plywood cover	2.10 to 2.80

Glass and Window Frames, per S.F. of glass

Glass only installed	
1/4" float	$6.60 to $8.90
1/4" float, tempered	8.90 to 11.20
1/4" float, colored	7.10 to 9.40
Store front, 1/4" glass in aluminum frame	
Anodized, 8' high	20.60 to 28.30
Anodized, 6' high	24.90 to 33.20
Anodized, 3' high	30.70 to 38.80
Satin bronze, 6' high	28.90 to 38.00
Satin bronze, 3' high	35.30 to 44.30

Satellite Receiver Systems

Satellite receiver systems are common in mountain and rural areas, where TV reception is limited. They are also often installed in residential or commercial areas, for homes, motels or hotels, restaurants and businesses. The cost of the unit varies with the size of the dish, amplifier, control unit, and type of mounting. A solid dish costs roughly 25% more than a mesh dish.

Installed cost for an all-automatic, motorized system, including wiring to one interior outlet:

6' dish	$2,000 to $2,200	
8' dish	2,200 to 2,800	
10' dish (mesh)	3,000 to 3,800	
10' dish (solid)	4,400 to 5,000	
12' dish	3,500 to 4,000	
13-1/2' dish (mesh)	4,000 to 4,500	

Deduct $500 for manual units.

Signs

Lighted Display Signs, Cost per S.F. of sign area

Painted sheet metal with floodlights	$65.00 to $85.00
Porcelain enamel with floodlights	71.00 to 95.00
Plastic with interior lights	80.00 to 125.00
Simple rectangular neon with painted sheet metal faces and a moderate amount of plain letters	85.20 to 150.00
Round or irregular neon with porcelain enamel faces and more elaborate lettering	120.00 to 180.00
Channel letters - individual neon illuminated metal letters with translucent plastic faces, (per upright inch, per letter)	10.50 to 15.75

All of the above sign costs are for double-faced signs. Use 2/3 of those costs for single-faced signs. Sign costs include costs of installation and normal electrical hookup. They do not include the cost of a post. If signs are mounted on separate posts, post mounting costs must be added. These costs are for custom-built signs (one-at-a-time orders). Mass-produced signs will have lower costs.

Post Mounting Costs for Signs

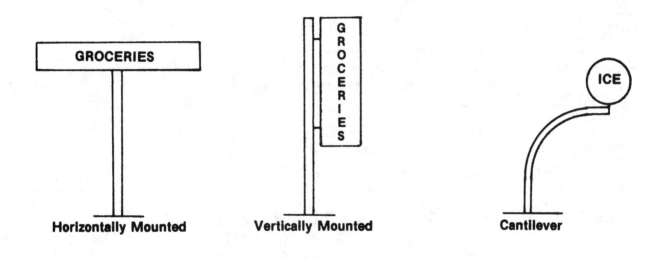

Horizontally Mounted   **Vertically Mounted**   **Cantilever**

Post Mounting Costs

Post Height	Pole Diameter at Base					
	4"	6"	8"	10"	12"	14"
15	$810	$960	$1,410	$1,910	$2,970	$3,560
20	960	1,130	1,640	1,980	3,540	3,980
25	1,050	1,300	1,700	2,360	3,940	4,350
30	1,150	1,540	1,760	2,580	4,160	4,820
35	--	1,675	1,980	2,850	4,710	5,180
40	--	1,780	2,410	3,040	4,890	5,700
45	--	--	2,830	3,460	5,340	5,970
50	--	--	3,040	3,940	5,660	6,540
55	--	--	--	4,160	5,970	7,120
60	--	--	--	4,390	6,390	7,440
65	--	--	--	--	6,860	7,910

If signs are mounted on separate posts, post mounting costs must be added. Post mounting costs include the installed cost of the post and foundation. On horizontally mounted signs, post height is the distance from the ground to the bottom of the sign. On vertically mounted signs, post height is the distance to the top of the post. For cantilevered posts, use one and one-half to two times the conventional post cost.

All of the above post costs are for single posts. Use 90% of the single post costs for each additional post.

Signs

If signs are mounted on buildings or canopies and if, because of the extra weight of the sign, extra heavy support posts or foundations are required, 125% of the post mounting cost should be used.

Sample calculations:

For example, the cost of a 4' x 25' plastic sign mounted on a 15' by 6" post shared by an adjacent canopy might be estimated as follows:

Sign cost 100 S.F. at $100.00	$10,000
Post cost $960 x 1/2	480

Total cost	$10,480

If this sign were mounted on a 8" post 20' high above the canopy with extra supports not needed, the cost might be estimated as follows:

Sign Cost	$10,000
Post Cost $1,640 x 1	1,640

Total Cost	$11,640

Sign Rotator

Small signs	Less than 50 S.F.	$1,700 to $1,850
Medium signs	50 to 100 S.F.	1,810 to 3,625
Large signs	100 to 200 S.F.	3,575 to 6,670
Extra large signs	Over 200 S.F.	$34.00 per S.F. of sign area

Yard Improvements

Bumpers pre-cast concrete or good quality painted wood.

Typical 4-foot lengths $9.40 to $12.50 per linear foot.

Asphaltic Concrete Paving, cost per square foot

	Under 1,000	1,000 to 2,000	2,000 to 5,000	5,000 to 12,000	12,000 or more
2", no base	$2.10	$1.05	$.88	$.78	$.71
3", no base	2.32	1.32	1.10	1.00	.94
4", no base	2.54	1.60	1.38	1.32	1.10
2", 4" base	3.58	2.04	1.66	1.54	1.38
3", 6" base	3.92	2.37	2.04	1.88	1.71

Concrete Paving, small quantities, per S.F.

4" concrete (without wire mesh) no base	$2.90
4" concrete (without wire mesh) 4" base	3.20
5" concrete (with wire mesh) no base	3.50
6" concrete (with wire mesh) no base	4.05
Add for color	.60

Gates, coin or card operated

Single gate, 10' arm	$3,000 to $5,000
Lane spikes, 6' wide	1,600 to 1,900

Striping

Parking spaces	
Single line between spaces	$10.00 per space
Double line between spaces	12.00 per space
Pavement line marking, 4"	.55 per LF

Curbs, per linear foot

Asphalt 6" high berm	$5.90 to $7.00
Concrete 6" wide x 12" high	8.60 to 10.75
Concrete 6" wide x 18" high	10.75 to 12.90
Wood bumper rail 6" x 6"	8.00 to 9.10

Yard Improvements

Lighting

Fluorescent arm-type fixtures	12 ft.	One arm unit, pole height			12 ft.	Two arm unit, pole height		Add for 3-fixture unit
		16 ft.	30 ft.			16 ft.	30 ft.	
2 tube, 60"	$1,140	$1,260	---		$1,470	$1,480	---	$600
4 tube, 48"	1,260	1,400	---		1,520	1,760	---	680
4 tube, 72"	1,340	1,490	$1,710		1,650	1,780	$2,040	790
6 tube, 72"	1,490	1,600	1,850		1,830	1,930	2,160	840
4 tube, 96"	1,540	1,670	1,930		1,890	2,040	2,350	920
6 tube, 96"	1,590	1,760	2,060		1,930	2,260	2,580	1,010

Mercury vapor	1-fixture unit, pole height			2-fixture unit, pole height			4-fixture unit, pole height		
	12 ft.	16 ft.	30 ft.	12 ft.	16 ft.	30 ft.	12 ft.	16 ft.	30 ft.
400 watt	$2,470	$2,560	$3,450	$3,450	$3,700	$4,570	$6,170	$6,780	$7,710
1,000 watt	---	3,200	3,590	---	4,320	5,430	---	---	8,930

Incandescent Scoop Reflector	1-fixture unit, pole height			2-fixture unit, pole height			Add for 3-fixture Unit
	12 ft.	16 ft.	30 ft.	12 ft.	16 ft.	30 ft.	
500 watt	$730	$860	$1,110	$970	$1,040	$1,340	$240
1,000 watt	920	1,040	1,340	1,220	1,280	1,590	300

Mall and Garden Lighting	1 fixture unit, pole height		
	3 ft.*	5 - 7 ft.	10 ft.
Incandescent 60-200 watt	$610	$ 875	$ 925
Mercury vapor 100-250 watt	680	1,110	2,360

*Area lights usually found in service stations.

Chain Link Fences, 9 gauge 2" mesh with top rail line posts 10' on center, per linear foot.

Height	Under 150 LF	150 to 1,000 LF	1,000 to 2,000 LF	Over 2,000 LF	3 strands barbed wire	Add for line posts 8' or less o.c.	Deduct For: No top rail	#11 gauge wire
4'	$10.90	$ 9.06	$ 8.63	$ 8.19	$2.18	$1.10	$1.64	$.93
5'	11.46	9.50	9.06	8.63	2.18	1.20	1.64	.98
6'	12.02	9.94	9.50	9.01	2.18	1.31	1.64	1.01
8'	13.10	10.92	10.38	9.83	2.18	1.42	1.64	1.10
10'	15.07	12.56	11.90	11.24	2.18	1.53	1.64	1.26
12'	17.47	14.53	13.76	13.00	2.18	1.64	1.64	1.47

Add for gates, per square foot of frame area $4.00 to $5.00
Add for gate hardware   per walkway gate $20.00 to $25.00
                            per driveway gate $40.00 to $48.00

Drainage, per linear foot, runs to 50'

6" non-reinforced concrete pipe	$9.10 to $9.70
8" reinforced concrete pipe	11.50 to 12.70
10" reinforced concrete pipe	12.70 to 14.20
12" reinforced concrete pipe	15.90 to 17.70

The cost varies with the depth of the pipe in the ground.

Drainage Items

Size and Type	Cost
Catch basin 2' x 2' x 2' deep	$475.00 to $540.00 ea.
For each additional foot in depth, add	90.00 ea.
Drop inlets	390.00 to 715.00 ea.
Manholes 4' diameter x 5' deep	1,050 to 1,585 ea.
For each additional foot in depth, add	115.00 per ft.

Wood Fences, per linear foot

5' high grapestake	$12.90
6' high grapestake	14.00
5' basketweave	26.30
6' basketweave	30.10
5' board and batt (simulated)	15.10
6' board and batt (simulated)	17.40
5' three rail diamond top	14.00
6' three rail diamond top	14.60
3' two rail picket	9.50
4' two rail picket	12.30
Typical 3' wide gate, 4' to 6' high	92.40

Agricultural Structures Section

Section Contents

General Purpose Barns
Quality Classification

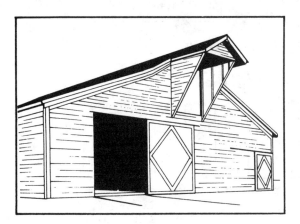

General Purpose Barn, Class 1　　　　**General Purpose Barn, Class 3**

Component	Class 1 Good Quality	Class 2 Average Quality	Class 3 Low Quality
Foundation	Continuous concrete.	Concrete or masonry piers.	Redwood or cedar mudsills.
Floor	Concrete.	Dirt.	Dirt.
Wall Structure	Good wood frame, 10' eave height.	Average wood frame, 10' eave height.	Light wood frame, 10' eave height.
Exterior Wall Cover	Good wood siding, painted.	Standard gauge corrugated iron, aluminum or average wood siding.	Light aluminum or low cost boards.
Roof Construction	Medium to high pitch, good wood trusses.	Medium to high pitch, average wood trusses.	Medium to high pitch, 2" x 4" rafters 24" to 36" o.c. or light wood trusses.
Roof Cover	Wood shingles.	Standard gauge corrugated iron or aluminum.	Light aluminum.
Electrical	Four outlets per 1,000 S.F.	Two outlets per 1,000 S.F.	None.
Plumbing	Two cold water outlets.	One cold water outlet.	None.

Square Foot Area

Quality Class	1,000	2,000	3,000	4,000	5,000	6,000	7,000	8,000	9,000	10,000	11,000
1, Good	26.42	22.78	21.22	20.28	19.76	19.24	18.93	18.62	18.41	18.20	18.82
2, Average	16.91	14.49	13.55	12.92	12.60	12.29	12.08	11.87	11.76	11.66	11.89
3, Low	12.40	10.71	9.95	9.53	9.24	9.05	8.88	8.75	8.64	8.57	8.48

Hay Storage Barns
Quality Classification

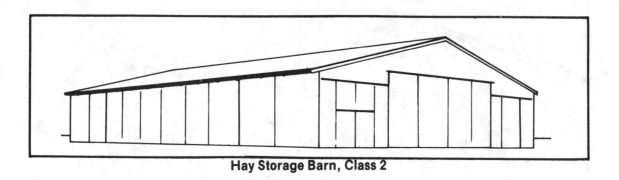

Hay Storage Barn, Class 2

Component	Class 1 Good Quality	Class 2 Average Quality	Class 3 Low Quality
Foundation	Continuous concrete.	Concrete or masonry piers.	Redwood or cedar mudsills.
Floor	Concrete.	Dirt.	Dirt.
Wall Structure	Good wood frame, 20' eave height.	Average wood frame, 20' eave height.	Light wood frame, 20' eave height.
Exterior Wall Cover	Good wood siding, painted.	Standard gauge corrugated iron or aluminum.	Light aluminum or low cost boards.
Roof Construction	Low to medium pitch, good wood trusses.	Low to medium pitch, average wood trusses.	Low to medium pitch, 2" x 4" rafters 24" to 36" o.c. or light wood trusses.
Roof Cover	Wood shingles.	Standard gauge corrugated iron or aluminum.	Light aluminum.
Electrical	None.	None.	None.
Plumbing	None.	None.	None.
Shape	Length between one and two times width.	Length between one and two times width.	Length between one and two times width.

Square Foot Area

Quality Class	1,000	2,000	3,000	4,000	5,000	6,000	7,000	8,000	9,000	10,000	11,000
1, Good	20.38	18.30	16.95	15.91	15.29	14.77	14.25	13.94	13.62	13.31	13.10
2, Average	11.97	10.71	9.98	9.45	9.03	8.72	8.40	8.19	7.98	7.88	7.77
3, Low	10.49	9.43	8.69	8.16	7.84	7.53	7.31	7.10	7.00	6.89	6.78

Feed Barns
Quality Classification

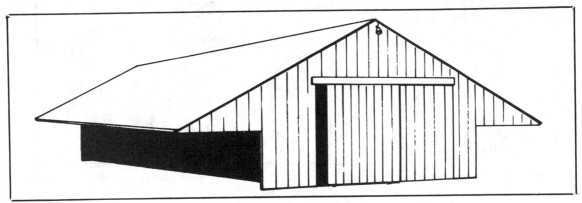

Feed Barn, Class 2

Component	Class 1 Good Quality	Class 2 Average Quality	Class 3 Low Quality
Foundation	Continuous concrete.	Concrete or masonry piers.	Redwood or cedar mudsills.
Floor	Concrete in center section.	Concrete in center section.	Dirt.
Wall Structure	Good wood frame, 8' eave height at drip line.	Average wood frame, 8' eave height at drip line.	Light wood frame, 8' eave height at drip line.
Exterior Wall Cover	Open sides, good siding painted on ends.	Open sides, standard gauge corrugated iron, aluminum or average wood siding on ends.	Open sides and ends.
Roof Construction	Medium to low pitch, good wood trusses.	Medium to low pitch, average wood trusses.	Medium to low pitch, 2" x 4" rafters 24" to 36" o.c. or light wood trusses.
Roof Cover	Wood shingles.	Standard gauge corrugated iron or aluminum.	Light gauge corrugated iron or aluminum.
Electrical	Four outlets per 1,000 S.F.	Two outlets per 1,000 S.F.	None.
Plumbing	Two cold water outlets.	One cold water outlet.	None.

Square Foot Area

Quality Class	1,000	2,000	3,000	4,000	5,000	6,000	7,000	8,000	9,000	10,000	11,000
1, Good	12.06	11.34	11.02	10.92	10.71	10.66	10.61	10.50	10.40	10.30	10.19
2, Average	10.40	9.82	9.56	9.40	9.29	9.24	9.19	9.14	9.08	9.03	8.98
3, Low	6.57	6.15	6.04	5.94	5.88	5.83	5.78	5.72	5.67	5.62	5.57

Shop Buildings

Quality Classification

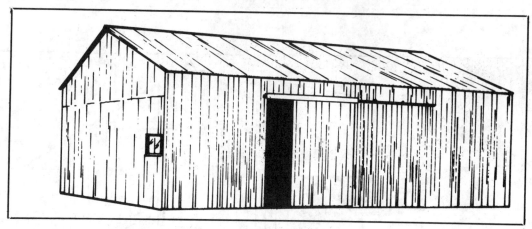

Shop, Class 2

Component	Class 1 Good Quality	Class 2 Average Quality	Class 3 Low Quality
Foundation	Continuous concrete.	Light concrete.	Light concrete.
Floor	Concrete.	Concrete.	Concrete.
Wall Structure	Good wood frame, 15' eave height.	Average wood frame, 15' eave height.	Light wood frame, 15' eave height.
Exterior Wall Cover	Good wood siding, painted.	Standard gauge corrugated iron, aluminum or average wood siding.	Light aluminum or low cost boards.
Roof Construction	Low to medium pitch, good wood trusses.	Low to medium pitch, average wood trusses.	Low to medium pitch, 2" x 4" rafters 24" to 36" o.c. or light wood trusses.
Roof Cover	Wood shingles.	Standard gauge corrugated iron or aluminum.	Light gauge corrugated iron or aluminum.
Electrical	Four outlets per 1,000 S.F.	Two outlets per 1,000 S.F.	Two outlets per 1,000 S.F.
Plumbing	Two cold water outlets.	One cold water outlet.	None.
Doors	One drive-thru door per 1,000 S.F. plus one walk-thru door.	One average sliding or swinging door per 2,000 S.F.	One light sliding or swinging door per 2,000 S.F.
Windows	Five percent of floor area.	None or few low cost.	None.
Shape	Length between one and two times width.	Length between one and two times width.	Length between one and two times width.

Square Foot Area

Quality Class	1,000	1,500	2,000	2,500	3,000	4,000	5,000	6,000	8,000	10,000
1, Good	21.32	19.45	18.41	17.68	17.16	16.33	15.81	15.39	14.87	14.46
2, Average	16.91	15.44	14.60	13.97	13.55	12.92	12.50	12.18	11.76	11.45
3, Low	13.36	12.19	11.45	11.02	10.71	10.18	9.86	9.65	9.33	9.01

Machinery and Equipment Sheds
Quality Classification

Equipment Shed, Class 3

Component	Class 1 Good Quality	Class 2 Average Quality	Class 3 Low Quality
Foundation	Continuous concrete.	Concrete or masonry piers.	Redwood or cedar mudsills.
Floor	Concrete.	Concrete.	Dirt.
Wall Structure	Good wood frame, 10' eave height.	Average wood frame, 10' eave height.	Light wood frame, 10' eave height.
Exterior Wall Cover	Good wood siding, painted.	Standard gauge corrugated iron, aluminum or average wood siding.	Light aluminum or low cost boards.
Roof Construction	Low to medium pitch, gable or shed type, good wood framing.	Low to medium pitch, gable or shed type, average wood framing.	Low to medium pitch, shed type, light wood framing.
Roof Cover	Wood shingles.	Standard gauge corrugated iron or aluminum.	Light aluminum.
Electrical	Four outlets per 1,000 S.F.	Two outlets per 1,000 S.F.	None.
Shape	Usually elongated, width between 15 and 30 feet, any length.	Usually elongated, width between 15 and 30 feet, any length.	Usually elongated, width between 15 and 30 feet, any length.

All Sides Closed
Square Foot Area

Quality Class	500	1,000	1,500	2,000	2,500	3,000	3,500	4,000	4,500	5,000	6,000
1, Good	16.74	14.66	13.94	13.52	13.31	13.10	13.00	12.90	12.79	12.74	12.69
2, Average	12.60	11.03	10.40	10.08	9.98	9.82	9.71	9.66	9.61	9.56	9.45
3, Low	8.27	7.21	6.89	6.68	6.57	6.52	6.47	6.36	6.31	6.25	6.20

One Side Open
Square Foot Area

Quality Class	500	1,000	1,500	2,000	2,500	3,000	3,500	4,000	4,500	5,000	6,000
1 Good	15.39	13.00	12.27	11.86	11.54	11.34	11.23	11.13	11.02	10.92	10.82
2, Average	11.45	9.66	9.08	8.77	8.56	8.45	8.35	8.30	8.24	8.19	8.14
3, Low	7.31	6.20	5.83	5.62	5.51	5.41	5.35	5.30	5.25	5.19	5.14

Small Sheds

Quality Classification

Component	Class 1 Good Quality	Class 2 Average Quality	Class 3 Low Quality
Foundation	Continuous concrete.	Concrete or masonry piers.	Redwood or cedar mudsills.
Floor	Concrete.	Boards.	Dirt.
Wall Structure	Good wood frame, 8' eave height.	Average wood frame, 8" eave height.	Light wood frame, 8' eave height.
Exterior Wall Cover	Good wood siding, painted.	Standard gauge corrugated iron, aluminum or average wood siding.	Light aluminum or low cost boards.
Roof Construction	Low to medium pitch, gable or shed type, good wood framing.	Low to medium pitch, gable or shed type, average wood framing.	Low to medium pitch, shed type, light wood framing.
Roof Cover	Wood shingles.	Standard gauge corrugated iron or aluminum.	Light aluminum.
Electrical	None.	None.	None.
Shape	Usually elongated, width between 6 and 12 feet, any length.	Usually elongated, width between 6 and 12 feet, any length.	Usually elongated, width between 6 and 12 feet, any length.

All Sides Closed
Square Foot Area

Quality Class	50	60	80	100	120	150	200	250	300	400	500
1, Good	21.11	19.24	16.95	15.60	14.66	13.83	12.90	12.27	11.96	11.44	11.13
2, Average	15.23	13.97	12.29	11.34	10.61	9.98	9.40	8.93	8.66	8.30	8.09
3, Low	10.07	8.06	8.16	7.47	7.05	6.63	6.15	5.88	5.72	5.51	5.35

One Side Open
Square Foot Area

Quality Class	50	60	80	100	120	150	200	250	300	400	500
1, Good	15.81	14.87	13.52	12.69	11.96	11.23	10.35	9.78	9.31	8.74	8.27
2, Average	11.55	10.92	9.92	9.24	8.77	8.19	7.61	7.14	6.83	6.41	6.09
3, Low	7.42	7.00	6.36	5.94	5.62	5.25	4.88	4.56	4.40	4.08	3.87

Pole Barns

These prices are for pole barns with a low pitch corrugated iron or aluminum covered roof supported by light wood trusses and poles 15' to 20' o.c. The gable end is enclosed and the roof overhangs about 2' on two sides. Wall height is 18 feet. Where sides are enclosed, the wall consists of a light wood frame covered with corrugated metal.

All Sides Open
Side Length

End Width	34	51	68	85	102	119	136	153	170	187
20	5.90	5.67	5.57	5.52	5.47	5.43	5.42	5.41	5.40	5.39
25	5.54	5.33	5.23	5.18	5.14	5.11	5.10	5.09	5.08	5.07
30	5.29	5.09	5.00	4.94	4.91	4.89	4.87	4.86	4.85	4.85
35	5.10	4.91	4.83	4.76	4.73	4.71	4.70	4.69	4.68	4.68
40	5.02	4.83	4.74	4.68	4.65	4.64	4.62	4.59	4.59	4.58
45	4.92	4.74	4.65	4.59	4.56	4.54	4.53	4.52	4.51	4.51
50	4.81	4.63	4.53	4.49	4.46	4.45	4.42	4.41	4.41	4.40
60	4.78	4.60	4.51	4.46	4.43	4.41	4.40	4.39	4.38	4.38
70	4.74	4.55	4.48	4.43	4.40	4.38	4.37	4.35	4.34	4.34
80	4.71	4.53	4.45	4.40	4.37	4.35	4.33	4.33	4.32	4.31

Ends and One Side Closed, One Side Open
Side Length

End Width	34	51	68	85	102	119	136	153	170	187
20	10.64	9.38	8.76	8.39	8.18	8.01	7.90	7.81	7.74	7.69
25	9.73	8.58	8.01	7.69	7.47	7.33	7.24	7.14	7.07	7.02
30	9.14	8.04	7.51	7.22	7.01	6.88	6.78	6.70	6.63	6.59
35	8.68	7.65	7.14	6.86	6.66	6.54	6.45	6.38	6.31	6.26
40	8.36	7.38	6.90	6.61	6.44	6.30	6.21	6.14	6.09	6.05
45	8.13	7.17	6.68	6.43	6.23	6.12	6.04	5.97	5.91	5.87
50	7.90	6.95	6.50	6.23	6.06	5.94	5.87	5.78	5.73	5.70
60	7.69	6.76	6.32	6.06	5.91	5.78	5.70	5.63	5.58	5.55
70	7.51	6.61	6.17	5.93	5.78	5.66	5.56	5.51	5.46	5.42
80	7.34	5.70	6.04	5.80	5.62	5.52	5.44	5.39	5.34	5.29

All Sides Closed
Side Length

End Width	34	51	68	85	102	119	136	153	170	187
20	12.76	11.45	10.82	10.43	10.18	10.00	9.86	9.77	9.69	9.62
25	11.39	10.24	9.65	9.32	9.08	8.91	8.81	8.70	8.64	8.58
30	10.44	9.39	8.86	8.55	8.35	8.19	8.09	8.00	7.93	7.89
35	9.82	8.82	8.33	8.02	7.84	7.68	7.59	7.52	7.45	7.41
40	9.33	8.40	7.92	7.63	7.45	7.33	7.22	7.14	7.07	7.03
45	8.94	8.03	7.58	7.32	7.14	7.01	6.92	6.85	6.79	6.73
50	8.60	7.73	7.29	7.02	6.85	6.74	6.65	6.58	6.52	6.48
60	8.21	7.36	6.96	6.71	6.55	6.43	6.36	6.28	6.22	6.18
70	7.94	7.14	6.75	6.51	6.34	6.22	6.15	6.08	6.04	6.00
80	7.68	6.88	6.51	6.28	6.11	6.02	5.94	5.87	5.83	5.79

Side sheds tying into one side of a pole barn are priced as follows. The shed consists of one row of poles 14' to 16' high, spaced 15' to 20' o.c. A light wood truss covered with a low pitch sheet metal roof spans the distance between the poles and the barn side. If the sides are open, the cost will be between $4.65 and $6.20 per square foot of area covered. If all sides are enclosed with sheet metal and a light wood frame, the square foot cost will be $7.25 to $9.50.

Low Cost Dairy Barns

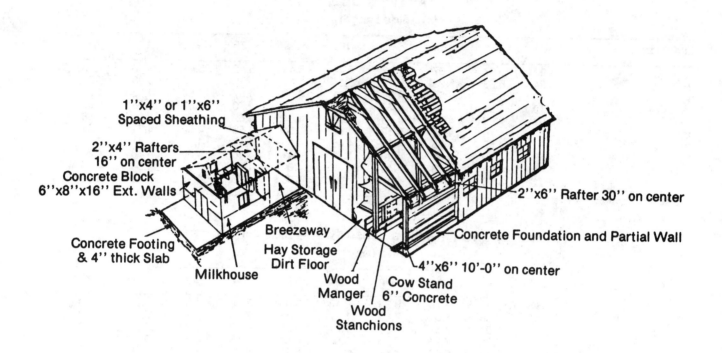

1''x4'' or 1''x6'' Spaced Sheathing

2''x4'' Rafters 16'' on center

Concrete Block 6''x8''x16'' Ext. Walls

Concrete Footing & 4'' thick Slab

Milkhouse

Breezeway

Hay Storage Dirt Floor

Wood Manger

Wood Stanchions

Cow Stand 6'' Concrete

4''x6'' 10'-0'' on center

Concrete Foundation and Partial Wall

2''x6'' Rafter 30'' on center

	Milk House	Dairy Barn
Foundation	Concrete.	Light concrete.
Floors	Concrete slab.	Concrete cow stands.
Walls	6" or 8' concrete block 36" high 2" x 4" @16" o.c. framing above.	Box frame, 4" x 6", 10' o.c.
Roof	Average wood frame, corrugated iron or aluminum cover.	Average wood frame, wood shingles, corrugated iron or aluminum cover.
Windows	Metal sash or metal louvers, on 5% of wall area.	Barn sash.
Interior	Smooth finish plaster.	Unfinished, wood stanchions.
Electrical	Minimum grade, fair fixtures.	None.
Plumbing	One wash basin.	None.
Square Foot Cost	$32.40 to $40.00 (including breezeway).	$12.40 to $14.90 (exclusive of milking equipment).

Stanchion Dairy Barns

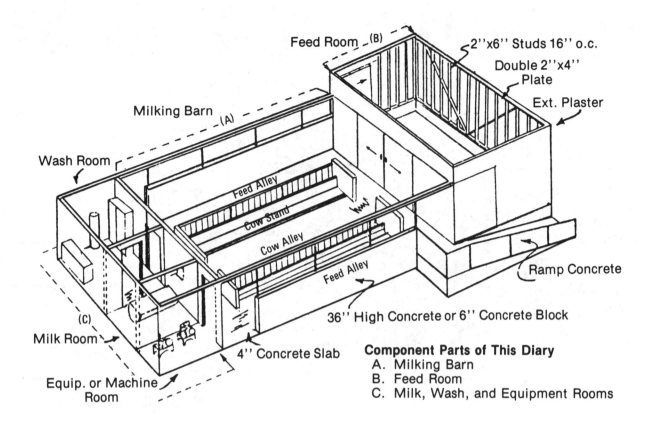

Feed Room (B)

2''x6'' Studs 16'' o.c.

Double 2''x4'' Plate

Ext. Plaster

Milking Barn (A)

Wash Room

Feed Alley

Cow Stand

Cow Alley

Feed Alley

Ramp Concrete

36'' High Concrete or 6'' Concrete Block

(C)

Milk Room

4'' Concrete Slab

Equip. or Machine Room

Component Parts of This Diary
A. Milking Barn
B. Feed Room
C. Milk, Wash, and Equipment Rooms

	Feed Room	Milk, Wash and Equipment Room	Milking Barn
Foundation	Reinforced concrete.	Reinforced concrete.	Reinforced concrete.
Floors	Concrete slab.	Concrete slab.	Concrete, well formed gutters and mangers.
Walls	2" x 4" or 2" x 6" @16" o.c. framing.	6" or 8" concrete block 36" high with 2" x 6" @16" o.c. framing above.	6" or 8" concrete block or reinforced concrete 36" high with 2" x 6" @16" o.c. framing above.
Roofs	Average wood frame, corrugated iron or aluminum cover.	Average wood frame, corrugated iron or aluminum cover.	Average wood frame, corrugated iron or aluminum cover.
Windows	None.	Metal sash or metal louvers on 10% of wall area.	Metal sash or metal louvers.
Interior	Unfinished.	Smooth finish plaster, cove base.	Smooth plaster 36" high, metal stanchions.
Electrical	Conduit, average fixtures.	Conduit, average fixtures.	Conduit, average fixtures.
Plumbing	None.	One wash basin with floor drains.	Floor drains and hose bibs.
Square Foot Costs	$13.00 to $21.60.	$45.40 to $58.30 (including breezeway).	$24.30 to $28.00 (exclusive of milking equipment).

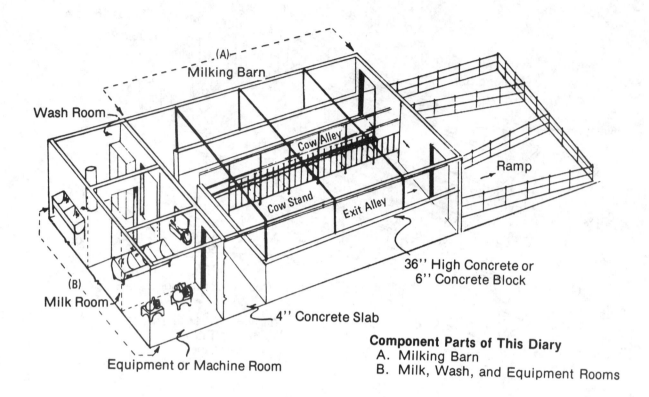

Component Parts of This Diary
A. Milking Barn
B. Milk, Wash, and Equipment Rooms

	Milking Barn	Milk, Wash, and Equipment Room
Foundation	Reinforced concrete.	Reinforced concrete.
Floors	Concrete, well-formed gutters and mangers.	Concrete slab
Walls	6" or 8" concrete block or reinforced concrete 36" high with 2" x 6" @16" o.c. framing above or all concrete block.	6" or 8" concrete block 36" high with 2" x 6" @16" o.c. framing above, or all concrete block.
Roofs.	Average wood frame, corrugated iron or aluminum cover.	Average wood frame, corrugated iron or aluminum cover.
Windows.	Metal sash or metal louvers. on 10% of wall area.	Metal sash or metal louvers
Interior.	Smooth plaster 36" high, base.	Smooth finish plaster, cove
Electrical.	Conduit, average fixtures.	Conduit, average fixtures.
Plumbing.	Floor drains and hose bibs.	One wash basin, floor drains.
Square Foot Costs	$28.10 (exclusive of milking equipment).	$29.20 (including breezeway).

Modern Herringbone Barns

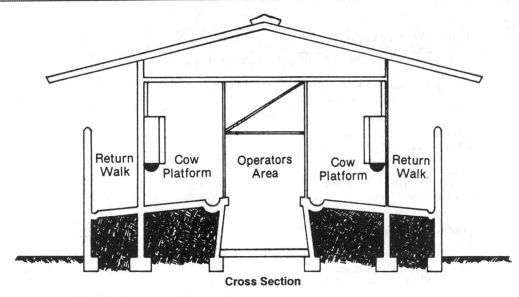

Cross Section

Milking Barn

Foundation	Reinforced concrete.
Floors	Concrete, wall-formed gutters and mangers.
Walls	6" or 8" concrete block or reinforced concrete 36" high with 2" x 6" @16" o.c. framing above or all concrete block.
Roof	Average wood frame, corrugated iron or aluminum cover.
Windows	Metal sash or metal louvers.
Interior	Smooth plaster 36" high.
Electrical	Conduit, average fixtures.
Plumbing	Floor drains and hose bibs.
Stanchions	Metal stanchions.
Costs Exclude	Stalls, feeding system and milking equipment.
Feed System & Stalls	Per double ten barn, $31,900.

Milk, Wash, and Equipment Room

	Average Quality	Good Quality
Foundation	Reinforced concrete.	Reinforced concrete.
Floors	Concrete slab.	Concrete slab.
Walls	8" concrete block with 2" x 6" @16" o.c. framing above or all concrete block.	8" concrete block with 2" x 6" @16" o.c. framing above or all concrete block.
Exterior	Stucco or concrete block.	Stucco and masonry veneer.
Roof	Corrugated iron or aluminum.	Wood shakes or mission tile, 3' to 4' overhang
Windows	Metal sash on 10% of wall area.	Metal sash on 10% of wall
Interior	Smooth finish plaster, cove base.	Smooth finish plaster, cove base.
Electrical	Conduit, average fixtures.	Conduit, average fixtures.
Plumbing	One wash basin, one water closet, one lavatory, floor drains.	One wash basin, one water closet, one lavatory, floor drains.
Square Foot Costs	$40.00 to $42.10.	$41.60 to $45.40.

Miscellaneous Dairy Costs

Holding Corral and Wash Area Costs

Components	Cost
Floor or Ramp	Sloping concrete with abrasive finish, $3.00 to $3.50 per square foot.
Wall	5' to 6' high plastered interior, $38.00 to $45.00 per linear foot.
Metal rail fence	Welded pipe, posts 10' o.c. in concrete, top rail and 3 cables, $9.20 per linear foot.
Cable fence	Pipe posts 10' o.c. set in concrete, $7.60 + .45 per linear foot per cable.
Gates	54" high, pipe with necessary bracing, $23.30 per linear foot.
Sprinklers	Hooded rainbird, $165.00 each, including plumbing and pump.
Roof	Pipe column supports, average wood frame, corrugated iron or aluminum cover, open sides, $4.00 to $5.80 per square foot.
Typical wash area (without sprinkler)	Sloped concrete floors, 5' concrete block exterior walls, welded pipe interior fences and gates, usual floor drains $19.50/S.F. with roof, $13.00/S.F. without roof.

Corral Costs

Components	Cost
4" concrete flatwork	$2.70 per square foot.
6" concrete flatwork	$3.00 per square foot.
12" curb	$8.70 per linear foot.
Cable fence	$8.70 per linear foot.
Water tank, concrete typical	$2,450 each.
Steel stanchions	$43.50 to $48.00 each. $17.90 to $22.00 per linear foot.
Steel lockable stanchions	$51.00 to $55.00 each. $20.00 to $24.00 per linear foot.
Light standards	$2,200.00 each.
Sump pumps, each 3 HP 4 HP 10 HP	 $2,400.00. $3,100.00. $5,200.00.
Hay shelters	$4.60 to $5.70 per square foot.
Loafing sheds	$4.90 to $6.00 per square foot.
Typical 80-cow corral	$26,000.00 to $30,300.00 total. $325.00 to $375.00 per cow.

Milk Line, 3" stainless steel

40-cow conventional barn	$17,000 total $425 per cow
Double ten herringbone barn	$10,200 total $510 per cow
Dual milk pump, add	$1,960
Incline filter, add	$2,100
Incline cold plate, add	$4,800 to $7,000

Refrigeration Compressors

7-1/2 HP	$5,500.00
10 HP	7,250.00
15 HP	9,100.00

Stainless Steel
Refrigerated Holding Tanks

5,000 gallon	$44,500
4,000 gallon	41,900
3,500 gallon	40,300
3,000 gallon	35,800
2,500 gallon	33,000
2,000 gallon	30,700
1,500 gallon	27,300

Vacuum Pumps (air pumps)

Units	Cost
8	5,500.00
12	9,000.00
20	12,400.00

Poultry Houses

Conventional Lay Cage Type

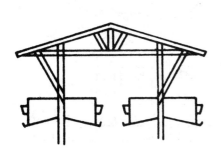
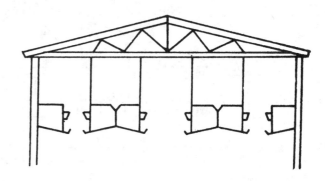

Basic Building

Component	Good Quality	Average Quality	Low Quality
Floors	2" concrete.	Dirt with 4' concrete walkways.	Dirt.
Foundations	Thickened slab.	Concrete piers.	Wood piers.
Frame	Light steel or average wood frame.	Average wood frame.	Light wood frame.
Roof Cover	Aluminum or corrugated iron.	Light aluminum or composition.	Light aluminum or composition.
Exterior	Plywood.	Vinyl curtains.	Wood lath.
Lighting	Good system, automatic controls. controls.	Average system, automatic controls.	Minimum system, manual
Plumbing	Good system.	Average system.	Fair system.
Insulation	Roof only.	None.	None.
Basic Building Cost Per S.F.	$5.00 to $6.00.	$3.30 to $4.40.	$2.80 to $3.30.

Equipment

Component	Best Quality	Good Quality	Average Quality	Low Quality
Cages	12" x 20" double deck.	12" x 20" single deck.	12" x 20" single deck.	12" x 12" single deck.
Water System	Automatic cup system.	Automatic cup system.	Simple "V" trough.	Simple "V" trough.
Feed System	Automatic system.	V trough.	V trough.	V trough.
Egg Gathering	Manual.	Manual.	Manual.	Manual.
Cooling	Pad and fan system.	Pad and fan system.	Simple fogging system.	Simple fogging system.
Cost Per S.F.	$13.70 to $15.40.	$8.25 to $9.80.	$6.30 to $7.80.	$5.30 to $7.20
One Bird Per S.F.	$4.50 to $5.20	$5.50 to $6.50	$4.10 to $5.30.	$3.50 to $4.80

Poultry Houses
Modern Controlled Environment Type

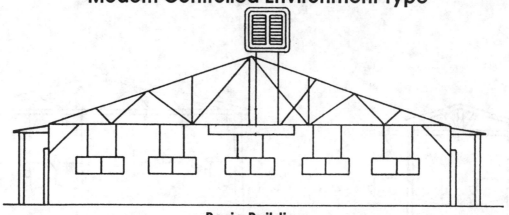

Basic Building

Foundation	Concrete.
Floor	3-1/2 concrete slab.
Wall Frame	2" x 4" @ 24" o.c.
Roof Frame	Wood trusses with 2" x 4" purlins @ 24" o.c.
Exterior	Two rib aluminum or corrugated iron.
Interior	4" fiberglass roll with aluminum foil facing or 3/4" insulation.
Lighting	Fluorescent or good automatic incandescent system.
Plumbing	Good basic system.
Basic Building Cost Per S.F.	$8.20 to $9.90.

Equipment

Component	Single Deck	Stair Step
Cages	12" x 20" single deck.	12" x 20" double deck.
Watering System	Automatic cup system.	Automatic cup system.
Feeding System	Manual.	Manual.
Egg Gathering System	Manual.	Manual.
Cooling	Evporative coolers.	Evaporative coolers.
Heating	None.	None.
Building & Equipment Square Foot Cost	$18.50 to $19.50.	$19.50 to $21.60.
Building & Equipment Cost per Bird	$12.30 to $13.20.	$6.60 to $7.20.

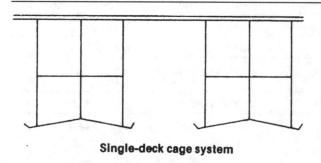

Single-deck cage system

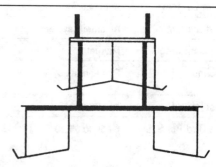

Stair-step cage system

Poultry Houses

High Rise Type

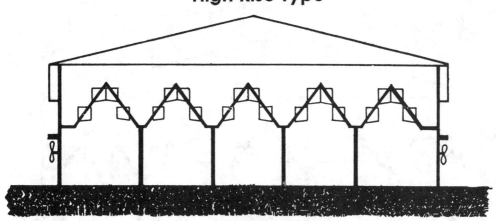

Basic Building

Foundation	Concrete piers.
Floors	Dirt.
Wall Frame	2" x 4" @24" o.c.
Roof Frame	Wood trusses with 2" x 4" purlins @24" o.c.
Exterior	Two rib aluminum or corrugated iron.
Interior	3/4" insulation.
Lighting	Fluorescent or good automatic incandescent system.
Plumbing	Good basic system.
Basic Building Cost Per S.F.	$8.80 to $10.90.

Equipment

Component	Flat Deck	Stair Step
Cages	12" x 20".	12" x 20".
Watering System	Automatic cup system.	Automatic cup system.
Feeding	Automatic system.	Automatic system.
Egg Gathering	Automatic system.	Automatic system.
Cooling	Negative pressure system.	Negative pressure system.
Heating	None.	None.
Building & Equipment Square Foot Cost	$17.00 to $18.80.	$20.40 to $22.10.
Building & Equipment Cost per Bird	$11.30 to $12.50.	$13.70 to $14.40.

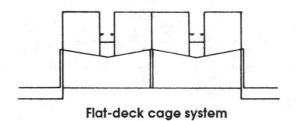

Flat-deck cage system

Poultry Houses

Deep Pit Type

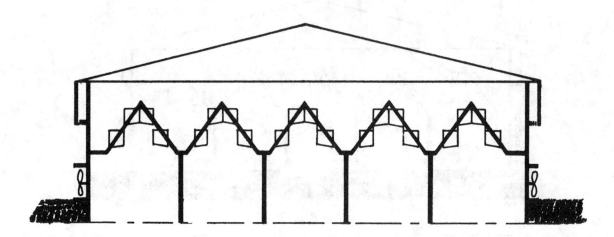

Basic Building

Foundation	Concrete piers.
Floors	Concrete with waterproof membrane.
Wall Frame	2" x 4" @24" o.c.
Roof Frame	Wood trusses with 2" x 4" purlins @24" o.c.
Exterior	Two rib aluminum or corrugated iron.
Interior	3/4" insulation.
Lighting	Fluorescent or good incandescent.
Plumbing	Good basic system.
Basic Building Cost Per S.F.	$9.30 to $11.20.

Equipment

Component	Flat Deck	Stair Step
Cages	12" x 20".	12" x 20".
Watering	Automatic cup system.	Automatic cup system.
Feeding	Automatic system.	Automatic system.
Egg Gathering	Automatic system.	Automatic system.
Cooling	Negative pressure system.	Negative pressure system.
Heating	None.	None.
Building & Equipment Square Foot Cost	$18.80 to $21.40.	$21.40 to $25.30.
Building & Equipment Cost per Bird	$12.50 to $14.40.	$14.40 to $17.40.

Poultry Houses

Equipment Costs

Add these costs to the basic building cost

Component	Serving One Row of Cages	Serving Two Rows of Cages
Automatic feeders	$1.55 per bird	$.78 per bird
Automatic egg gathering	1.20 per bird	.57 per bird
Automatic water cup system	1.20 per bird	.57 per bird
	4.00 per cup	2.70 per cup
"V" water trough	.25 per bird	.18 per bird
16" feed trough	.35 per bird	.23 per bird

Foggers	1/2" galvanized pipe	$1.60/linear foot
	3/4" galvanized pipe	1.65/linear foot
	1" galvanized pipe	1.75/linear foot

Roof sprinklers $1.75 per linear foot

Evaporative coolers $910.00 each. $1.15 per S.F. of building

Fans
30"	$510.00 each
36"	710.00 each
48"	770.00 each

Negative pressure air conditioning system $1.00 to $1.10 per S.F. of building.
Cooling pads in walls $.60 per S.F. of surface.

Heating systems $1,550.00 per unit.

Cages, 12" x 20" or 18" $6.00 each. $1.50 per bird.

Labor Camps

Quality Classification

	Class 1 Best Quality	Class 2 Prefabricated	Class 3 Good Quality	Class 4 Average Quality	Class 5 Low Quality
Slab Foundation	Spread footing around perimeter and thickened slab at partitions.	Spread footing around perimeter.	Thickened around perimeter.	Thickened around perimeter.	Thickened around perimeter.
Floor	4" concrete slab reinforced.	4" concrete slab reinforced.	4" concrete slab.	4" concrete slab.	4" concrete slab.
Walls	Masonry exterior walls, wood frame interior partitions and ceiling.	Metal building, prefabricated.	2" x 4" studs @16" o.c., 2" x 4" stud partitions.	Box construction 4" x 4" at 48" o.c.	Box construction 2" x 4" at 48" o.c.
Exterior Cover	Natural blocks.	Prefabricated metal building.	Average grade redwood board and batt or horizontal siding or stucco finish.	Fair grade redwood or fir board and batt or horizontal board.	Poor grade of redwood or fir, vertical or horizontal.
Interior Finish	Gypsum wallboard.	None.	Gypsum wallboard.	Plywood partitions.	None.
Roof Framing	Rafters, collar beams and ceiling joists.	Prefabricated metal building.	Rafters, collar beams and ceiling joists.	Very simple truss.	Rafters and tie at plate line.
Roofing	Composition shingles.	Metal.	Aluminum or wood.	Composition or sheet metal.	Composition or used sheet metal.
Doors	1 metal door and metal frame each room.	Metal doors.	1 average door each room.	3 or 4 average doors.	2 or 3 cheap doors.
Windows	1 steel sash or aluminum window in each room.	Approximately 20% of floor area.	1 steel or aluminum window in each room.	1 window each room.	Few and small.
Electrical	1 good light and plug in each room.	1 light, 1 plug for each 300 S.F.	1 light, 1 plug in each room.	1 pull chain light and plug for each 400 S.F.	1 pull chain light for each 500 S.F.

Labor Camp, Class 1

Labor Camp, Class 4

Square Foot Area

Quality Class	400	600	800	1,000	1,200	1,500	2,000	2,500	3,000
1, Best	33.76	31.18	29.78	28.90	28.27	27.63	26.93	26.49	26.16
2, Prefabricated	30.63	28.30	27.03	26.23	25.66	25.05	24.44	24.03	23.73
3, Good	28.19	26.03	24.86	24.13	23.62	23.07	22.49	22.10	21.86
4, Average	23.38	21.60	20.64	20.03	19.60	19.15	18.66	18.34	18.12
5, Low	19.17	17.68	16.91	16.40	16.04	15.68	15.29	15.03	14.84

Costs do not include any plumbing. Add $625 to $750 per fixture.

Miscellaneous Agricultural Structures

Livestock Scales

Type	Size	Capacity	In-Place
Full-capacity beam	16' x 8'	5 ton	$9,400
Printing beam	16' x 8'	5 ton	11,700
Full-capacity beam	22' x 8'	10 ton	12,400
Printing beam	22' x 8'	10 ton	15,000

Additional Costs for Livestock Scales

Types and Size	Cost
Each foot arm is removed from scale	$105.00/L.F.
Angle iron stock rack for 16' x 8' scale	3,100 ea.
Angle iron stock rack (wood) for 16' x 8'	4,600 ea.
Angle iron stock rack (wood) for 22' x 8'	5,400 ea.

Scale pit has 4" concrete walls and slab poured in place. May be poured in or on top of the ground. If on top, compacted ramps and steps to the scale beam are included.

Motor Truck Scales

Five inch reinforced concrete platform. All-steel structure and scale mechanism. Reinforced concrete pit. Motor truck scales are of two general types, the beam type (either manual or type registering) and the full-automated dial type. The construction of both, insofar as the weight-carrying mechanism is concerned is very similar. The method of recording and weight capacity make the cost vary.

Capacity	Platform Size	Total Cost
20 tons	24' x 10'	$18,600
30 tons	34' x 10'	23,300
40 tons	40' x 10'	25,700
50 tons	45' x 10'	29,200
50 tons	50' x 10'	30,500
50 tons	60' x 10'	34,800
50 tons	70' x 10'	38,000

Above costs are for full-capacity beam scales. Add $850 for the registering type beam.

Septic Tanks

2 bedroom home with 1,200 gallon tank	$3,600
3 bedroom home with 1,500 gallon tank	4,800
4 bedroom home with 2,000 gallon tank	5,700

Bulk Feed Tanks

Size and Type	Cost
3 to 4 ton	$1,150
5 to 6 ton	1,750
6 to 7 ton	1,850
7 to 8 ton	2,200
9 to 10 ton	2,300
10 to 12 ton	2,700
3 to 4 ton twin	1,300
4 to 5 ton twin	1,500
8 ton dairy feed	2,200
12 ton dairy feed	2,950
15 ton dairy feed	3,300

Tanks are equipped with a scissor-type opening chute

Domestic Water Systems Submersible pump, installed at 105' depth excluding well and casing

	Typical Installation		
	1/2 HP	3/4 HP	1 HP
Total cost	$1,650	$1,850	$2,200
Pressure tank size	82 gal.	82 gal.	120 gal.
Cost per ft. above or below 105' depth	$2.10	$2.30	$2.50
	1-1/2 HP	2 HP	3 HP
Total cost	$2,480	$3,150	$3,800
Pressure tank size	220 gal.	220 gal.	315 gal.
Cost per ft. above or below 105' depth	$2.60	$3.90	$5.30

6" wells average $27.00 per foot of depth
8" wells average $32.60 per foot or depth.

Pressure Tank Sizes and Installed Costs

42 gal. 16" dia. x 48" depth	$200 to $ 280
82 gal. 20" dia. x 60" depth	280 to 330
120 gal. 24" dia. x 60" depth	330 to 500
220 gal. 30" dia. x 72" depth	830 to 930
315 gal. 36" dia. x 72" depth	1,150 to 1,300
525 gal. 36" dia. x 120" depth	1,500 to 1,700

Typical Physical Lives in years for agricultural structures

Building Type	Good	Average	Low
Barns	40	30	20
Dairy barns	25	25	--
Dairy barns, low cost	--	20	20
Storage sheds	40	30	20
Poultry houses, modern	30	25	--
Poultry houses, conventional	--	25	20

To determine the useful life remaining, use the percent good table for residential structures on page 40.

Military Construction Costs

The Office of the Secretary of Defense has prepared the following square foot guidelines to reflect the cost of military construction for fiscal 1995 and 1996. The costs are based on construction of permanent facilities built on military bases worldwide. Use the "Construction Cost Indices" at the end of this section to adapt the square foot costs to any other area. The "Size Cost Adjustment Chart" should be used to determine the approximate cost of a structure larger or smaller than the typical size shown.

Included in these costs are all items of equipment which are permanently built-in or attached to the structure, including items with fixed utility connections.

The costs include items such as the following:

Furniture, cabinets and shelving, built-in.

Venetian blinds and shades.

Window screens and screen doors.

Elevators and escalators.

Drinking water coolers.

Telephone, fire alarm and intercom systems.

Theater seats.

Pneumatic tube systems.

Heating, ventilating and air conditioning installations.

Electrical generators and auxiliary gear.

Waste disposers such as incinerators.

Food preparation & serving equipment, built-in.

Raised flooring.

Hoods and vents.

Chapel pews and pulpit.

Refrigerators, built-in.

Laboratory furniture, built-in.

Cranes and hoists, built-in.

Dishwashers.

The costs listed are the estimated contract award costs, excluding contingencies, supervision, and administration. They include construction to the five-foot line only, but do not include the cost of outside utilities or other site improvements.

Figures listed do not include the cost of piles or other special foundations which are considered as an additional supporting item. The cost of air conditioning is included to the extent authorized by the Construction Criteria Manual.

The costs of equipment such as furniture and furnishings which are loose, portable, or can be detached from the structure without tools are excluded from the unit costs. The cost of permanently attached equipment related directly to the operating function for which the structure is being provided, such as technical, scientific, production and processing equipment, is normally excluded from these costs. The following items are excluded from the costs on page 235.

Furniture, loose.

Furnishings, including rugs, loose.

Filing cabinets and portable safes.

Office machines, portable.

Wall clocks, plug-in

Food preparation and serving equipment, including appliances, portable.

Training aids and equipment, including simulators.

Shop equipment.

Bowling lanes, including automatic pin spotting equipment, score table and players' seating.

Automatic data processing equipment.

Communications equipment.

Photographic equipment, portable.

Any operational equipment for which installation, mounting and connections are provided in building design and which are detachable without damage to the building or equipment.

Estimating Procedure

Determine the area relationship of the proposed building by dividing the gross area by the typical size as shown in the Square Foot Cost Table. Locate the quotient on the Area Relationship scale and trace vertically to the Factor Line, then trace horizontally to the Cost Relationship scale. This value is then multiplied by the unit cost in the Square Foot Cost Table and factored by the Construction Cost Index to determine the adjusted unit cost for the building.

Military Construction Costs

Facilities	Typical size Gross S.F.	Unit cost per S.F. FY 95	Unit cost per S.F. FY 96
Administrative office			
Multi-purpose	25,000	94.00	96.00
Data processing	21,000	115.00	118.00
Aircraft operations, no tower	20,000	115.00	118.00
Airfield control tower	3,000	189.00	193.00
Applied instruction building	25,000	95.00	97.00
Barracks, dormitory	100,000	94.00	96.00
Bowling alley, 8 pin spotters	7,800	128.00	131.00
Chapel center	10,000	122.00	125.00
Commissary (includes storage)	60,000	65.00	68.00
Enlisted service club	16,000	123.00	126.00
Exchange store, with snack bar	20,000	72.00	74.00
Family housing, U.S.	None	65.00	66.00
Family housing, outside U.S.	None	69.00	70.00
Family support			
Child development center	15,000	97.00	99.00
Education center	10,000	94.00	96.00
Youth center	15,000	97.00	99.00
Family service center	5,000	108.00	110.00
Fire station, community	3,500	110.00	112.00
Hangars maintenance			
General purpose	38,000	90.00	92.00
High bay	35,000	132.00	134.00
Medical facility			
Station hospital	None	159.00	162.00
Regional medical center	None	189.00	192.00
Troop clinic	12,000	129.00	131.00
Outpatient clinic	30,000	132.00	136.00
Dental clinic	24,000	213.00	219.00
Messhall, enlisted (with kitchen equipment)	16,000	147.00	150.00
Operations building			
General purpose	20,000	98.00	100.00
Headquarters	36,000	155.00	158.00
Squadron	30,000	80.00	82.00

Facilities	Typical size Gross S.F.	Unit cost per S.F. FY 95	Unit cost per S.F. FY 96
Physical fitness training center	20,000	108.00	110.00
Recreation center	20,000	93.00	96.00
Reserve facility			
Center	20,000	81.00	83.00
Vehicle maintenance	20,000	94.00	96.00
Communications center	6,000	168.00	172.00
Library	12,000	94.00	96.00
School for dependents			
Elementary, overseas	None	95.00	98.00
High school, overseas	None	100.00	102.00
Elementary, U.S.	None	84.00	87.00
Jr. high, middle, U.S.	None	87.00	89.00
High school, U.S.	None	95.00	98.00
Shops			
Vehicle maint. (wheeled)	30,000	92.00	95.00
Vehicle maint. (tracked)	20,000	94.00	96.00
Aircraft avionics	23,000	92.00	94.00
Installation maintenance	31,000	91.00	93.00
Parachute and dinghy	8,000	93.00	85.00
Aircraft machine shop	20,000	105.00	108.00
Storage facility			
Cold storage warehouse, w/processing	11,000	99.00	101.00
General purpose storage shed	20,000	34.00	35.00
General purpose warehouse low bay	40,000	52.00	53.00
General purpose warehouse high bay	40,000	64.00	66.00
Hazardous & flamable storage	10,000	115.00	118.00
Operational storehouse	4,000	62.00	63.00
Temporary lodging facility	25,000	66.00	68.00
Unmarried officers quarters	44,000	97.00	99.00

Size Cost Adjustment

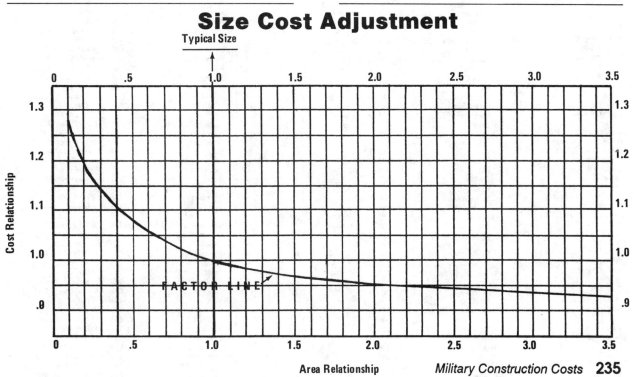

State	Index
Alabama	0.90
Birmingham	0.90
Mobile	1.00
Montgomery	0.79
Anniston AD	0.88
Fort Rucker	0.85
Fort McClellan	0.87
Maxwell AFB	0.79
Phosphate Dev.Works	0.85
Redstone ARS	0.80
Alabama AAP	0.79
Alaska	2.46
Anchorage	1.79
Adak	3.52
Fairbanks	2.06
Aleutian Is	3.48
Clear AFS	2.78
Eielson AFB	2.06
Elmendorf AFB	1.79
Fort Greeley	2.66
Fort Richardson	1.79
Ft Wainwright	2.06
Galena	3.35
Kodiak	2.85
Arizona	1.00
Flagstaff	1.03
Phoenix	0.94
Tucson	0.98
Fort Huachuca	1.00
Navajo AD	1.03
Yuma PG	1.19
Arkansas	0.91
Pine Bluff	0.90
Little Rock	0.87
Fort Smith	0.96
Fort Chaffee	0.96
California	1.20
Los Angeles	1.19
San Diego	1.21
San Francisco	1.19
29 Palms	1.32
Beale AFB	1.19
Bridgeport NWTC	1.24
Castle AFB	1.08
Centerville Beach	1.18
Camp Pendleton	1.21
Desert Area	1.25
Edwards	1.23
El Centro	1.19
Fort Hunter Liggett	1.34
Fort Irwin	1.36
Fort Ord	1.25
George AFB	1.20
Lemoore AFB	1.14
March AFB	1.22
Mather AFB	1.15
McClellan AFB	1.15
Monterey Area	1.25
Norton AFB	1.16
Oakland AB	1.29
Port Hueneme Area	1.18
River Bank AAP	1.07
Sacramento AD	1.13
San Clemente Is	1.46
San Nicolas Is	2.20
Sharpe AD	1.05
Sierra AD	1.34
Stockton	1.07
Travis AFB	1.22
Vandenberg AFB	1.27
Colorado	0.97
Colorado Springs	0.95
Denver	0.97
Pueblo	1.00
Air Force Academy	1.01
Cheyenne Mountain	1.07
Flacon AFS	1.01
Fitzsimmons AMC	0.99
Fort Carson	1.01
Peterson AFB	1.01
Pueblo AD	0.98
Rock Mountain ARS	0.97
Connecticut	1.13
Bridgeport	1.12
Hartford	1.10
New London	1.20
Stratford Eng. PLA	1.12
Delaware	0.99
Dover AFB	0.97
Lewes NF	0.97
Milford	1.02
Florida	0.93
Miami	0.96
Panama City	0.87
Tampa	0.97
Cape Canaveral	0.91
Cape Kennedy	0.91
Elgin AFB	0.82

State	Index
Florida	0.93
Jacksonville Area	0.87
Key West NAS	1.05
McDill AFB	0.96
Orlando	0.84
Panama City Area	0.87
Pensacola Area	0.82
Tyndall AFB	0.87
Georgia	0.84
Albany	0.85
Atlanta	0.86
Macon	0.81
Athens	0.89
Marietta	0.87
Fort Benning	0.80
Fort Gillem	0.86
Fort Gordon	0.92
Fort McPherson	0.86
Fort Stewart	0.82
Kings Bay	0.98
Warner Robbins AFB	0.81
Hawaii	1.33
Honolulu	1.39
Maui	1.25
Aliamanu	1.39
Barbers Pt NAS	1.44
Barking Sand	1.80
Ewa Beach	1.44
Ford Island	1.53
Fort DeRussy	1.39
Fort Shafter	1.39
Helemand	1.44
Hickam AFB	1.39
Kaneohe MCAS	1.44
Lualualei	1.39
Moanalua	1.39
Pearl City	1.39
Pearl Harbor	1.39
Pohakuloa	1.35
Schofield Barracks	1.39
Tripler AMC	1.39
Wheeler AFB	1.44
Idaho	1.06
Boise	1.03
Idaho Falls	1.00
Mtn Home AFB	1.11
Illinois	1.01
Bellville	1.00
Chicago	1.06
Rock Island	0.97
Chanute AFB	1.14
Fort Sheridan	1.07
Great Lakes	1.07
Rock Island ARS	1.00
Savannah AD	1.06
Scott AFB	1.06
Indiana	1.02
Indianapolis	1.05
Logansport	1.05
Madison	0.98
Crane AAP	1.12
Fort Ben Harrison	1.09
Grissom AFB	1.09
Indiana AAP	1.00
Jefferson PG	0.98
Newport AAP	1.03
Iowa	1.01
Burlington	1.04
Cedar Rapids	0.99
Des Moines	0.99
Iowa AAP	1.06
Kansas	0.92
Manhattan	0.94
Topeka	0.90
Wichita	0.91
Fort Leavenworth	0.99
Fort Riley	0.89
Kansas AAP	0.95
Sunflower AAP	0.96
McConnell AFB	0.91
Kentucky	0.94
Bowling Green	0.93
Lexington	0.95
Louisville	0.95
Fort Campbell	0.98
Fort Knox	1.03
Lexington/Bluegrass AD	1.02
Louisiana	0.89
Alexandria	0.84
New Orleans	0.94
Shreveport	0.88
Barksdale AFB	0.88
England AFB	0.84
Fort Polk	0.96
Lake Charles	0.96
Louisiana AAP	0.88
Maine	0.98
Bangor	0.88

State	Index
Maine	0.98
Caribou	1.12
Portland	0.93
Brunswick	1.12
Cutler	0.96
Loring AFB	1.11
Prospect Harbor	0.95
Winter Harbor	0.95
Maryland	0.94
Baltimore	0.92
Fredrick	0.95
Indian Head	1.03
Lexington Park	0.96
Aberdeen PG	0.91
Andrews PG	1.03
Annapolis	1.03
Fort Detrick	0.95
Fort Holabird	0.92
Fort Meade	0.92
Fort Ritchie	0.95
Harry Diamond Lab	1.03
Patuxent River Area	1.03
Massachusetts	1.10
Boston	1.08
Fitchburg	1.11
Springfield	1.11
Army Mat. & Mech	1.15
Fort Devens	1.18
Hanscomb AFB	1.15
Natick	1.18
South Weymouth	1.24
Michigan	1.05
Bay City	1.02
Detroit	1.07
Marquette	1.06
Republic Elfcom	1.22
Sawyer AFB	1.06
Wurfsmith AFB	1.03
Minnesota	1.13
Duluth	1.12
Minneapolis	1.15
St. Cloud	1.12
Rosemont	1.19
Twin City AAP	1.19
Mississippi	0.82
Biloxi	0.85
Columbus	0.82
Jackson	0.80
Columbus AFB	0.82
Gulf Port Area	0.85
Keesler AFB	0.85
Meridian	0.85
Mississippi AAP	0.85
Pascagoula	0.84
Missouri	0.93
Kansas City	1.00
St. Louis	1.04
Rolla	0.89
Fort Leonard Wood	0.92
Kansas City	0.94
Lake City AAP	0.93
Kirksville	0.95
St. Louis AAP	1.04
Whitman AFB	1.04
Montana	1.15
Billings	1.08
Butte	1.19
Great Falls	1.19
Malmstrom AFB	1.19
Nebraska	1.03
Grand Island	1.01
Lincoln	1.03
Omaha	1.05
Cornhuster AAP	0.82
Offcutt AFB	1.05
Nevada	1.17
Hawthorn	1.27
Las Vegas	1.09
Reno	1.15
Fallon	1.34
Hawthorn AAP	1.20
Nellis AFB	1.09
New Hampshire	1.06
Concord	1.07
Nashua	1.04
Portsmouth	1.08
Portsmouth Area	1.10
New Jersey	1.10
Newark	1.10
Red Bank	1.11
Trenton	1.08
Atlantic City	1.25
Bayonne Mot	1.12
Earle	1.11
Fort Dix	1.00
Fort Monmouth	1.07
North Bergen	1.30
Lakehurst	1.07
McGuire AFB	1.10

State	Index
New Jersey	1.10
Picatinny ARS	1.18
New Mexico	1.01
Alamagordo	1.00
Albuquerque	1.00
Gallup	1.02
Santa Fe	0.87
Cannon AFB	0.99
Fort Wingate	1.02
Holloman	1.08
Kirtland AFB	1.00
Wh.Sands Missile Rng	1.08
New York	1.14
Albany	1.07
N.Y. City	1.27
Syracuse	1.08
Brooklyn	1.27
Fort Drum	1.08
Garden City	1.29
Griffiss AFB	1.08
Plattsburg AFB	1.12
Seneca AD	1.11
Staten Island	1.40
USMA	1.14
Watervliet AFS	1.07
North Carolina	0.85
Fayetteville	0.84
Greensboro	0.88
Wilmington	0.83
Camp Lejeune Area	0.92
Cherry Point	0.96
Fort Bragg	0.87
Pope AFB	0.87
Seymour AFB	0.78
Sunny Point Mt	0.89
North Dakota	0.98
Bismarck	0.96
Grand Forks	0.95
Minot AFB	1.04
Grand Forks AFB	0.95
Ohio	1.01
Columbus	0.97
Dayton	1.03
Youngstown	1.02
Cleveland	1.10
Ravenna AAP	1.05
Wright-Patterson AFB	1.04
Oklahoma	0.89
Lawton	0.92
McAlester	0.87
Oklahoma City	0.88
Altus AFB	0.99
Fort Sill	0.92
Tinker AFB	0.87
Oregon	1.02
Pendleton	1.09
Portland	0.99
Salem	0.97
Astoria	1.13
Charleston	1.09
Coos Head	1.06
Umatilla AD	1.15
Pennsylvania	0.99
Harrisburg	0.88
Philadelphia	1.08
Pittsburgh	1.01
Carlisle Barracks	0.88
Indiantown Gap Missle Range	0.94
Letterkenny AD	1.01
Mechanicsburg Area	0.88
New Cumberland AD	0.88
Scranton AAP	1.02
Tobyhanna AD	1.09
Warminster Area	1.02
Rhode Island	1.11
Bristol	1.11
Newport	1.20
Providence	1.07
Davisville	1.13
Newport Area	1.19
South Carolina	0.85
Charleston	0.89
Columbia	0.80
Myrtle Beach	0.87
Beaufort Area	0.93
Charleston AFB	0.89
Fort Jackson	0.80
Shaw AFB	0.78
South Dakota	0.99
Aberdeen	0.96
Sioux Falls	0.99
Rapid City	1.02
Ellsworth AFB	1.03
Tennessee	0.87
Chattanooga	0.85
Kingsport	0.81
Memphis	0.94

State	Index
Tennessee	0.87
Arnold AFB	0.92
Holston AAP	0.81
Milan AAP	0.92
Volunteer AAP	0.90
Texas	0.86
San Angelo	0.79
San Antonio	0.88
Forth Worth	0.91
Bergstrom AFB	0.87
Brooks AFB	0.88
Camp Bullis	0.73
Carswell AFB	0.91
Chase Field, Beeville	0.93
Corpus Christi AD	0.88
Dallas	0.93
Dyess AFB	0.96
Fort Bliss	0.96
Fort Hood	0.84
Fort Sam Houston	0.88
Galveston	0.89
Goodfellow AFB	0.79
Ingleside	0.89
Kelly AFB	0.88
Kingville	0.96
Lackland AFB	0.88
Laughlin AFB	0.90
Lonestar AAP	0.88
Longhorn AAP	0.82
Randolph AFB	0.88
Red River AD	0.88
Reese AFB	0.84
Sheppard AFB	0.97
Utah	1.02
Ogden	0.99
Salt Lake City	1.00
Tooele	1.06
Dugway PG	1.15
Fort Douglas	1.00
Hill AFB	1.03
Vermont	0.95
Burlington	0.99
Montpelier	0.90
Rutland	0.96
Virginia	0.92
Norfolk	0.90
Radford	0.92
Richmond	0.93
Arlington	1.04
Cameron Station	1.05
Dahlgren	1.02
Fort Belvoir	1.05
Fort Eustis	0.93
Fort Hill	0.96
Fort Lee	0.90
Fort Monroe	0.92
Fort Myer	1.04
Fort Pickett	0.97
Fort Story	0.92
Norfolk Area	0.92
Quantico	0.96
Radford AAP	0.91
Vint Hill Farms	0.96
Washington	1.15
Spokane	1.14
Aberdeen	1.06
Tacoma	1.10
Yakima	1.21
Everett	1.19
Fairchild AFB	1.19
Fort Lewis	1.10
Indian Head	0.99
Jim Creek	1.20
McChord AFB	1.26
Pacific Beach	1.13
Seattle Area	1.14
Whidbey Island	1.16
Yakima FC	1.27
West Virginia	0.95
Bluefield	0.89
Clarksburg	0.99
Charleston	0.98
Sugar Grove	1.29
Wisconsin	1.04
LaCrosse	1.06
Madison	1.00
Milwaukee	1.06
Badger AAP	1.00
Clam Lake	1.15
Fort McCoy	1.17
Wyoming	1.04
Casper	1.00
Cheyenne	1.04
Laramie	1.08
F.E.Warren AFB	1.04
Washington DC Area	1.04
Fort McNair	1.04
Walter Reed AMC	1.04

Index

Other Practical References

• National Repair & Remodeling Estimator

The complete pricing guide for dwelling reconstruction costs. Reliable, specific data you can apply on every repair and remodeling job. Up-to-date material costs and labor figures based on thousands of jobs across the country. Provides recommended crew sizes; average production rates; exact material, equipment, and labor costs; a total unit cost and a total price including overhead and profit. Separate listings for high- and low-volume builders, so prices shown are accurate for any size business. Estimating tips specific to repair and remodeling work to make your bids complete, realistic, and profitable. Includes an electronic version of the book on computer disk with a stand-alone *Windows* estimating program **FREE** on a 3½" high-density (1.44 Mb) disk.* **416 pages, 11 x 8½, $32.50. Revised annually**

• National Electrical Estimator

This year's prices for installation of all common electrical work: conduit, wire, boxes, fixtures, switches, outlets, loadcenters, panelboards, raceway, duct, signal systems, and more. Provides material costs, manhours per unit, and total installed cost. Explains what you should know to estimate each part of an electrical system. Includes an electronic version of the book on computer disk with a stand-alone *Windows* estimating program **FREE** on a 3½" high-density (1.44 Mb) disk.* **464 pages, 8½ x 11, $31.75. Revised annually**

• National Painting Cost Estimator

A complete guide to estimating painting costs for just about any type of residential, commercial, or industrial painting, whether by brush, spray, or roller. Shows typical costs and bid prices for fast, medium, and slow work, including material costs per gallon; square feet covered per gallon; square feet covered per manhour; labor, material, overhead, and taxes per 100 square feet; and how much to add for profit. Includes an electronic version of the book on computer disk with a stand-alone *Windows* estimating program **FREE** on a 3½" high-density (1.44 Mb) disk.* **464 pages, 8½ x 11, $32.00. Revised annually**

• Contractor's Guide to the Building Code Rev.

This completely revised edition explains in plain English exactly what the Uniform Building Code requires. Based on the most recent code, it covers many changes made since then. Also covers the Uniform Mechanical Code and the Uniform Plumbing Code. Shows how to design and construct residential and light commercial buildings that'll pass inspection the first time. Suggests how to work with an inspector to minimize construction costs, what common building shortcuts are likely to be cited, and where exceptions are granted. **544 pages, 5½ x 8½, $28.00**

• Construction Estimating Reference Data

Provides the 300 most useful manhour tables for practically every item of construction. Labor requirements are listed for sitework, concrete work, masonry, steel, carpentry, thermal and moisture protection, door and windows, finishes, mechanical and electrical. Each section details the work being estimated and gives appropriate crew size and equipment needed. This new edition contains *DataEst*, a computer estimating program on a high density disk. This fast, powerful program and complete instructions are yours free on high density 5¼" disk when you buy the book. **432 pages, 11 x 8½, $39.50**

• Construction Forms & Contracts

125 forms you can copy and use — or load into your computer (from the **FREE** disk enclosed). Then you can customize the forms to fit your company, fill them out, and print. Loads into Word for Windows, Lotus 1-2-3, WordPerfect, or Excel programs. You'll find forms covering accounting, estimating, fieldwork, contracts, and general office. Each form comes with complete instructions on when to use it and how to fill it out. These forms were designed, tested and used by contractors, and will help keep your business organized, profitable and out of legal, accounting and collection troubles. Includes a 3½" high-density disk for your PC. Add $15.00 if you need double-density or Macintosh disks. **432 pages, 8½ x 11, $39.75**

• Daily Job Log

This handy builder's calendar will help you stay on schedule, on budget and on top of what's happening at each of your jobs for the whole year. Here you'll keep a daily log of appointments, change orders, deliveries, submittal deadlines, job visits, weather delays and due dates — everything that affects your work schedule — in one handy place. There's a calendar page for each day, each with a humorous construction cartoon to help start your morning right, as well as monthly summary pages, estimating forms, cost reference data, and advice from a lawyer on how the data you enter in this log can serve as protection against unfounded claims by others and as the foundation for admissible evidence should you have a case end up in court. **352 pages, 8½ x 11, $24.50**

• National Construction Estimator

Current building costs for residential, commercial, and industrial construction. Estimated prices for every common building material. Manhours, recommended crew, and labor cost for installation. Includes an electronic version of the book on computer disk with a stand-alone *Windows* estimating program **FREE** on a 3½" high-density (1.44 Mb) disk.* **592 pages, 8½ x 11, $31.50. Revised annually**

• Basic Engineering for Builders

If you've ever been stumped by an engineering problem on the job, yet wanted to avoid the expense of hiring a qualified engineer, you should have this book. Here you'll find engineering principles explained in non-technical language and practical methods for applying them on the job. With the help of this book you'll be able to understand engineering functions in the plans and how to meet the requirements, how to get permits issued without the help of an engineer, and anticipate requirements for concrete, steel, wood and masonry. See why you sometimes have to hire an engineer and what you can undertake yourself: surveying, concrete, lumber loads and stresses, steel, masonry, plumbing, and HVAC systems. This book is designed to help the builder save money by understanding engineering principles that you can incorporate into the jobs you bid. **400 pages, 8½ x 11, $34.00**

• National Plumbing & HVAC Estimator

Manhours, labor and material costs for all common plumbing and HVAC work in residential, commercial, and industrial buildings. You can quickly work up a reliable estimate based on the pipe, fittings and equipment required. Every plumbing and HVAC estimator can use the cost estimates in this practical manual. Sample estimating and bidding forms and contracts also included. Explains how to handle change orders, letters of intent, and warranties. Describes the right way to process submittals, deal with suppliers and subcontract specialty work. Includes an electronic version of the book on computer disk with a stand-alone *Windows* estimating program **FREE** on a 3½" high-density (1.44 Mb) disk. **352 pages, 8½ x 11, $32.25. Revised annually**

• Encyclopedia of Construction Terms

Almost everything you could possibly want to know about any word or technique in construction. Hundreds of up-to-date construction terms, materials, drawings and pictures with detailed, illustrated articles describing equipment and methods. Terms and techniques are explained or illustrated in vivid detail. Use this valuable reference to check spelling, find clear, concise definitions of construction terms used on plans and construction documents, or learn about little-known tools, equipment, tests and methods used in the building industry. It's all here. **416 pages, 8½ x 11, $36.00**

• Profits in Buying & Renovating Homes

Step-by-step instructions for selecting, repairing, improving, and selling highly profitable "fixer-uppers." Shows which price ranges offer the highest profit-to-investment ratios, which neighborhoods offer the best return, practical directions for repairs, and tips on dealing with buyers, sellers, and real estate agents. Shows you how to determine your profit before you buy, what "bargains" to avoid, and how to make simple, profitable, inexpensive upgrades. **304 pages, 8½ x 11, $19.75**

• Wood-Frame House Construction

Step-by-step construction details, from the layout of the outer walls, excavation and formwork, to finish carpentry and painting. Contains all new, clear illustrations and explanations updated for construction in the '90s. Everything you need to know about framing, roofing, siding, interior finishings, floor covering and stairs — your complete book of wood-frame homebuilding. **320 pages, 8½ x 11, $19.75. Revised edition**

• Rough Framing Carpentry

If you'd like to make good money working outdoors as a framer, this is the book for you. Here you'll find shortcuts to laying out studs; speed cutting blocks, trimmers and plates by eye; quickly building and blocking rake walls; installing ceiling backing, ceiling joists, and truss joists; cutting and assembling hip trusses and California fills; arches and drop ceilings — all with production line procedures that save you time and help you make more money. Over 100 on-the-job photos of how to do it right and what can go wrong. **304 pages, 8½ x 11, $26.50**

• Profits in Building Spec Homes

If you've ever wanted to make big profits in building spec homes yet were held back by the risks involved, you should have this book. Here you'll learn how to do a market study and feasibility analysis to make sure your finished home will sell quickly, and for a good profit. You'll find tips that can save you thousands in negotiating for land, learn how to impress bankers and get the financing package you want, how to nail down cost estimating, schedule realistically, work effectively yet harmoniously with subcontractors so they'll come back for your next home, and finally, what to look for in the agent you choose to sell your finished home. Includes forms, checklists, worksheets, and step-by-step instructions. **208 pages, 8½ x 11, $27.25**

Order Toll-free
24 Hours A Day
1-800-829-8123

We accept phone orders charged to your MasterCard, Visa, Discover or American Express.
Or, use the postage paid cards on the facing page.

Craftsman Book Company
6058 Corte del Cedro
P. O. Box 6500
Carlsbad, CA 92018

In a hurry?

We accept phone orders
charged to your MasterCard,
Visa, Discover or American Express

Call 1-800-829-8123
FAX (619) 438-0398

Name (Please print)

Company

Address

City/State/Zip

☐ Send check or money order and we pay postage
If you prefer, use your ☐ Visa, ☐ MasterCard,
☐ Discover or ☐ American Express
Card#
Expiration date _____ Initials _____

10-Day Money Back Guarantee

☐ 34.00 Basic Engineering for Builders
☐ 39.50 Construction Estimating Reference Data with **FREE** *DataEst* estimating program on a 5¼" high-density disk.**
☐ 39.75 Construction Forms & Contracts with **FREE** 3½" high-density disk for PC. Add $15 if you need ☐ *double-density* or ☐ *Macintosh* disk.
☐ 28.00 Contractor's Guide to the Building Code, Revised
☐ 24.50 Daily Job Log
☐ 36.00 Encyclopedia of Construction Terms
☐ 31.50 National Construction Estimator with **FREE** *Windows* estimating program on a 3½" high-density disk.*
☐ 31.75 National Electrical Estimator with **FREE** *Windows* estimating program on a 3½" high-density disk.*
☐ 32.00 National Painting Cost Estimator with **FREE** *Windows* estimating program on a 3½" high-density disk.*
☐ 32.25 National Plumbing & HVAC Estimator with **FREE** *Windows* estimating program on a 3½" high-density disk.*
☐ 32.50 National Repair & Remodeling Estimator with **FREE** *Windows* estimating program on a 3½" high-density disk.*
☐ 27.25 Profits in Building Spec Homes
☐ 19.75 Profits in Buying & Renovating Homes
☐ 26.50 Rough Framing Carpentry
☐ 19.75 Wood-Frame House Construction
☐ 18.00 National Building Cost Estimator
☐ **FREE** Full Color Catalog
*Add $10 if you need extra ☐ 3½" double-density or ☐ 5¼" high -density disks.
**Add $10 if you need extra ☐ 5¼" double-density disks or ☐ 3½" high

---

Receive Fresh Cost Data Every Year -- Automatically

Join Craftsman's Standing Order Club and automatically receive special membership discounts on annual cost books!

Qty. **National Construction Estimator** Standing Order price $26.78	Qty. **National Repair & Remodeling Estimator** Standing Order price $27.63	Qty. **National Electrical Estimator** Standing Order price $26.99	Qty **National Building Cost Manual** Standing Order price $15.30	Qty. **National Painting Cost Estimator** Standing Order price $27.20	Qty. **National Plumbing & HVAC Estimator** Standing Order price $27.41

How many times have you missed one of Craftsman's pre-publication discounts, or procrastinated about buying the updated book and ended up bidding your jobs using obsolete cost data that you "updated" on your own? As a Standing Order Member you never have to worry about ordering every year. Instead, you will receive a confirmation of your standing order each September. If the order is correct, do nothing, and your order will be shipped and billed as listed on the confirmation card. If you wish to change your address, or your order, you can do so at that time. Your standing order for the books you selected will be shipped as soon as the publications are printed. You'll automatically receive any pre-publication discount price.

Bill to:
Name

Company

Address

City/State/Zip

Ship to:
Name

Company

Address

City/State/Zip

Purchase order #

Phone # (   )

Signature

Mail This Card Today For a Free Full Color Catalog

Over 100 books, videos, and audios at your fingertips with information that can save you time and money. Here you'll find information on carpentry, contracting, estimating, remodeling, electrical work, and plumbing.

All items come with an unconditional 10-day money-back guarantee. If they don't save you money, mail them back for a full refund.

Name (please print)

Company

Address

City   State   Zip

Craftsman Book Company, 6058 Corte del Cedro, P. O. Box 6500, Carlsbad, CA 92018

BUSINESS REPLY MAIL
FIRST CLASS MAIL PERMIT NO. 271 CARLSBAD CA

POSTAGE WILL BE PAID BY ADDRESSEE

Craftsman Book Company
6058 Corte del Cedro
P. O. Box 6500
Carlsbad, CA 92018-9974

BUSINESS REPLY MAIL
FIRST CLASS MAIL PERMIT NO. 271 CARLSBAD CA

POSTAGE WILL BE PAID BY ADDRESSEE

Craftsman Book Company
6058 Corte del Cedro
P. O. Box 6500
Carlsbad, CA 92018-9974

BUSINESS REPLY MAIL
FIRST CLASS MAIL PERMIT NO. 271 CARLSBAD CA

POSTAGE WILL BE PAID BY ADDRESSEE

Craftsman Book Company
6058 Corte del Cedro
P. O. Box 6500
Carlsbad, CA 92018-9974